The Statistical Foundations of Entropy

The Statistical Foundations of Entropy

Editors

Petr Jizba
Jan Korbel

MDPI • Basel • Beijing • Wuhan • Barcelona • Belgrade • Manchester • Tokyo • Cluj • Tianjin

Editors
Petr Jizba
Czech Technical University in Prague
Czech Republic
+420-224-358-295

Jan Korbel
Medical University of Vienna & Complexity Science Hub Vienna
Austria
+43-1-40160-36253

Editorial Office
MDPI
St. Alban-Anlage 66
4052 Basel, Switzerland

This is a reprint of articles from the Special Issue published online in the open access journal *Entropy* (ISSN 1099-4300) (available at: https://www.mdpi.com/journal/entropy/special_issues/Statistical_Foundations).

For citation purposes, cite each article independently as indicated on the article page online and as indicated below:

LastName, A.A.; LastName, B.B.; LastName, C.C. Article Title. *Journal Name* **Year**, *Volume Number*, Page Range.

ISBN 978-3-0365-3557-9 (Hbk)
ISBN 978-3-0365-3558-6 (PDF)

© 2022 by the authors. Articles in this book are Open Access and distributed under the Creative Commons Attribution (CC BY) license, which allows users to download, copy and build upon published articles, as long as the author and publisher are properly credited, which ensures maximum dissemination and a wider impact of our publications.
The book as a whole is distributed by MDPI under the terms and conditions of the Creative Commons license CC BY-NC-ND.

Contents

About the Editors vii

Petr Jizba and Jan Korbel
The Statistical Foundations of Entropy
Reprinted from: *Entropy* **2021**, *23*, 1367, doi:10.3390/e23101367 1

Ariel Caticha
Entropy, Information, and the Updating of Probabilities
Reprinted from: *Entropy* **2021**, *23*, 895, doi:10.3390/e23070895 5

Velimir M. Ilić and Ivan B. Djordjević
On the α-q-Mutual Information and the α-q-Capacities
Reprinted from: *Entropy* **2021**, *23*, 702, doi:10.3390/e23060702 25

Carlos Medel-Portugal, Juan Manuel Solano-Altamirano and José Luis E. Carrillo-Estrada
Classical and Quantum H-Theorem Revisited: Variational Entropy and Relaxation Processes
Reprinted from: *Entropy* **2021**, *23*, 366, doi:10.3390/e23030366 49

Petr Jizba, Jacob Dunningham and Martin Prokš
From Rényi Entropy Power to Information Scan of Quantum States
Reprinted from: *Entropy* **2021**, *23*, 334, doi:10.3390/e23030334 65

Xiaolin Shi, Yimin Shi and Kuang Zhou
Estimation for Entropy and Parameters of Generalized Bilal Distribution under Adaptive Type II Progressive Hybrid Censoring Scheme
Reprinted from: *Entropy* **2021**, *23*, 206, doi:10.3390/e23020206 89

Jan Korbel
Calibration Invariance of the MaxEnt Distribution in the Maximum Entropy Principle
Reprinted from: *Entropy* **2021**, *23*, 96, doi:10.3390/e23010096 111

Marek Czachor
Unifying Aspects of Generalized Calculus
Reprinted from: *Entropy* **2020**, *22*, 1180, doi:10.3390/e22101180 121

Miroslav Grmela, Václav Klika and Michal Pavelka
Dynamic and Renormalization-Group Extensions of the Landau Theory of Critical Phenomena
Reprinted from: *Entropy* **2020**, *22*, 978, doi:10.3390/e22090978 141

Rosa Bernardini Papalia and Esteban Fernandez Vazquez
Entropy-Based Solutions for Ecological Inference Problems: A Composite Estimator
Reprinted from: *Entropy* **2020**, *22*, 781, doi:10.3390/e22070781 161

About the Editors

Petr Jizba is an assistant professor at the Faculty of Nuclear Sciences and Physical Engineering of the Czech Technical University in Prague. He obtained his Ph.D. in theoretical physics at the University of Cambridge. After his graduation, he spent several years as a researcher at the Tsukuba University in Japan and as a Humboldt fellow at the Freie Universität in Berlin. His main interests are quantum field theory and quantum mechanics, statistical physics, and econophysics. His main research topics are the applications of the path and functional integral in quantum field theory and quantum mechanics, the applications of generalized entropies in complex systems, information transfer in a complex time series, and the applications of superstatistics in econophyiscs. He is co-author of the book "Quantum Field Theory and its Macroscopic Manifestations: Boson condensation, ordered patterns and topological defects".

Jan Korbel is a Postdoc at the Section of Science for Complex Systems of the Medical University of Vienna and the Complexity Science Hub Vienna. He obtained his Ph.D. in mathematical physics at the Faculty of Nuclear Sciences and Physical Engineering of the Czech Technical University in Prague under the supervision of Petr Jizba. During his doctoral studies, he spent a year as a research intern at the Max Planck Institute for the History of Science. He also spent a year as a postdoc at Zhejiang University, Hangzhou, China. His main interests are non-equilibrium statistical physics, complex systems, and econophysic. His main research topics are the applications of non-standard entropies in non-equilibrium thermodynamics, financial time series analysis (multifractals, superstatistics, fractional calculus), and the applications of information theory in statistical physics and econophysics.

Editorial

The Statistical Foundations of Entropy

Petr Jizba [1] and Jan Korbel [1,2,3,*]

1. Faculty of Nuclear Sciences and Physical Engineering, Czech Technical University, 11519 Prague, Czech Republic; petr.jizba@fjfi.cvut.cz
2. Section for the Science of Complex Systems, Center for Medical Statistics, Informatics and Intelligent Systems (CeMSIIS), Medical University of Vienna, Spitalgasse 23, 1090 Vienna, Austria
3. Complexity Science Hub Vienna, Josefstädterstrasse 39, 1080 Vienna, Austria
* Correspondence: jan.korbel@meduniwien.ac.at

Citation: Jizba, P.; Korbel, J. The Statistical Foundations of Entropy. *Entropy* **2021**, *23*, 1367. https://doi.org/10.3390/e23101367

Received: 13 October 2021
Accepted: 15 October 2021
Published: 19 October 2021

Publisher's Note: MDPI stays neutral with regard to jurisdictional claims in published maps and institutional affiliations.

Copyright: © 2021 by the authors. Licensee MDPI, Basel, Switzerland. This article is an open access article distributed under the terms and conditions of the Creative Commons Attribution (CC BY) license (https://creativecommons.org/licenses/by/4.0/).

During the last few decades, the notion of entropy has become omnipresent in many scientific disciplines, ranging from traditional applications in statistical physics and chemistry, information theory, and statistical estimation to more recent applications in biology, astrophysics, geology, financial markets, or social networks. All these examples belong to the large family of complex dynamical systems that is typified by phase transitions, scaling behavior, multiscale and emerging phenomena, and many other non-trivial phenomena. Frequently, it turns out that in these systems, the usual Boltzmann–Gibbs–Shannon entropy and ensuing statistical physics are not adequate concepts. This especially happens in cases where the sample space in question does not grow exponentially.

This Special Issue, "The Statistical Foundations of Entropy", is dedicated to discussing solutions and delving into concepts, methods, and algorithms for improving our understanding of the statistical foundations of entropy in complex systems, with a particular focus on the so-called generalized entropies that go beyond the usual Boltzmann–Gibbs–Shannon framework. The nine high-quality articles included in this Special Issue propose and discuss new tools and concepts derived from information theory, non-equilibrium statistical physics, and the theory of complex dynamical systems to investigate various non-conventional aspects of entropy with assorted applications. They illustrate the potential and pertinence of novel conceptual tools in statistical physics that, in turn, help us to *shed fresh light* on the statistical foundations of entropy.

In the first contribution [1], the authors discuss the important topic of ecological inference used to estimate individual behavior from the knowledge-aggregated data. The authors discuss two popular approaches: the Generalized Maximum Entropy approach and the Distributionally Weighted Regression equation. Moreover, the authors show that it is possible to obtain a combined solution with the so-called Generalized Cross-Entropy solution for the matrix adjustment problem.

The theory of critical phenomena is one of the essential parts of statistical physics with important overlaps to other scientific fields. The authors of the second paper [2] extend the classic Landau theory of critical phenomena in the context of multiscale dynamics. By using renormalization group theory, the authors can describe critical points through the inseparability of levels at certain points. These findings allow experimentalists more precise measurements of critical points.

The third contribution [3] describes a general approach toward generalized entropies based on the non-Newtonian calculus. The main idea is to redefine the usual arithmetic operations (e.g., addition and multiplication) and related calculus operators (e.g., differentiation and integration) in such a way that some crucial properties remain valid. Entropy and the corresponding thermodynamic quantities are then defined analogously to the Boltzmann–Gibbs–Shannon case in terms of these deformed operations. This approach incorporates many popular generalized entropies and gives a general recipe for the thermodynamics of such generalized entropies. Since the potential of this general approach is immense, the editors decided to promote this article by choosing the paper as the Editor's Choice.

The aim of the fourth paper [4] was to show that for each distribution obtained from the principle of maximum entropy, there exists freedom in the choice of entropic functions and constraints, which make the reverse identification of entropic function and constraints from the form of MaxEnt distribution impossible. The paper consists of two simple examples of such invariance, where the different choices of entropy and constraints lead to the same MaxEnt distribution, which differs only in the Lagrange parameters. Since the Lagrange parameters may have some thermodynamic interpretation, it is essential to identify some additional properties of the system to decide which function plays the role of thermodynamic entropy.

In the fifth paper [5], the authors focused on the problem of estimation for entropy and the parameters of generalized Bilal distribution under an adaptive type II progressive hybrid censoring scheme. To this end, they used the maximum likelihood estimator and Newton–Raphson iteration method. In addition, some other estimators, such as the Bayesian estimator and confidence intervals, were also provided. Finally, the study contains an illustrative example that applies the obtained results.

The main result of the sixth contribution [6] was to introduce a generalization the concept of Shannon's entropy power based on Rényi entropy. Consequently, the authors could generalize several popular identities, including the de Bruijn identity, isoperimetric inequality, or Stam inequality. Moreover, this enables one to introduce a new class of the one-parameter family of Rényi entropy power-based quantum mechanical uncertainty relations. These relations turn out to be very useful in many applications in quantum mechanics, including entanglement and quantum metrology.

The seventh paper [7] revisited the Boltzmann H theorem for both the classical and quantum systems. The authors considered a spatially inhomogeneous system initially out of equilibrium and studied its relaxation toward equilibrium. They accomplished this by considering small local cells and assuming the local equilibrium hypothesis. The global H-function is a sum of the local H functions. The authors recovered the H theorem for the case of spatially inhomogeneous gasses for both the classical and quantum cases. The correspondence principle connects the classical and quantum H functions.

In the eighth paper [8], the authors introduced a generalization of mutual information based on two-parameter Sharma–Mittal entropy. They also discussed the distinction between mutual information and capacity for the case of generalized entropies based on Sharma–Mittal entropy. They accomplished this by considering a proper axiomatic framework. Finally, the authors showed that the proper definition of both Sharma–Mittal mutual information and Sharma–Mittal capacity solves the issue of non-physical behavior.

Finally, in the last paper of this Special Issue [9], the authors reviewed the maximum entropy principle from the point of view of a general inference framework. The authors discussed a general procedure of updating the prior distribution to a posterior probability distribution through an eliminative induction process. The authors showed that under the assumption of subsystem independence, the logarithmic relative entropy (also called Kullback–Leibler divergence) is the unique solution that fulfills the prescribed axiomatic framework. Furthermore, the authors showed that this general framework contains the MaxEnt principle and Bayes' rule as special cases and unifies the entropic and Bayesian inference methods.

Entropy is presumably one of the most intricate and complex scientific concepts. Its comprehension is an open challenge that requires new abstractions and methodological approaches. Information theory, statistical physics, and estimation theory methods provide a versatile and flexible framework with the potential to move research in this field forward. In this Special Issue, various conceptual and methodological approaches were provided to further deepen our understanding of the statistical foundations of entropy. They provided relevant pieces of research that addressed timely, current topics associated with the entropy paradigm. It is our hope that the reader will enjoy the articles included in this Special Issue and will find them helpful.

Author Contributions: Both authors contributed equally to writing and revising the manuscript. All authors have read and agreed to the published version of the manuscript.

Acknowledgments: As guest editors, we would like to thank to all the anonymous peer reviewers who revised and assessed the submissions and the authors that contributed to the Special Issue "The Statistical Foundations of Entropy".

Conflicts of Interest: The authors declare no conflict of interest.

References

1. Bernardini Papalia, R.; Fernandez Vazquez, E. Entropy-Based Solutions for Ecological Inference Problems: A Composite Estimator. *Entropy* **2020**, *22*, 781. [CrossRef] [PubMed]
2. Grmela, M.; Klika, V.; Pavelka, M. Dynamic and Renormalization-Group Extensions of the Landau Theory of Critical Phenomena. *Entropy* **2020**, *22*, 978. [CrossRef] [PubMed]
3. Czachor, M. Unifying Aspects of Generalized Calculus. *Entropy* **2020**, *22*, 1180. [CrossRef] [PubMed]
4. Korbel, J. Calibration Invariance of the MaxEnt Distribution in the Maximum Entropy Principle. *Entropy* **2021**, *23*, 96. [CrossRef] [PubMed]
5. Shi, X.; Shi, Y.; Zhou, K. Estimation for Entropy and Parameters of Generalized Bilal Distribution under Adaptive Type II Progressive Hybrid Censoring Scheme. *Entropy* **2021**, *23*, 206. [CrossRef] [PubMed]
6. Jizba, P.; Dunningham, J.; Prokš, M. From Rényi Entropy Power to Information Scan of Quantum States. *Entropy* **2021**, *23*, 334. [CrossRef] [PubMed]
7. Medel-Portugal, C.; Solano-Altamirano, J.M.; Carrillo-Estrada, J.L.E. Classical and Quantum H-Theorem Revisited: Variational Entropy and Relaxation Processes. *Entropy* **2021**, *23*, 366. [CrossRef] [PubMed]
8. Ilić, V.M.; Djordjević, I.B. On the α-q-Mutual Information and the α-q-Capacities. *Entropy* **2021**, *23*, 702. [CrossRef] [PubMed]
9. Caticha, A. Entropy, Information, and the Updating of Probabilities. *Entropy* **2021**, *23*, 895. [CrossRef] [PubMed]

Article

Entropy, Information, and the Updating of Probabilities

Ariel Caticha

Physics Department, University at Albany-SUNY, Albany, NY 12222, USA; acaticha@albany.edu;
Tel.: +1-(518)-442-4592

Abstract: This paper is a review of a particular approach to the method of maximum entropy as a general framework for inference. The discussion emphasizes pragmatic elements in the derivation. An epistemic notion of information is defined in terms of its relation to the Bayesian beliefs of ideally rational agents. The method of updating from a prior to posterior probability distribution is designed through an eliminative induction process. The logarithmic relative entropy is singled out as a unique tool for updating (a) that is of universal applicability, (b) that recognizes the value of prior information, and (c) that recognizes the privileged role played by the notion of independence in science. The resulting framework—the ME method—can handle arbitrary priors and arbitrary constraints. It includes the MaxEnt and Bayes' rules as special cases and, therefore, unifies entropic and Bayesian methods into a single general inference scheme. The ME method goes beyond the mere selection of a single posterior, and also addresses the question of how much less probable other distributions might be, which provides a direct bridge to the theories of fluctuations and large deviations.

Keywords: maximum entropy; Bayesian inference; updating probabilities

1. Introduction

Inductive inference is a framework for coping with uncertainty, for reasoning with incomplete information. The framework must include a means to represent a state of partial knowledge—this is handled through the introduction of probabilities—and it must allow us to change from one state of partial knowledge to another when new information becomes available. Indeed, any inductive method that recognizes that a situation of incomplete information is in some way unfortunate—by which we mean that it constitutes a problem in need of a solution—would be severely deficient if it failed to address the question of how to proceed in those fortunate circumstances when new information becomes available. The theory of probability, if it is to be useful at all, demands a method for assigning and updating probabilities.

The challenge is to develop updating methods that are both systematic, objective and practical. When information consists of data and a likelihood function, Bayesian updating is the uniquely natural method of choice. Its foundation lies in recognizing the value of prior information: whatever was learned in the past is valuable and should not be disregarded, which amounts to requiring that beliefs ought to be revised but only to the extent required by the new data. This immediately raises a number of questions: How do we update when the information is not in the form of data? If the information is not data, what else could it possibly be? Indeed, what, after all, is "information"? On a separate line of development, the method of Maximum Gibbs–Shannon Entropy (MaxEnt) allows one to process information in the form of constraints on the allowed probability distributions. This provides a partial answer to one of our questions: in addition to data, information can also take the form of constraints. However, it immediately raises several other questions: What is the interpretation of entropy? Is there unique entropy? Are Bayesian and entropic methods mutually compatible?

The purpose of this paper is to review one particular approach to entropic updating. The presentation below, which is meant to be pedagogical and self-contained, is based on work presented in a sequence of papers [1–5] and in the sets of lectures [6–8]. As we shall see below, we adopt a pragmatic approach in which entropy is a tool designed for the specific purpose of updating probabilities.

Historically, the method of maximum relative entropy (ME) is a direct descendant of the MaxEnt method, pioneered by Jaynes [9,10]. In the MaxEnt framework, entropy is interpreted through the Shannon axioms as a measure of the amount of information that is missing in a probability distribution. This approach has its limitations. The Shannon axioms refer to probabilities of discrete variables; for continuous variables, the Shannon entropy is not defined. A more serious objection is that even if we grant that the Shannon axioms do lead to a reasonable expression for entropy, to what extent do we believe the axioms themselves? Shannon's third axiom, the grouping property, is indeed sort of reasonable, but is it necessary? Is entropy the only consistent measure of uncertainty or of information? Indeed, there exist examples in which the Shannon entropy does not seem to reflect one's intuitive notion of information [8,11]. One could introduce other entropies justified by different choices of axioms (e.g., [12–14]), but this move raises problems of its own: Which entropy should one adopt? If different systems are handled using different entropies, how does one handle composite systems?

From our perspective, the problem can be traced to the fact that neither Shannon nor Jaynes were concerned with the task of updating probabilities. Shannon's communication theory aimed to characterize the sources of information, to measure the capacity of the communication channels, and to learn how to control the degrading effects of noise. On the other hand, Jaynes conceived MaxEnt as a method to assign probabilities on the basis of constraint information and a fixed underlying measure and not from an arbitrary prior distribution.

Considerations such as these motivated several attempts to develop ME directly as a method for updating probabilities without invoking questionable measures of information [1,5,15–17]. The important contribution by Shore and Johnson was the realization that one could axiomatize the updating method itself rather than the information measure. Their axioms have, however, raised criticisms [11,18–20] and counter-criticisms [2,6,8,21,22]. Despite the controversies, Shore and Johnson's pioneering papers have had an enormous influence: they identified the correct goal to be achieved.

The concept of relative entropy is introduced as a tool for updating probabilities. Hereinafter, we drop the qualifier "relative" and adopt the simpler term "entropy". The reasons for the improved nomenclature are the following: (1) The general concept should receive the general name "entropy", while the more specialized concepts should be the ones receiving a qualifier, such as "thermodynamic" or "Clausius" entropy, and "Gibbs–Shannon" entropy. (2) All entropies are relative, even if they happen to be relative to an implicit uniform prior. Making this fact explicit has tremendous pedagogical value. (3) The practice is already in use with the concept of energy: all energies are relative too, but there is no advantage in constantly referring to a "relative energy". Accordingly, ME will be read as "maximum entropy"; additional qualifiers are redundant.

As with all tools, entropy too is *designed* to perform a certain function, and its performance must meet certain *design criteria* or *specifications*. There is no implication that the method is "true", or that it succeeds because it achieves some special contact with reality. Instead, the claim is that the method succeeds in the pragmatic sense that it works as designed—this is satisfactory because when properly deployed, it leads to empirically adequate models. In this approach, *entropy needs no interpretation* whether it be in terms of heat, multiplicity of states, disorder, uncertainty, or even in terms of an amount of information. Incidentally, this may explain why the search for the meaning of entropy has proved so elusive: we need not know what "entropy" means—we only need to know how to use it.

Since our topic is the updating of probabilities when confronted with new information, our starting point is to address the question, "what is information?". In Section 2, we develop a concept of information that is both pragmatic and Bayesian. "Information" is defined in terms of its effects on the beliefs of rational agents. The design of entropy as a tool for updating is the topic of Section 3. There, we state the design specifications that define what function entropy is supposed to perform, and we derive its functional form. To streamline the presentation, some of the mathematical derivations are left to the appendices.

To conclude, we present two further developments. In Section 4, we show that Bayes' rule can be derived as a special case of the ME method. An earlier derivation of this important result following a different line of argument was given by Williams [23] before a sufficient understanding of entropy as an updating tool had been achieved. It is not, therefore, surprising that Williams' achievement has not received the widespread appreciation it deserves. Thus, within the ME framework, entropic and Bayesian methods are unified into a single consistent theory of inference. One advantage of this insight is that it allows a number of generalizations of Bayes' rule [2,8]. Another is that it provides an important missing piece for the old puzzles of quantum mechanics concerning the so-called collapse of the wave function and the quantum measurement problem [24,25].

There is yet another function that the ME method must perform in order to fully qualify as a method of inductive inference. Once we have decided that the distribution of maximum entropy is to be preferred over all others, the following question arises immediately: the maximum of the entropy functional is never infinitely sharp, so are we confident that distributions that lie very close to the maximum are completely ruled out? In Section 5, the ME method is deployed to assess quantitatively the extent to which distributions with lower entropy are ruled out. The significance of this result is that it provides a direct link to the theories of fluctuations and large deviations. Concluding remarks are given in Section 6.

2. What Is Information?

The term "information" is used with a wide variety of different meanings [10,26,27]. There is the Shannon notion of information that is meant to measure an amount of information and is quite divorced from semantics. There is also an algorithmic notion of information that captures a notion of complexity and originates in the work of Solomonov, Kolmogorov, and Chaitin [26]; there is a related notion of entropy as a minimum description length [28]. Furthermore, in the general context of the thermodynamics of computation, it is said that "information is physical" because systems "carry" or "contain" information about their own physical state [29–31] (see also [32,33]).

Here, we follow a different path [3,4]. We seek an epistemic notion of information that is closer to the everyday colloquial use of the term—roughly, information is what we request when we ask a question. In a Bayesian framework, this requires an explicit account of the relation between information and the beliefs of ideally rational agents. We emphasize that our concern here is with *idealized rational agents*. Our subject is not the psychology of actual humans who often change their beliefs by processes that are neither fully rational nor fully conscious. We adopt a Bayesian interpretation of probability as a degree of credibility: the degree to which we *ought* to believe that a proposition is true if only we were ideally rational. For a discussion of a decision theory that might be relevant to the economics and psychology of partially rational agents see [34–36]. An entropic framework for modelling economies that bypasses all issues of bounded rationality is described in [37].

It is implicit in the recognition that most of our beliefs are held on the basis of incomplete information that not all probability assignments are equally good; some beliefs are preferable to others in the very pragmatic sense that they enhance our chances to successfully navigate this world. Thus, a theory of probability demands a theory of updating probabilities in order to improve our beliefs.

We are now ready to address the question: What, after all, is "information"? The answer is pragmatic. *Information is what information does*. Information is defined by its effects: (a) it restricts our options as to what we are honestly and rationally allowed to believe; and (b) it induces us to update from prior beliefs to posterior beliefs. This, I propose, is a defining characteristic of information:

Information is that which induces a change from one state of rational belief to another.

One aspect of this notion is that for a rational agent, the identification of what constitutes information—as opposed to mere noise—already involves a judgment, an evaluation. Another aspect is that the notion that information is directly related to changing our minds does not involve any reference to *amounts* of information, but it nevertheless allows precise quantitative calculations. Indeed, constraints on the acceptable posterior probabilities are precisely the kind of information that the method of maximum entropy is designed to handle. In short,

Information constrains probability distributions. The constraints are the information.

To the extent that the probabilities are Bayesian, this definition captures the Bayesian notion that information is directly related to changing our minds, that it is the driving force behind the process of learning. It also incorporates an important feature of rationality: being rational means accepting that "not everything goes", and that our beliefs must be constrained in very specific ways. However, the indiscriminate acceptance of any arbitrary constraint does not qualify as rational behavior. To be rational, an agent must exercise some judgment before accepting a particular piece of information as a reliable basis for the revision of its beliefs, which raises questions about what judgments might be considered sound. Furthermore, there is no implication that the information must *be true*; only that we *accept it as true*. False information is information too, at least to the extent that we are prepared to accept it and allow it to affect our beliefs.

The paramount virtue of the definition above is that it is useful; it allows precise quantitative calculations. The constraints that constitute information can take a wide variety of forms. They can be expressed in terms of expected values, they can specify the functional form of a distribution, or be imposed through various geometrical relations. Examples are given in Section 5 and in [38].

Concerning the act of updating, it may be worthwhile to point out an analogy with dynamics. In Newtonian mechanics, the state of motion of a system is described in terms of momentum, and the change from one state to another is said to be "caused" by an applied force or impulse. Bayesian inference is analogous in that a state of belief is described in terms of probabilities, and the change from one state to another is "caused" by information. Just as a force is that which induces a change from one state of motion to another, so *information is that which induces a change from one state of belief to another*. Updating is a form of dynamics. In [39], the analogy is taken seriously: the logic is reversed and quantum mechanics is derived as an example of the entropic updating of probabilities.

3. The Pragmatic Design of Entropic Inference

Once we have decided, as a result of the confrontation of new information with old beliefs, that our beliefs require revision, the problem becomes one of deciding how precisely this ought to be done. First, we identify some general features of the kind of belief revision that one might count as rational. Then, we design a method—a systematic procedure—that implements those features. To the extent that the method performs as desired, we can claim success. The point is not that success derives from our method having achieved some intimate connection to the inner wheels of reality; success simply means that the method seems to be working.

The one obvious requirement is that the updated probabilities ought to agree with the newly acquired information. Unfortunately, this requirement, while necessary, is not sufficiently restrictive: we can update in many ways that preserve both internal consistency and consistency with the new information. Additional criteria are needed. What rules would an ideally rational agent choose?

3.1. General Criteria

The rules are motivated by the same pragmatic criteria that motivate the design of probability theory itself [8]—universality, consistency, and practical utility. However, this is admittedly too vague; we must be very specific about the precise way in which the criteria are implemented.

3.1.1. Universality

In principle, different systems and different situations could require different problem-specific induction methods. However, in order to be useful in practice, the method we seek must be of *universal* applicability. Otherwise, it would fail us when most needed, for we would not know which method to choose when not much is known about the system. To put in different words, what we want to design is a general-purpose method that captures what all the other problem-specific methods might have in common. The idea is that the peculiarities of a particular problem will be captured by the specific constraints that describe the information that is relevant to the problem at hand.

The analogy with mechanics can be found here as well. The possibility of a science of mechanics hinges on identifying a law of motion of universal applicability (e.g., the Schrödinger equation), while the specifics of each system are introduced through initial conditions and the choice of potentials or forces. Here, we shall design an entropy of universal applicability, while the specifics of each problem are introduced through prior probabilities and the choice of constraints.

3.1.2. Parsimony

To specify the updating, we adopt a very conservative criterion that recognizes the value of information: what has been laboriously learned in the past is valuable and should not be disregarded unless rendered obsolete by new information. The only aspects of one's beliefs that should be updated are those for which new evidence has been supplied. Thus, we adopt the following.

Principle of Minimal Updating (PMU): *Beliefs should be updated only to the minimal extent required by the new information.*

The special case of updating in the absence of new information deserves a comment. The PMU states that when there is no new information, ideally, rational agents should not change their minds. In fact, it is difficult to imagine any notion of rationality that would allow the possibility of changing one's mind for no apparent reason.

Minimal updating offers yet another pragmatic advantage. As we shall see below, rather than identifying what features of a distribution are singled out for updating and then specifying the detailed nature of the update, we will adopt design criteria that stipulate what is not to be updated. The practical advantage of this approach is that it enhances objectivity—there are many ways to change something but only one way to keep it the same. The analogy with mechanics can be pursued even further: if updating is a form of dynamics, then minimal updating is the analogue of inertia. Rationality and objectivity demand a considerable amount of inertia.

3.1.3. Independence

The next general requirement turns out to be crucially important because without it, the very possibility of scientific theories would be compromised. The point is that every scientific model, whatever the topic, if it is to be useful at all, must assume that all relevant variables have been taken into account and that whatever was left out—the rest of the universe—should not matter. To put it another way, in order to do scientific work, we must be able to understand parts of the universe without having to understand the universe as a whole. Granted, a pragmatic understanding need not be complete and exact; it must be merely adequate for our purposes.

The assumption, then, is that it is possible to focus our attention on a suitably chosen system of interest and neglect the rest of the universe because the system and the rest of the universe are "sufficiently independent". Thus, in any form of science, the notion of statistical independence must play a central and privileged role. This idea—that some things can be neglected and that not everything matters—is implemented by imposing a criterion that tells us how to handle independent systems. The chosen criterion is quite natural: *whenever two systems are a priori believed to be independent and we receive information about just one, it should not matter if the other is included in the analysis or not*. This is an example of the PMU in action; it amounts to requiring that independence to be preserved unless information about correlations is explicitly introduced.

Again, we emphasize that none of these criteria are imposed by nature. They are desirable for pragmatic reasons; they are imposed by design.

3.2. Entropy as a Tool for Updating Probabilities

Consider a set of propositions $\{x\}$ about which we are uncertain. The proposition x can be discrete or continuous, in one or in several dimensions. It could, for example, represent the microstate of a physical system, a point in phase space, or an appropriate set of quantum numbers. The uncertainty about x is described by a probability distribution $q(x)$. The goal is to update from the prior distribution $q(x)$ to a posterior distribution $p(x)$ when new information—by which we mean a set of constraints—becomes available. The question is, which distribution among all those that satisfy the constraints should we select?

Our goal is to design a method that allows a systematic search for the preferred posterior distribution. The central idea, first proposed by Skilling [16], is disarmingly simple: to select the posterior, first rank all candidate distributions in increasing order of "preference" and then pick the distribution that ranks the highest. Irrespective of what it is that makes one distribution "preferable" over another (we will get to that soon enough), it is clear that any such ranking must be transitive: if distribution p_1 is preferred over distribution p_2, and p_2 is preferred over p_3, then p_1 is preferred over p_3. Transitive rankings are implemented by assigning to each p a real number $S[p]$, which is called the entropy of p in such a way that if p_1 is preferred over p_2, then $S[p_1] > S[p_2]$. The selected distribution (one or possibly many, for *there may be several equally preferred distributions*) is that which maximizes the entropy functional.

The importance of Skilling's strategy of ranking distributions cannot be overestimated: it answers the questions "why entropy?" and "why a maximum?". The strategy implies that the updating method will take the form of a variational principle—the method of maximum entropy (ME)—involving a certain functional that maps distributions to real numbers. These features are not imposed by nature; they are all imposed by design. They are dictated by the function that the ME method is supposed to perform. (Thus, it makes no sense to seek a generalization in which entropy is a complex number or a vector; such generalized entropies would just not perform the desired function.)

Next, we specify the ranking scheme, that is, we choose a specific functional form for the entropy $S[p]$. Note that the purpose of the method is to update *from priors to posteriors* so the ranking scheme must depend on the particular prior q and therefore, the entropy S must be a functional of both p and q. The entropy $S[p,q]$ describes a ranking of the distributions p *relative* to the given prior q. $S[p,q]$ is the entropy of p *relative* to q, and accordingly, $S[p,q]$ is commonly called *relative entropy*. This is appropriate and sometimes we will follow this practice. However, since all entropies are relative, even when relative to a uniform distribution, the qualifier "relative" is redundant and can be dropped.

The functional $S[p,q]$ is designed by a process of elimination—this is a process of *eliminative induction*. First, we state the desired design criteria; this is the crucial step that defines what makes one distribution preferable over another. Candidate functionals that fail to satisfy the criteria are discarded—hence, the qualifier "eliminative". As we shall see, the criteria adopted below are so constraining that there is a single entropy functional $S[p, q]$ that survives the process of elimination.

This approach has a number of virtues. First, to the extent that the design criteria are universally desirable, the single surviving entropy functional will also be of universal applicability. Second, the reason why alternative entropy candidates are eliminated is quite explicit—at least one of the design criteria is violated. Thus, *the justification behind the single surviving entropy is not that it leads to demonstrably correct inferences, but rather, that all other candidates demonstrably fail to perform as desired.*

3.3. Specific Design Criteria

Consider a lattice of propositions generated by a set $\mathcal{X}$ of atomic propositions that are mutually exclusive and exhaustive and are labeled by a discrete index $i = 1, 2, \ldots, n$. The extension to infinite sets and to continuous labels turns out to be straightforward. The index i might, for example, label the microstates of a physical system but, since the argument below is supposed to be of general validity, we shall not assume that the labels themselves carry any particular significance. We can always permute labels; this should have no effect on the updating of probabilities.

We adopt design criteria that reflect the structure of the lattice of propositions—the propositions are related to each other by disjunctions (OR) and conjunctions (AND) and the consistency of the web of beliefs is implemented through the sum and product rules of the probability theory. Our criteria refer to the two extreme situations of propositions that are mutually exclusive and of propositions that are mutually independent. At one end, we deal with the probabilities of propositions that are highly correlated (if one proposition is true, the other is false and vice versa); at the other end, we deal with the probabilities of propositions that are totally uncorrelated (the truth or falsity of one proposition has no effect on the truth or falsity of the other). One extreme is described by the simplified sum rule, $p(i \vee j) = p(i) + p(j)$, and the other extreme by the simplified product rule, $p(i \wedge j) = p(i)p(j)$. (For an alternative approach to the foundations of inference that exploits the various symmetries of the lattice of propositions see [40,41].

The two design criteria and their consequences for the functional form of entropy are given below. Detailed proofs are deferred to the appendices.

3.3.1. Mutually Exclusive Subdomains

DC1: *Probabilities that are conditioned on one subdomain are not affected by information about other non-overlapping subdomains.*

Consider a subdomain $\mathcal{D} \subset \mathcal{X}$ composed of atomic propositions $i \in \mathcal{D}$ and suppose the information to be processed refers to some other subdomain $\mathcal{D}' \subset \mathcal{X}$ that does not overlap with $\mathcal{D}$, $\mathcal{D} \cap \mathcal{D}' = \emptyset$. In the absence of any new information about $\mathcal{D}$, the PMU demands, we do not change our minds about probabilities that are conditional on $\mathcal{D}$. Thus, we design the inference method so that $q(i|\mathcal{D})$, the prior probability of i conditioned on $i \in \mathcal{D}$, is not updated. Thus, the selected conditional posterior is

$$P(i|\mathcal{D}) = q(i|\mathcal{D}) . \tag{1}$$

We adopt the following notation: priors are denoted by q, candidate posteriors by lower case p, and the selected posterior by upper case P. We shall write either $p(i)$ or p_i. Furthermore, we adopt the notation, standard in physics where the probabilities of x and θ are written $p(x)$ and $p(\theta)$ but there is no implication that p refers to the same mathematical function.

We emphasize that the point is not that we make the unwarranted assumption that keeping $q(i|\mathcal{D})$ unchanged is guaranteed to lead to correct inferences. It need not; induction is risky. The point is, rather, that in the absence of any evidence to the contrary, there is no reason to change our minds and the prior information takes priority.

The consequence of DC1 is that non-overlapping domains of i contribute additively to the entropy,
$$S(p,q) = \sum_i F(p_i, q_i), \qquad (2)$$
where F is some unknown function of two arguments. The proof is given in Appendix A.

Comment 1: It is essential that DC1 refers to *conditional* probabilities—local information about a domain $\mathcal{D}'$ can (via normalization) have a non-local effect on the probability of another domain $\mathcal{D}$.

Comment 2: An important special case is the "update" from a prior $q(i)$ to a posterior $P(i)$ in a situation in which no new information is available. The criterion DC1 applied to a situation where the subdomain $\mathcal{D}$ covers the whole space of is, $\mathcal{D} = \mathcal{X}$, requires that *in the absence of any new information, the prior conditional probabilities are not to be updated*: $P(i|\mathcal{X}) = q(i|\mathcal{X})$ or $P(i) = q(i)$.

Comment 3: The criterion DC1 implies Bayesian conditionalization as a special case. Indeed, if the information is given through the constraint $p(\tilde{\mathcal{D}}) = 0$, where $\tilde{\mathcal{D}}$ is the complement of $\mathcal{D}$, then $P(i|\mathcal{D}) = q(i|\mathcal{D})$, which is referred to as Bayesian conditionalization. More explicitly, if θ is the variable to be inferred on the basis of prior information about a likelihood function $q(i|\theta)$ and observed data i', then the update from the prior q to the posterior P, is
$$q(i,\theta) = q(i)q(\theta|i) \rightarrow P(i,\theta) = P(i)P(\theta|i), \qquad (3)$$
consists of updating $q(i) \rightarrow P(i) = \delta_{ii'}$ to agree with the new information and invoking the PMU so that $P(\theta|i') = q(\theta|i')$ remains unchanged. Therefore,
$$P(i,\theta) = \delta_{ii'} q(\theta|i') \quad \text{so that} \quad P(\theta) = q(\theta|i') = q(\theta) \frac{q(i'|\theta)}{q(i')}, \qquad (4)$$
which is Bayes' rule. Thus, *entropic inference is designed to include Bayesian inference as a special case*. Note, however, that imposing DC1 is not identical to imposing Bayesian conditionalization: DC1 is not restricted to information in the form of absolute certainties, such as $p(\mathcal{D}) = 1$.

Comment 4: If the label i is turned into a continuous variable x, the criterion DC1 requires that information that refers to points infinitely close but just outside the domain $\mathcal{D}$ will have no influence on probabilities conditional on $\mathcal{D}$. This may seem surprising, as it may lead to updated probability distributions that are discontinuous, but it is not a problem. In situations where we have explicit reasons to believe that conditions of continuity or differentiability hold, then such conditions should be imposed explicitly. The inference process should not be expected to discover and replicate information with which it was not supplied.

3.3.2. Subsystem Independence

DC2: *When two systems are a priori believed to be independent and the information we receive about one of them makes no reference to the other, then it should not matter whether the latter is included in the analysis of the former or not.*

Consider a system of propositions labeled by a composite index, $i = (i_1, i_2) \in \mathcal{X} = \mathcal{X}_1 \times \mathcal{X}_2$. For example, $\{i_1\} = \mathcal{X}_1$ and $\{i_2\} = \mathcal{X}_2$ might describe the microstates of two separate physical systems. Assume that all prior evidence led us to believe the two subsystems are independent, that is, any two propositions $i_1 \in \mathcal{X}_1$ and $i_2 \in \mathcal{X}_2$ are believed to be independent. This belief is reflected in the prior distribution: if the individual subsystem priors $q_1(i_1)$ and $q_2(i_2)$, then the prior for the whole system is $q_1(i_1)q_2(i_2)$. Next, suppose that new information is acquired such that $q_1(i_1)$ would by itself be updated to $P_1(i_1)$, and that $q_2(i_2)$ would by itself be updated to $P_2(i_2)$. DC2 requires that $S[p,q]$ be such that the joint prior $q_1(i_1)q_2(i_2)$ updates to the product $P_1(i_1)P_2(i_2)$ so that inferences about one subsystem do not affect inferences about the other.

The consequence of DC2 is to fully determine the unknown function F in (2) so that probability distributions $p(i)$ should be ranked relative to the prior $q(i)$ according to the relative entropy,

$$S[p,q] = -\sum_i p(i) \log \frac{p(i)}{q(i)}. \tag{5}$$

Comment 1: We emphasize that the point is not that when we have no evidence for correlations, we draw the firm conclusion that the systems must necessarily be independent. Induction involves risk; the systems might, in actual fact, be correlated through some unknown interaction potential. The point is rather that if the joint prior reflected independence and the new evidence is silent on the matter of correlations, then the evidence we actually have—namely, the prior—takes precedence, and there is no reason to change our minds. As before, the PMU requires that a feature of the probability distribution—in this case, independence—will not be updated unless the evidence requires it.

Comment 2: We also emphasize that DC2 *is not a consistency requirement*. The argument we deploy is *not* that both the prior *and* the new information tell us the systems are independent in which case consistency requires that it should not matter whether the systems are treated jointly or separately. DC2 refers to a situation where the new information does not say whether the systems are independent or not. Rather, the updating is being *designed*— through the PMU—so that the independence reflected in the prior is maintained in the posterior by default.

Comment 3: The generalization to continuous variables $x \in \mathcal{X}$ is approached as a Riemann limit from the discrete case. A continuous probability density $p(x)$ or $q(x)$ can be approximated by the discrete distributions. Divide the region of interest $\mathcal{X}$ into a large number N of small cells. The probabilities of each cell are as follows:

$$p_i = p(x_i)\Delta x_i \quad \text{and} \quad q_i = q(x_i)\Delta x_i, \tag{6}$$

where Δx_i is an appropriately small interval. The discrete entropy of p_i relative to q_i is as follows:

$$S_N = -\sum_{i=1}^{N} \Delta x_i\, p(x_i) \log\left[\frac{p(x_i)\Delta x_i}{q(x_i)\Delta x_i}\right], \tag{7}$$

and in the limit as $N \to \infty$ and $\Delta x_i \to 0$ we get the Riemann integral

$$S[p,q] = -\int dx\, p(x) \log\left[\frac{p(x)}{q(x)}\right]. \tag{8}$$

(To simplify the notation, we include multi-dimensional integrals by writing $d^n x = dx$.) It is easy to check that the ranking of distributions induced by $S[p,q]$ is invariant under coordinate transformations. The insight that coordinate invariance could be derived as a consequence of the requirement of subsystem independence first appeared in [5].

3.4. The ME Method

We can now summarize the overall conclusion.

The ME method: *The goal is to update from a prior distribution q to a posterior distribution when there is new information in the form of constraints $\mathcal{C}$ that specify a family $\{p\}$ of candidate posteriors. The preferred posterior P is that which maximizes the relative entropy,*

$$S[p,q] = -\sum_i p_i \log \frac{p_i}{q_i} \quad \text{or} \quad S[p,q] = -\int dx\, p(x) \log\left[\frac{p(x)}{q(x)}\right], \tag{9}$$

within the family $\{p\}$ specified by the constraints $\mathcal{C}$.

This extends the method of maximum entropy beyond its original purpose as a rule to assign probabilities from a given underlying measure (MaxEnt) to a method for updating probabilities from any arbitrary prior (ME). Furthermore, the logic behind the updating procedure does not rely on any particular meaning assigned to the entropy whether in terms of information, or heat, or disorder. Entropy is merely a tool for inductive inference. *No interpretation for $S[p,q]$ is given and none is needed.*

The derivation above has singled out *a unique $S[p,q]$ to be used in inductive inference.* Other "entropies" (such as the one-parameter families of entropies proposed in [12–14] might turn out to be useful for other purposes—perhaps as measures of some kind of "information", as measures of discrimination or distinguishability among distributions, of ecological diversity, or for some altogether different function—but they are unsatisfactory for the purpose of updating because they fail to perform the functions stipulated by the design criteria DC1 and DC2. They induce correlations that are unwarranted by the information in the priors or the constraints.

4. Bayes' Rule as a Special Case of ME

Back in Section 3.3.1, we saw that ME is designed to include Bayes' rule as a special case. Here, we wish to verify this explicitly [2]. The goal is to update our beliefs about $\theta \in \Theta$ (θ represents one or many parameters) on the basis of three pieces of information: (1) the prior information codified into a prior distribution $q(\theta)$; (2) the new information conveyed by data $x \in \mathcal{X}$ (obtained in one or many experiments); and (3) the known relation between θ and x given by a model defined by the sampling distribution or likelihood, $q(x|\theta)$. The updating will result in replacing the *prior* probability distribution $q(\theta)$ by a *posterior* distribution $P(\theta)$ that applies after the data information has been processed.

The crucial element that will allow the Bayes' rule to be smoothly integrated into the ME scheme is the realization that before the data are collected, not only do we not know θ, but we do not know x either. Thus, the relevant space for inference is not the space Θ but the product space $\Theta \times \mathcal{X}$, and the relevant joint *prior* is $q(x,\theta) = q(\theta)q(x|\theta)$. Let us emphasize two points: first, the likelihood function is an integral part of the *prior* distribution; second, the prior information about how x is related to θ is contained in the *functional form* of the distribution $q(x|\theta)$ and not in the numerical values of the arguments x and θ, which, at this point, are still unknown.

Next, data are collected and the observed values turn out to be x'. We must update to a posterior that lies within the family of distributions $p(x, \theta)$ that reflect the fact that the previously unknown x is now known to be x', that is,

$$p(x) = \int d\theta\, p(\theta, x) = \delta(x - x'). \tag{10}$$

The information in this data constrains but is not sufficient to fully determine the joint distribution,

$$p(x,\theta) = p(x)p(\theta|x) = \delta(x-x')p(\theta|x'). \tag{11}$$

Any choice of $p(\theta|x')$ is, in principle, possible. So far, the formulation of the problem parallels Section 3.3.1 exactly. We are, after all, solving the same problem. The next step is to apply the ME method.

According to the ME method, the selected joint posterior $P(x,\theta)$ is that which maximizes the entropy,

$$S[p,q] = -\int dx d\theta\, p(x,\theta) \log \frac{p(x,\theta)}{q(x,\theta)}, \tag{12}$$

subject to the data constraints. Note that Equation (10) represents an *infinite* number of constraints on the family $p(x,\theta)$: there is one constraint and one Lagrange multiplier $\lambda(x)$ for each value of x. Maximizing S, (12), subject to (10) and normalization,

$$\delta\{S + \alpha[\int dx d\theta\, p(x,\theta) - 1] + \int dx\, \lambda(x)[\int d\theta\, p(x,\theta) - \delta(x-x')]\} = 0, \tag{13}$$

yields the joint posterior

$$P(x,\theta) = q(x,\theta)\frac{e^{\lambda(x)}}{Z}, \qquad (14)$$

where Z is a normalization constant, and the multiplier $\lambda(x)$ is determined from (10) as follows:

$$\int d\theta\, q(x,\theta)\frac{e^{\lambda(x)}}{Z} = q(x)\frac{e^{\lambda(x)}}{Z} = \delta(x-x'), \qquad (15)$$

so that the joint posterior is

$$P(x,\theta) = q(x,\theta)\frac{\delta(x-x')}{q(x)} = \delta(x-x')q(\theta|x). \qquad (16)$$

The corresponding marginal posterior probability $P(\theta)$ is

$$P(\theta) = \int dx\, P(\theta,x) = q(\theta|x') = q(\theta)\frac{q(x'|\theta)}{q(x')}, \qquad (17)$$

which is Bayes' rule. Thus, Bayes' rule is derivable from, and therefore consistent with, the ME method.

To summarize, the prior $q(x,\theta) = q(x)q(\theta|x)$ is updated to the posterior $P(x,\theta) = P(x)P(\theta|x)$, where $P(x) = \delta(x-x')$ is fixed by the observed data while $P(\theta|x') = q(\theta|x')$ remains unchanged. Note that in accordance with the PMU philosophy that drives the ME method, *one only updates those aspects of one's beliefs for which corrective new evidence has been supplied*. In [2,8,42], further examples are given that show how ME allows generalizations of Bayes' rule to situations where the data itself are uncertain, there is information about moments of x or moments of θ, or even in situations where the likelihood function is unknown. In conclusion, the ME method of maximum entropy can fully reproduce and then go beyond the results obtained by the standard Bayesian methods.

5. Deviations from Maximum Entropy

The basic ME problem is to update from a prior $q(x)$ given information specified by certain constraints. The constraints specify a family of candidate distributions as follows:

$$p_\theta(x) = p(x|\theta) \qquad (18)$$

which can be conveniently labeled with a finite number of parameters θ^a, $a = 1\ldots n$. (The generalization to an infinite number of parameters poses technical but not insurmountable difficulties.) Thus, the parameters θ are coordinates on the statistical manifold specified by the constraints. The distributions in this manifold are ranked according to their entropy,

$$S[p_\theta,q] = -\int dx\, p(x|\theta)\log\frac{p(x|\theta)}{q(x)} = S(\theta), \qquad (19)$$

and the selected posterior is the distribution $p(x|\theta_0)$ that maximizes the entropy $S(\theta)$. (The notation indicates that $S[p_\theta,q]$ is a functional of p_θ while $S(\theta)$ is a function of θ.)

The question we now address concerns the extent to which $p(x|\theta_0)$ should be preferred over other distributions with lower entropy or, to put it differently, to what extent is it rational to believe that the selected value ought to be the entropy maximum θ_0 rather than any other value θ [1]? This is a question about the probability $p(\theta)$ of various values of θ. The original problem which led us to design the maximum entropy method was to assign a probability to the quantity x; we now see that the full problem is to assign probabilities to both x and θ. We are concerned not just with $p(x)$, but rather with the joint distributions which we denote as $\pi(x,\theta)$; the universe of discourse has been expanded from $\mathcal{X}$ (the space of xs) to the product space $\mathcal{X}\times\Theta$ (Θ is the space of parameters θ).

To determine the joint distribution $\pi(x,\theta)$, we make use of essentially the only (universal) method at our disposal—the ME method itself—but this requires that we address the standard two preliminary questions: First, what is the prior distribution? What do we know about x and θ before we receive information about the constraints? Second, what is the new information that constrains the allowed joint distributions $\pi(x,\theta)$?

This first question is the more subtle one: when we know absolutely nothing about the θs, we know neither their physical meaning nor whether there is any relation to the xs. A joint prior that reflects this lack of correlations is a product, $q(x,\theta) = q(x)q(\theta)$. We will assume that the prior $q(x)$ is known—it is the same prior we had used when we updated from $q(x)$ to $p(x|\theta_0)$ using (19).

However, we are not totally ignorant about the θs: we know that they label distributions $\pi(x|\theta)$ on some as yet unspecified statistical manifold Θ. Then there exists a natural measure of distance in the space Θ. It is given by the information metric $d\ell^2 = g_{ab}d\theta^a d\theta^b$ [8,43], where

$$g_{ab} = \int dx\, p(x|\theta) \frac{\partial \log p(x|\theta)}{\partial \theta^a} \frac{\partial \log p(x|\theta)}{\partial \theta^b}, \tag{20}$$

and the corresponding volume elements are given by $g^{1/2}(\theta)d^n\theta$, where $g(\theta)$ is the determinant of the metric. The uniform prior for θ, which assigns equal probabilities to equal volumes, is proportional to $g^{1/2}(\theta)$, and therefore we choose $q(\theta) = g^{1/2}(\theta)$. Therefore, the joint prior is $q(x,\theta) = q(x)g^{1/2}(\theta)$.

Next, we tackle the second question: what are the constraints on the allowed joint distributions $\pi(x,\theta)$? Consider the space of all joint distributions. To each choice of the functional form of $\pi(x|\theta)$ (for example, whether we talk about Gaussians, Boltzmann–Gibbs distributions, or something else), there corresponds a different subspace defined by distributions of the form $\pi(x,\theta) = \pi(\theta)\pi(x|\theta)$. The crucial constraint is that which specifies the subspace by imposing that $\pi(x|\theta)$ takes the particular functional form given by the constraint (18), $\pi(x|\theta) = p(x|\theta)$. This defines the meaning to the θs and also fixes the prior $g^{1/2}(\theta)$ on the relevant subspace.

The preferred joint distribution, $P(x,\theta) = P(\theta)p(x|\theta)$, is the distribution, $\pi(x,\theta) = \pi(\theta)p(x|\theta)$, that maximizes the joint entropy,

$$\begin{aligned}\mathcal{S}[\pi,q] &= -\int dx\, d\theta\, \pi(\theta)p(x|\theta) \log \frac{\pi(\theta)p(x|\theta)}{g^{1/2}(\theta)q(x)} \\ &= -\int d\theta\, \pi(\theta) \log \frac{\pi(\theta)}{g^{1/2}(\theta)} + \int d\theta\, \pi(\theta)S(\theta),\end{aligned} \tag{21}$$

where $S(\theta)$ is given in (19). Varying (21) with respect to $\pi(\theta)$ with $\int d\theta\, \pi(\theta) = 1$ and $p(x|\theta)$ fixed yields the posterior probability that the value of θ lies within the small volume $g^{1/2}(\theta)d^n\theta$,

$$P(\theta)d^n\theta = \frac{1}{\zeta} e^{S(\theta)} g^{1/2}(\theta) d^n\theta \quad \text{with} \quad \zeta = \int d^n\theta\, g^{1/2}(\theta)\, e^{S(\theta)}. \tag{22}$$

Equation (22) is the result we seek. It tells us that, as expected, the preferred value of θ is the value θ_0 that maximizes the entropy $S(\theta)$, Equation (19), because this maximizes the scalar density $\exp S(\theta)$. However, it also tells us the degree to which values of θ away from the maximum are ruled out. (Note that the density $\exp S(\theta)$ is a scalar function and the presence of the Jacobian factor $g^{1/2}(\theta)$ makes Equation (22) manifestly invariant under changes of the coordinates θ in the space Θ.)

This discussion allows us to refine our understanding of the ME method. ME is not an all-or-nothing recommendation to pick the single distribution that maximizes entropy and reject all others. The ME method is more nuanced: in principle, all distributions within the constraint manifold ought to be included in the analysis; they contribute in proportion to the exponential of their entropy and this turns out to be significant in situations where the entropy maximum is not particularly sharp.

Going back to the original problem of updating from the prior $q(x)$, given information that specifies the manifold $\{p(x|\theta)\}$, the preferred update within the family $\{p(x|\theta)\}$ is $p(x|\theta_0)$, but to the extent that other values of θ are not totally ruled out, a better update is obtained marginalizing the joint posterior $P(x,\theta) = P(\theta)p(x|\theta)$ over θ,

$$P(x) = \int d^n\theta \, P(\theta)p(x|\theta) = \int d^n\theta \, g^{1/2}(\theta) \frac{e^{S(\theta)}}{\zeta} p(x|\theta) \,. \tag{23}$$

In situations where the entropy maximum at θ_0 is very sharp, we recover the old result,

$$P(x) \approx p(x|\theta_0) \,. \tag{24}$$

When the entropy maximum is not very sharp a more honest update is Equation (23), which, incidentally, is a form of superstatistics.

One of the limitations of the standard MaxEnt method is that it selects a single "posterior" $p(x|\theta_0)$ and strictly rules out all other distributions. The result (22) overcomes this limitation and finds many applications. For example, it extends the Einstein theory of thermodynamic fluctuations beyond the regime of small fluctuations; it provides a bridge to the theory of large deviations; and, suitably adapted for Bayesian data analysis, it leads to the notion of entropic priors [44].

6. Discussion

Consistency with the law of large numbers.

Entropic methods of inference are of general applicability but there exist special situations—for example, those involving large numbers of independent subsystems—where inferences can be made by purely probabilistic methods without ever invoking the concept of entropy. In such cases, one can check (see, for example, [6,45]) that the two methods of calculation are consistent with each other. It is significant, however, that alternative entropies, such as those proposed in [12–14], do not pass this test [46,47], which rules them out as tools for updating. Some probability distributions obtained by maximizing the alternative entropies have, however, turned out to be physically relevant. It is, therefore, noteworthy that those successful distributions can also be derived through a more standard application of MaxEnt or ME, as advocated in this review [8,48–51]. In other words, what is being ruled out are not the distributions themselves, but the alternative entropies from which they were inferred.

On priors.

Choosing the prior density $q(x)$ can be tricky. Sometimes, symmetry considerations can be useful but otherwise, there is no fixed set of rules to translate information into a probability distribution except, of course, for Bayes' rule and the ME method themselves.

What if the prior $q(x)$ vanishes for some values of x? $S[p,q]$ can be infinitely negative when $q(x)$ vanishes within some region $\mathcal{D}$. This means that the ME method confers an infinite preference on those distributions $p(x)$ that vanish whenever $q(x)$ does. One must emphasize that this is as it should be. A similar situation also arises in the context of Bayes' theorem, where assigning a vanishing prior represents a tremendously serious commitment because no amount of data to the contrary would allow us to revise it. In both ME and Bayes updating, we should recognize the implications of assigning a vanishing prior. Assigning a very low but non-zero prior represents a safer and possibly less prejudiced representation of one's prior beliefs.

Commuting and non-commuting constraints.

The ME method allows one to process information in the form of constraints. When we are confronted with several constraints, we must be particularly cautious. Should they be processed simultaneously or sequentially? And, if the latter, in what order? The answer depends on the problem at hand [42].

We refer to constraints as *commuting* when it makes no difference whether they are handled simultaneously or sequentially. The most common example is that of Bayesian updating on the basis of data collected in several independent experiments. In this case, the order in which the observed data $x' = \{x'_1, x'_2, \ldots\}$ are processed does not matter for the purpose of inferring θ. In general, however, constraints need not commute and when this is the case, the order in which they are processed is critical.

To decide whether constraints are to be handled sequentially or simultaneously, one must be clear about how the ME method handles constraints. The ME machinery interprets a constraint in a very mechanical way: all distributions satisfying the constraint are, in principle, allowed, while all distributions violating it are ruled out. Therefore, sequential updating is appropriate when old constraints become obsolete and are superseded by new information, while simultaneous updating is appropriate when old constraints remain valid. The two cases refer to different states of information, and therefore, it is to be expected that they will result in different inferences. These comments are meant to underscore the importance of understanding what information is and how it is processed by the ME method; failure to do so will lead to errors that do not reflect a shortcoming of the ME method but rather a misapplication of it.

Pitfalls?

Entropy is a tool for reasoning and—as with all tools for reasoning or otherwise—it can be misused, leading to unsatisfactory results [52]. Should that happen, the inevitable questions are "what went wrong?" and "how do we fix it?" It helps to first ask what components of the analysis can be trusted so that the possible mistakes can be looked for elsewhere. The answers proposed by the ME method are radically conservative: problems always arise through a wrong choices of variables, priors, or constraints. Indeed, one should not blame the entropic method for not having discovered and taken into account relevant information that was not explicitly introduced into the analysis. Indeed, just as one would be very reticent about questioning the basic rules of arithmetic, or the basic rules of calculus, one should not question the basic sum and product rules of the probability calculus and, taking this one step farther, one should not question the applicability of entropy as the updating tool. The adoption of this conservative approach leads us to reject alternative entropies and quantum probabilities. Fortunately, those constructs are not actually needed—as mentioned above, those Tsallis distributions that have turned out be useful can be derived with standard entropic methods [8,48–51], and quantum mechanics can be handled within standard probability theory without invoking exotic probabilities [39,53].

Funding: This research received no external funding.

Institutional Review Board Statement: Not applicable.

Informed Consent Statement: Not applicable.

Data Availability Statement: Not applicable.

Acknowledgments: I would like to acknowledge many valuable discussions on probability and entropy with N. Caticha, A. Giffin, K. Knuth, R. Preuss, C. Rodríguez, J. Skilling, and K. Vanslette.

Conflicts of Interest: The author declares no conflict of interest.

Appendix A. DC1—Mutually Exclusive Subdomains

In these appendices, we establish the consequences of the two criteria DC1 and DC2, leading to the final result: Equation (9). The details of the proofs are important not just because they lead to our final conclusions, but also because the translation of the verbal statement of the criteria into precise mathematical form is a crucial part of unambiguously specifying what the criteria actually say.

First, we prove that criterion DC1 leads to the expression Equation (2) for $S[p,q]$. Consider the case of a discrete variable, p_i with $i = 1 \ldots n$, so that $S[p,q] = S(p_1 \ldots p_n, q_1 \ldots q_n)$. Suppose the space of states $\mathcal{X}$ is partitioned into two non-overlapping domains $\mathcal{D}$ and $\tilde{\mathcal{D}}$ with $\mathcal{D} \cup \tilde{\mathcal{D}} = \mathcal{X}$, and that the information to be processed is in the form of a constraint that refers to the domain $\tilde{\mathcal{D}}$,

$$\sum_{j \in \tilde{\mathcal{D}}} a_j p_j = A . \tag{A1}$$

DC1 states that the constraint on $\tilde{\mathcal{D}}$ does not have an influence on the *conditional* probabilities $p_{i|\mathcal{D}}$. It may, however, influence the probabilities p_i within $\mathcal{D}$ through an overall multiplicative factor. To deal with this complication, consider then a special case where the overall probabilities of $\mathcal{D}$ and $\tilde{\mathcal{D}}$ are also constrained:

$$\sum_{i \in \mathcal{D}} p_i = P_{\mathcal{D}} \quad \text{and} \quad \sum_{j \in \tilde{\mathcal{D}}} p_j = P_{\tilde{\mathcal{D}}}, \tag{A2}$$

with $P_{\mathcal{D}} + P_{\tilde{\mathcal{D}}} = 1$. Under these special circumstances, constraints on $\tilde{\mathcal{D}}$ will not influence p_is within $\mathcal{D}$, and vice versa.

To obtain the posterior, maximize $S[p,q]$ subject to these three constraints,

$$0 = \left[\delta S - \lambda \left(\sum_{i \in \mathcal{D}} p_i - P_{\mathcal{D}} \right) + \right. \\ \left. - \tilde{\lambda} \left(\sum_{j \in \tilde{\mathcal{D}}} p_j - P_{\tilde{\mathcal{D}}} \right) + \mu \left(\sum_{j \in \tilde{\mathcal{D}}} a_j p_j - A \right) \right],$$

leading to

$$\frac{\partial S}{\partial p_i} = \lambda \quad \text{for} \quad i \in \mathcal{D}, \tag{A3}$$

$$\frac{\partial S}{\partial p_j} = \tilde{\lambda} + \mu a_j \quad \text{for} \quad j \in \tilde{\mathcal{D}} . \tag{A4}$$

Equations (A1)–(A4) are $n+3$ equations; we must solve for the p_is and the three Lagrange multipliers, λ, $\tilde{\lambda}$, and μ. Since $S = S(p_1 \ldots p_n, q_1 \ldots q_n)$ its derivative

$$\frac{\partial S}{\partial p_i} = f_i(p_1 \ldots p_n, q_1 \ldots q_n)$$

could, in principle, also depend on all $2n$ variables. However, this violates the DC1 criterion because any arbitrary change in a_j within $\tilde{\mathcal{D}}$ would influence the p_is within $\mathcal{D}$. The only way that probabilities conditioned on $\mathcal{D}$ can be shielded from arbitrary changes in the constraints pertaining to $\tilde{\mathcal{D}}$ is that for any $i \in \mathcal{D}$, the function f_i depends only on p_js with $j \in \mathcal{D}$. Furthermore, this must hold not just for one particular partition of $\mathcal{X}$ into domains $\mathcal{D}$ and $\tilde{\mathcal{D}}$, but it must hold for *all conceivable partitions*, including the partition into atomic propositions. Therefore, f_i can depend only on p_i,

$$\frac{\partial S}{\partial p_i} = f_i(p_i, q_1 \ldots q_n) . \tag{A5}$$

The power of the criterion DC1 is not exhausted yet. The information that affects the posterior can enter not just through constraints, but also through the prior. Suppose that the local information about domain $\tilde{\mathcal{D}}$ is altered by changing the prior within $\tilde{\mathcal{D}}$. Let $q_j \to q_j + \delta q_j$ for $j \in \tilde{\mathcal{D}}$. Then (A5) becomes

$$\frac{\partial S}{\partial p_i} = f_i(p_i, q_1 \ldots q_j + \delta q_j \ldots q_n),$$

which shows that p_i with $i \in \mathcal{D}$ will be influenced by information about $\tilde{\mathcal{D}}$ unless f_i with $i \in \mathcal{D}$ is independent of all the q_js for $j \in \tilde{\mathcal{D}}$. Again, this must hold for all possible partitions into $\mathcal{D}$ and $\tilde{\mathcal{D}}$, and therefore,

$$\frac{\partial S}{\partial p_i} = f_i(p_i, q_i) \quad \text{for all} \quad i \in \mathcal{X}.$$

The choice of the functions $f_i(p_i, q_i)$ can be restricted further. If we maximize $S[p, q]$, subject to constraints

$$\sum_i p_i = 1 \quad \text{and} \quad \sum_i a_i p_i = A$$

we obtain

$$\frac{\partial S}{\partial p_i} = f_i(p_i, q_i) = \lambda + \mu a_i \quad \text{for all} \quad i \in \mathcal{X},$$

where λ and μ are Lagrange multipliers. Solving for p_i gives a posterior,

$$P_i = g_i(q_i, \lambda, \mu, a_i)$$

for some functions g_i. As stated in Section 3.3 we do not assume that the labels i themselves carry any particular significance. This means, in particular, that for any proposition labeled i, we want the selected posterior P_i to depend only on the numbers q_i, λ, μ, and a_i. We do not want to have different updating rules for different propositions: two different propositions i and i' with the same $q_i = q_{i'}$ and the same $a_i = a_{i'}$ should be updated to the same posteriors, $P_i = P_{i'}$. In other words, the functions g_i and f_i must be independent of i. Therefore,

$$\frac{\partial S}{\partial p_i} = f(p_i, q_i) \quad \text{for all} \quad i \in \mathcal{X}. \tag{A6}$$

Integrating, one obtains

$$S[p, q] = \sum_i F(p_i, q_i) + \text{constant}.$$

for some still undetermined function F. The constant has no effect on the entropy maximization and can be dropped.

The corresponding expression for a continuous variable x is obtained replacing i by x, and the sum over i by an integral over x leading to Equation (2),

$$S[p, q] = \int dx \, F(p(x), q(x)).$$

Appendix B. DC2—Independent Subsystems

Here, we show that DC2 leads to Equation (9). Let the microstates of a composite system be labeled by $(i_1, i_2) \in \mathcal{X} = \mathcal{X}_1 \times \mathcal{X}_2$. We shall consider two special cases.

Case (a)

First, we treat the two subsystems separately. Suppose that for subsystem 1, we have the extremely constraining information that updates $q_1(i_1)$ to be $P_1(i_1)$, and for subsystem 2 we have no new information at all. For subsystem 1, we maximize $S_1[p_1, q_1]$ subject to the constraint $p_1(i_1) = P_1(i_1)$ and the selected posterior is, of course, $p_1(i_1) = P_1(i_1)$. For subsystem 2, we maximize $S_2[p_2, q_2]$ subject only to normalization and there is no update, $P_2(i_2) = q_2(i_2)$.

When the systems are treated jointly, however, the inference is not nearly as trivial. We want to maximize the entropy of the joint system,

$$S[p, q] = \sum_{i_1, i_2} F(p(i_1, i_2), q_1(i_1) q_2(i_2)),$$

subject to the constraint on subsystem 1,

$$\sum_{i_1} p(i_1, i_2) = P_1(i_1) .$$

Notice that this is not just one constraint: we have one constraint for each value of i_1, and each constraint must be supplied with its own Lagrange multiplier, $\lambda_1(i_1)$. Then,

$$\delta\left[S - \sum_{i_1} \lambda_1(i_1)\left(\sum_{i_2} p(i_1, i_2) - P_1(i_1)\right)\right] = 0 .$$

The independent variations $\delta p(i_1, i_2)$ yield the following:

$$f(p(i_1, i_2), q_1(i_1)q_2(i_2)) = \lambda_1(i_1) ,$$

where f is given in (A6),

$$\frac{\partial S}{\partial p} = \frac{\partial}{\partial p} F(p, q_1 q_2) = f(p, q_1 q_2) .$$

Next, we impose that the selected posterior is the product $P_1(i_1)q_2(i_2)$. The function f must be such that

$$f(P_1 q_2, q_1 q_2) = \lambda_1 .$$

Since the RHS is independent of the argument i_2, the f function must be such that the i_2-dependence cancels out, and this cancellation must occur for all values of i_2 and all choices of the prior q_2. Therefore, we impose that for any value of x the function $f(p, q)$ must satisfy

$$f(px, qx) = f(p, q) .$$

Choosing $x = 1/q$, we obtain

$$f\left(\frac{p}{q}, 1\right) = f(p, q) \quad \text{or} \quad \frac{\partial F}{\partial p} = f(p, q) = \phi\left(\frac{p}{q}\right). \tag{A7}$$

Thus, the function $f(p, q)$ has been reduced to a function $\phi(p/q)$ of a single argument.

Case (b)

Next, we consider a situation in which both subsystems are updated by extremely constraining information: when the subsystems are treated separately, $q_1(i_1)$ is updated to $P_1(i_1)$ and $q_2(i_2)$ is updated to $P_2(i_2)$. When the systems are treated jointly, we require that the joint prior for the combined system $q_1(i_1)q_2(i_2)$ be updated to $P_1(i_1)P_2(i_2)$.

First we treat the subsystems separately. Maximize the entropy of subsystem 1,

$$S[p_1, q_1] = \sum_{i_1} F(p_1(i_1), q_1(i_1)) \quad \text{subject to} \quad p_1(i_1) = P_1(i_1) .$$

To each constraint—one constraint for each value of i_1—we must supply one Lagrange multiplier, $\lambda_1(i_1)$. Then, we obtain

$$\delta\left[S - \sum_{i_1} \lambda_1(i_1)(p(i_1) - P_1(i_1))\right] = 0 .$$

Using Equation (A7),

$$\frac{\partial S}{\partial p_1} = \frac{\partial}{\partial p_1} F(p_1, q_1) = \phi\left(\frac{p_1}{q_1}\right),$$

and, imposing that the selected posterior be $P_1(i_1)$, we find that the function ϕ must obey

$$\phi\left(\frac{P_1(i_1)}{q_1(i_1)}\right) = \lambda_1(i_1) . \tag{A8}$$

Similarly, for system 2 we find the following:

$$\phi\left(\frac{P_2(i_2)}{q_2(i_2)}\right) = \lambda_2(i_2). \tag{A9}$$

Next, we treat the two subsystems jointly. Maximize the entropy of the joint system as follows:

$$S[p,q] = \sum_{i_1,i_2} F\left(p(i_1,i_2), q_1(i_1)q_2(i_2)\right),$$

subject to the following constraints on the joint distribution $p(i_1, i_2)$:

$$\sum_{i_2} p(i_1, i_2) = P_1(i_1) \quad \text{and} \quad \sum_{i_1} p(i_1, i_2) = P_2(i_2).$$

Again, there is one constraint for each value of i_1 and of i_2 and we introduce Lagrange multipliers, $\eta_1(i_1)$ or $\eta_2(i_2)$. Then,

$$\delta\left[S - \sum_{i_1}\eta_1(i_1)\left(\sum_{i_2} p(i_1,i_2) - P_1(i_1)\right) - \{1 \leftrightarrow 2\}\right] = 0,$$

where $\{1 \leftrightarrow 2\}$ indicates a third term, similar to the second, with 1 and 2 interchanged. The independent variations $\delta p(i_1, i_2)$ yield

$$\phi\left(\frac{p(i_1, i_2)}{q_1(i_1)q_2(i_2)}\right) = \eta_1(i_1) + \eta_2(i_2),$$

and we impose that the selected posterior be the product $P_1(i_1)P_2(i_2)$. Therefore, the function ϕ must be such that

$$\phi\left(\frac{P_1 P_2}{q_1 q_2}\right) = \eta_1 + \eta_2.$$

To solve this equation, we take the exponential of both sides, let $\xi = \exp\phi$, and rewrite as

$$\xi\left(\frac{P_1 P_2}{q_1 q_2}\right) e^{-\eta_2(i_2)} = e^{\eta_1(i_1)}. \tag{A10}$$

This shows that for any value of i_1, the dependences of the LHS on i_2 through P_2/q_2 and η_2 must cancel each other out. In particular, if for some subset of i_2s, the subsystem 2 is updated so that $P_2 = q_2$, which amounts to no update at all, the i_2 dependence on the left is eliminated but the i_1 dependence remains unaffected,

$$\xi\left(\frac{P_1}{q_1}\right) e^{-\eta_2'} = e^{\eta_1(i_1)}.$$

where η_2' is some constant independent of i_2. A similar argument with $\{1 \leftrightarrow 2\}$ yields

$$\xi\left(\frac{P_2}{q_2}\right) e^{-\eta_1'} = e^{\eta_2(i_2)},$$

where η_1' is a constant. Taking the exponential of (A8) and (A9) leads to the following:

$$\xi\left(\frac{P_1}{q_1}\right) e^{-\eta_2'} = e^{\lambda_1 - \eta_2'} = e^{\eta_1} \quad \text{and} \quad \xi\left(\frac{P_2}{q_2}\right) e^{-\eta_1'} = e^{\lambda_2 - \eta_1'} = e^{\eta_2}.$$

Substituting back into (A10), we obtain

$$\xi\left(\frac{P_1 P_2}{q_1 q_2}\right) = \xi\left(\frac{P_1}{q_1}\right) \xi\left(\frac{P_2}{q_2}\right),$$

where a constant factor $e^{-(\eta_1' + \eta_2')}$ is absorbed into a new function ξ. The general solution of this functional equation is a power,

$$\xi(xy) = \xi(x)\xi(y) \implies \xi(x) = x^a ,$$

so that

$$\phi(x) = a \log x + b ,$$

where a and b are constants. Finally, integrate (A7),

$$\frac{\partial F}{\partial p} = \phi\left(\frac{p}{q}\right) = a \log \frac{p}{q} + b ,$$

to obtain

$$F[p,q] = ap \log \frac{p}{q} + b'p + c$$

where b' and c are constants.

At this point, the entropy takes the general form

$$S[p,q] = \sum_i \left(ap_i \log \frac{p_i}{q_i} + b'p_i + c \right).$$

The additive constant c may be dropped: it contributes a term that does not depend on the probabilities and has no effect on the ranking scheme. Furthermore, since $S[p,q]$ will be maximized subject to constraints that include normalization, the b' term has no effect on the selected distribution and can also be dropped. Finally, the multiplicative constant a has no effect on the overall ranking, except in the trivial sense that inverting the sign of a will transform the maximization problem to a minimization problem or vice versa. We can, therefore, set $a = -1$ so that maximum S corresponds to maximum preference, which gives us Equation (9) and concludes our derivation.

References and Notes

1. Caticha, A. Relative Entropy and Inductive Inference. *AIP Conf. Proc.* **2004**, *707*, 75.
2. Caticha, A.; Giffin, A. Updating Probabilities. *AIP Conf. Proc.* **2006**, *872*, 31.
3. Caticha, A. Information and Entropy. *AIP Conf. Proc.* **2007**, *954*, 11.
4. Caticha, A. Towards an Informational Pragmatic Realism. *Mind Mach.* **2014**, *24*, 37. [CrossRef]
5. Vanslette, K. Entropic Updating of Probabilities and Density Matrices. *Entropy* **2017**, *19*, 664. [CrossRef]
6. Caticha, A. *Lectures on Probability, Entropy, and Statistical Physics*; MaxEnt: São Paulo, Brazil, 2008.
7. Caticha, A. *Entropic Inference and the Foundations of Physics*; EBEB: São Paulo, Brazil, 2012. Available online: https://www.albany.edu/physics/faculty/ariel-caticha (accessed on 12 July 2021).
8. Caticha, A. Entropic Physics: Probability, Entropy, and the Foundations of Physics. 2021. Available online: https://www.albany.edu/physics/faculty/ariel-caticha (accessed on 12 July 2021).
9. Jaynes, E.T. Information Theory and Statistical Mechanics, I and II. *Phys. Rev.* **1957**, *106*, 620. [CrossRef]
10. Rosenkrantz, R.D. (Ed.) *E. T. Jaynes: Papers on Probability, Statistics and Statistical Physics*; Reidel: Dordrecht, The Netherlands, 1983.
11. Uffink, J. Can the Maximum Entropy Principle be explained as a consistency requirement? *Stud. Hist. Philos. Mod. Phys.* **1995**, *26*, 223. [CrossRef]
12. Renyi, A. On measures of entropy and information. In Proceedings of the 4th Berkeley Symposium on Mathematical Statistics and Probability, Berkeley, CA, USA, 20 June–30 July 1961; p. 547.
13. Aczél, J.; Daróczy, Z. *On Measures of Information and Their Characterizations*; Academic Press: New York, NY, USA, 1975.
14. Tsallis, C. Possible Generalization of Boltzmann-Gibbs Statistics. *J. Stat. Phys.* **1988**, *52*, 479. [CrossRef]
15. Shore, J.E.; Johnson, R.W. Axiomatic derivation of the Principle of Maximum Entropy and the Principle of Minimum Cross-Entropy. *IEEE Trans. Inf. Theory* **1980**, *26*, 26–37. [CrossRef]
16. Skilling, J. The Axioms of Maximum Entropy. In *Maximum-Entropy and Bayesian Methods in Science and Engineering*; Erickson, G.J., Smith, C.R., Eds.; Kluwer: Dordrecht, The Netherlands, 1988; pp. 173–187.
17. Skilling, J. Classic Maximum Entropy. In *Maximum Entropy and Bayesian Methods*; Skilling, J., Ed.; Kluwer: Dordrecht, The Netherlands, 1989; pp. 45–52.

18. Karbelkar, S.N. On the axiomatic approach to the maximum entropy principle of inference. *Pramana J. Phys.* **1986**, *26*, 301–310. [CrossRef]
19. Tsallis, C. Conceptual Inadequacy of the Shore and Johnson Axioms for Wide Classes of Complex systems. *Entropy* **2015**, *17*, 2853. [CrossRef]
20. Jizba, P.; Korbel, J. Maximum Entropy Principle in Statistical Inference: Case for Non-Shannonian Entropies. *Phys. Rev. Lett.* **2019**, *122*, 120601. [CrossRef]
21. Pressé, S.; Ghosh, K.; Lee, J.; Dill, K.A. Nonadditive Entropies Yield Probability Distributions with Biases not Warranted by the Data. *Phys. Rev. Lett.* **2013**, *111*, 180604. [CrossRef]
22. Pressé, S.; Ghosh, K.; Lee, J.; Dill, K.A. Reply to Tsallis' Conceptual inadequacy of the Shore and Johnson axioms for wide classes of complex systems. *Entropy* **2015**, *17*, 5043. [CrossRef]
23. Williams, P.M. Bayesian Conditionalization and the Principle of Minimum Relative Information. *Br. J. Philos. Sci.* **1980**, *31*, 131. [CrossRef]
24. Johnson, D.T.; Caticha, A. Entropic dynamics and the quantum measurement problem. *AIP Conf. Proc.* **2012**, *1443*, 104.
25. Vanslette, K.; Caticha, A. Quantum measurement and weak values in entropic quantum dynamics. *AIP Conf. Proc.* **2017**, *1853*, 090003.
26. Cover, T.; Thomas, J. *Elements of Information Theory*; Wiley: New York, NY, USA, 1991.
27. Golan, A. *Foundations of Info-Metrics: Modeling, Inference, and Imperfect Information*; Oxford University Press: New York, NY, USA, 2018.
28. Rissanen, J. Modeling by shortest data description. *Automatica* **1978**, *14*, 465–471. [CrossRef]
29. Landauer, R. Information is Physical. *Phys. Today* **1991**, *44*, 23–29. [CrossRef]
30. Bennett, C. The thermodynamics of computation—A review. *Int. J. Theor. Phys.* **1982**, *21*, 905–940. [CrossRef]
31. Bennett, C. Notes on Landauer's principle, reversible computation, and Maxwell's demon. *Stud. Hist. Philos. Mod. Phys.* **2003**, *34*, 501–510. [CrossRef]
32. Norton, J.D. Waiting for Landauer. *Stud. Hist. Philos. Mod. Phys.* **2011**, *36*, 184–198. [CrossRef]
33. Norton, J.D. The End of the Thermodynamics of Computation: A No-Go Result. *Philos. Sci.* **2013**, *80*, 1182–1192. [CrossRef]
34. Binmore, K. On the foundations of decision theory. *Homo Oecon.* **2017**, *34*, 259. [CrossRef]
35. Harre, M.S. Information Theory for Agents in Artificial Intelligence, Psychology, and Economics. *Entropy* **2021**, *23*, 310. [CrossRef] [PubMed]
36. Evans, B.P.; Prokopenko, M. A Maximum Entropy Model of Bounded Rational Decision-Making with Prior Beliefs and Market Feedback. *Entropy* **2021**, *23*, 669. [CrossRef]
37. Caticha, A.; Golan, A. An Entropic framework for Modeling Economies. *Physica A* **2014**, *408*, 149. [CrossRef]
38. Pessoa, P.; Costa, F.X.; Caticha, A. Entropic dynamics on Gibbs statistical manifolds. *Entropy* **2021**, *23*, 494. [CrossRef]
39. Caticha, A. The Entropic Dynamics approach to Quantum Mechanics. *Entropy* **2019**, *21*, 943. [CrossRef]
40. Knuth, K.H. Lattice duality: The origin of probability and entropy. *Neurocomputing* **2005**, *67*, 245–274. [CrossRef]
41. Knuth, K.H.; Skilling, J. Foundations of Inference. *Axioms* **2012**, *1*, 38–73. [CrossRef]
42. Giffin, A.; Caticha, A. Updating Probabilities with Data and Moments. *AIP Conf. Proc.* **2007**, *954*, 74.
43. Amari, S. *Differential-Geometrical Methods in Statistics*; Springer: Berlin, Germnay, 1985.
44. Caticha, A.; Preuss, R. Maximum entropy and Bayesian data analysis: Entropic prior distributions. *Phys. Rev. E* **2004**, *70*, 046127. [CrossRef]
45. Grendar, M., Jr.; Grendar, M. Maximum Probability and Maximum Entropy Methods: Bayesian interpretation. *AIP Conf. Proc.* **2004**, *707*, 490.
46. La Cour, B.R.; Schieve, W.C. Tsallis maximum entropy principle and the law of large numbers. *Phys. Rev. E* **2000**, *62*, 7494. [CrossRef] [PubMed]
47. Nauenberg, M. Critique of q-entropy for thermal statistics. *Phys. Rev. E* **2003**, *67*, 036114. [CrossRef] [PubMed]
48. Plastino, A.R.; Plastino, A. From Gibbs microcanonical ensemble to Tsallis generalized canonical distribution. *Phys. Lett. A* **1994**, *193*, 140–143. [CrossRef]
49. Beck, C. Dynamical Foundations of nonextensive Statistical Mechanics. *Phys. Rev. Lett.* **2001**, *87*, 180601. [CrossRef]
50. Beck, C.; Cohen, E.G.D. Superstatistics. *Physica A* **2003**, *322*, 267–275. [CrossRef]
51. Crooks, G.E. Beyond Boltzmann-Gibbs statistics: Maximum entropy hyperensembles out of equilibrium. *Phys. Rev. E* **2007**, *75*, 041119. [CrossRef]
52. Caticha, A. Entropic inference: Some pitfalls and paradoxes we can avoid. *AIP Conf. Proc.* **2013**, *1553*, 176.
53. Bohm, D.; Hiley, B.J. *The Undivided Universe: An Ontological Interpretation on Quantum Theory*; Routledge: New York, NY, USA, 1993.

Article

On the α-q-Mutual Information and the α-q-Capacities

Velimir M. Ilić [1],* and Ivan B. Djordjević [2]

[1] Mathematical Institute of the Serbian Academy of Sciences and Arts, Kneza Mihaila 36, 11000 Beograd, Serbia
[2] Department of Electrical and Computer Engineering, University of Arizona, 1230 E. Speedway Blvd., Tucson, AZ 85721, USA; ivan@email.arizona.edu
* Correspondence: velimir.ilic@gmail.com

Citation: Ilić, V.M.; Djordjević, I.B. On the α-q-Mutual Information and the α-q-Capacities. *Entropy* **2021**, *23*, 702. https://doi.org/10.3390/e23060702

Academic Editors: Petr Jizba and Jan Korbel

Received: 12 February 2021
Accepted: 26 May 2021
Published: 1 June 2021

Publisher's Note: MDPI stays neutral with regard to jurisdictional claims in published maps and institutional affiliations.

Copyright: © 2021 by the authors. Licensee MDPI, Basel, Switzerland. This article is an open access article distributed under the terms and conditions of the Creative Commons Attribution (CC BY) license (https://creativecommons.org/licenses/by/4.0/).

Abstract: The measures of information transfer which correspond to non-additive entropies have intensively been studied in previous decades. The majority of the work includes the ones belonging to the Sharma–Mittal entropy class, such as the Rényi, the Tsallis, the Landsberg–Vedral and the Gaussian entropies. All of the considerations follow the same approach, mimicking some of the various and mutually equivalent definitions of Shannon information measures, and the information transfer is quantified by an appropriately defined measure of mutual information, while the maximal information transfer is considered as a generalized channel capacity. However, all of the previous approaches fail to satisfy at least one of the ineluctable properties which a measure of (maximal) information transfer should satisfy, leading to counterintuitive conclusions and predicting nonphysical behavior even in the case of very simple communication channels. This paper fills the gap by proposing two parameter measures named the α-q-mutual information and the α-q-capacity. In addition to standard Shannon approaches, special cases of these measures include the α-mutual information and the α-capacity, which are well established in the information theory literature as measures of additive Rényi information transfer, while the cases of the Tsallis, the Landsberg–Vedral and the Gaussian entropies can also be accessed by special choices of the parameters α and q. It is shown that, unlike the previous definition, the α-q-mutual information and the α-q-capacity satisfy the set of properties, which are stated as axioms, by which they reduce to zero in the case of totally destructive channels and to the (maximal) input Sharma–Mittal entropy in the case of perfect transmission, which is consistent with the maximum likelihood detection error. In addition, they are non-negative and less than or equal to the input and the output Sharma–Mittal entropies, in general. Thus, unlike the previous approaches, the proposed (maximal) information transfer measures do not manifest nonphysical behaviors such as sub-capacitance or super-capacitance, which could qualify them as appropriate measures of the Sharma–Mittal information transfer.

Keywords: rényi entropy; tsallis entropy; landsberg—vedral entropy; gaussian entropy; sharma—mittal entropy; α-mutual information; α-channel capacity

1. Introduction

In the past, extensive work has been written on defining the information measures which generalize the Shannon entropy [1], such as the one-parameter Rényi entropy [2], the Tsallis entropy [3], the Landsberg–Vedral entropy [4], the Gaussian entropy [5], and the two-parameter Sharma–Mittal entropy [5,6], which reduces to former ones for special choices of the parameters. The Sharma–Mittal entropy can axiomatically be founded as the unique q-additive measure [7,8] which satisfies generalized Shannon–Kihinchin axioms [9,10] and which has widely been explored in different research fields starting from statistics [11] and thermodynamics [12,13] to quantum mechanics [14,15], machine learning [16,17] and cosmology [18,19]. The Sharma–Mittal entropy has also been recognized in the field of information theory, where the measures of conditional Sharma–Mittal entropy [20], Sharma–Mittal divergences [21] and Sharma–Mittal entropy rate [22] have been established and analyzed.

Considerable research has also been done in the field of communication theory in order to analyze information transmission in the presence of noise if, instead of Shannon's entropy, the information is quantified with (instances of) Sharma–Mittal entropy and, in general, the information transfer is quantified by an appropriately defined measure of mutual information, while the maximal information transfer is considered as a generalized channel capacity. Thus, after Rényi's proposal for the additive generalization of Shannon entropy [2], several different definitions for Rényi information transfer were proposed by Sibson [23], Arimoto [24], Augustin [25], Csiszar [26], Lapidoth and Pfister [27] and Tomamichel and Hayashi [28]. These measures have been explored thoroughly and their operational characterization in coding theory, hypothesis testing, cryptography and quantum information theory was established, which qualifies them as a reasonable measure of Rényi information transfer [29]. Similar attempts have also been made in the case of non-additive entropies. Thus, starting from the work of Daroczy [30], who introduced a measure for generalized information transfer related to the Tsallis entropy, several attempts followed for the measures which correspond to non-additive particular instances of the Sharma–Mittal entropy, so the definitions for the Rényi information transfer were considered in [24,31], for the Tsallis information transfer in [32] and for the Landsber–Vedral information transfer in [4,33].

In this paper we provide a general treatment of the Sharma–Mittal entropy transfer and a detailed analysis of existing measures, showing that all of the definitions related to non-additive entropies fail to satisfy at least one of the ineluctable properties common to the Shannon case, which we state as axioms, by which the information transfer has to be non-negative, less than the input and output uncertainty, equal to the input uncertainty in the case of perfect transmission and equal to zero, in the case of a totally destructive channel. Thus, breaking some of these axioms implies unexpected and counterintuitive conclusions about the channels, such as achieving super-capacitance or sub-capacitance [4], which could be treated as nonphysical behavior. As an alternative, we propose the α-q-mutual information as a measure of Sharma–Mittal information transfer, maximized with the α-q-capacity. The α-q mutual information generalizes the α-mutual information by Arimoto [24], which is defined as a q-difference between the input Sharma–Mittal entropy and the appropriately defined conditional Sharma–Mittal entropy if the output is given, while the α-q-capacity represents a generalization of Arimoto's α-capacity in the case of $q = 1$. In addition, several other instances can be obtained by specifying the values of parameters α and q, which includes the information transfer measures for the Tsallis, the Landsber–Vedral and the Shannon entropy, as well as the case of the Gaussian entropy which was not considered before in the context of information transmission.

The paper is organized as follows. The basic properties and special instances of the Sharma–Mittal entropy are listed in Section 2. Section 3 reviews the basics of communication theory, introduces the basic communication channels and establishes the set of axioms which information transfer measures should satisfy. The information transfer measures which are defined by Arimoto are introduced in Section 4, and the alternative definitions for Rényi information transfer measures are discussed in Section 5. Finally, the α-q-mutual information and the α-q-capacities are proposed and their properties analyzed in Section 6 while the previously proposed measures of Sharma–Mittal entropy transfer are discussed in Section 7.

2. Sharma–Mittal Entropy

Let the sets of positive and nonnegative real numbers be denoted with $\mathbb{R}^+$ and $\mathbb{R}_0^+$, respectively, and let the mapping $\eta_q : \mathbb{R} \to \mathbb{R}$ be defined in

$$\eta_q(x) = \begin{cases} x, & \text{for } q = 1 \\ \dfrac{2^{(1-q)x} - 1}{(1-q)\ln 2}, & \text{for } q \neq 1 \end{cases} \qquad (1)$$

so that its inverse is given in

$$\eta_q^{-1}(x) = \begin{cases} x, & \text{for } q = 1 \\ \dfrac{1}{1-q} \log((1-q)x \ln 2 + 1), & \text{for } q \neq 1 \end{cases}. \quad (2)$$

The mapping η_q and its inverse are increasing continuous (hence invertible) functions such that $\eta(0) = 0$. The q-logarithm is defined in

$$\text{Log}_q(x) = \eta_q(\log x) = \begin{cases} \log x, & \text{for } q = 1 \\ \dfrac{x^{(1-q)} - 1}{(1-q) \ln 2}, & \text{for } q \neq 1 \end{cases}, \quad (3)$$

and its inverse, the q-exponential, is defined in

$$\text{Exp}_q(y) = \begin{cases} 2^y, & \text{for } q = 1 \\ (1 + (1-q)y \ln 2)^{\frac{1}{1-q}}, & \text{for } q \neq 1 \end{cases}, \quad (4)$$

for $1 + (1-q)y \ln 2 > 0$. Using η_q, we can define the pseudo-addition operation $\oplus_q$ [7,8]

$$x \oplus_q y = \eta_q\left(\eta_q^{-1}(x) + \eta_q^{-1}(y)\right) = x + y + (1-q)xy; \quad x, y \in \mathbb{R}, \quad (5)$$

and its inverse operation, the pseudo substraction

$$x \ominus_q y = \eta_q\left(\eta_q^{-1}(x) - \eta_q^{-1}(y)\right) = \frac{x - y}{1 + (1-q)y \ln 2}; \quad x, y \in \mathbb{R}. \quad (6)$$

The $\oplus_q$ can be rewritten in terms of the generalized logarithm by settings $x = \log u$ and $y = \log v$ so that

$$\text{Log}_q(u \cdot v) = \text{Log}_q(u) \oplus_q \text{Log}_q(v); \quad u, v \in \mathbb{R}_+. \quad (7)$$

Let the set of all n-dimensional distributions be denoted with

$$\Delta_n \equiv \left\{ (p_1, \ldots, p_n) \mid p_i \geq 0, \sum_{i=1}^n p_i = 1 \right\}; \quad n > 1. \quad (8)$$

Let the function $H_n : \Delta_n \to \mathbb{R}_0^+$ satisfy the following the Shannon–Khinchin axioms, for all $n \in \mathbb{N}, n > 1$.

GSK1 H_n is continuous in Δ_n;
GSK2 H_n takes its largest value for the uniform distribution, $U_n = (1/n, \ldots, 1/n) \in \Delta_n$, i.e., $H_n(P) \leq H_n(U_n)$, for any $P \in \Delta_n$;
GSK3 H_n is expandable: $H_{n+1}(p_1, p_2, \ldots, p_n, 0) = H_n(p_1, p_2, \ldots, p_n)$ for all $(p_1, \ldots, p_n) \in \Delta_n$;
GSK4 Let $P = (p_1, \ldots, p_n) \in \Delta_n$, $PQ = (r_{11}, r_{12}, \ldots, r_{nm}) \in \Delta_{nm}$, $n, m \in \mathbb{N}, n, m > 1$ such that $p_i = \sum_{j=1}^m r_{ij}$, and $Q_{|k} = (q_{1|k}, \ldots, q_{m|k}) \in \Delta_m$, where $q_{i|k} = r_{ik}/p_k$ and $\alpha \in \mathbb{R}_0^+$ are some fixed parameters. Then,

$$H_{nm}(PQ) = H_n(P) \oplus_q H_m(Q|P), \quad \text{where} \quad H_m(Q|P) = f^{-1}\left(\sum_{k=1}^n p_k^{(\alpha)} f(H_m(Q_{|k})) \right), \quad (9)$$

where f is an invertible continuous function and $P^{(\alpha)} = (p_1^{(\alpha)}, \ldots, p_n^{(\alpha)}) \in \Delta_n$ is the α-escort distribution of distribution $P \in \Delta_n$ defined in

$$p_k^{(\alpha)} = \frac{p_k^\alpha}{\sum_{i=1}^n p_i^\alpha}, \quad k = 1, \ldots, n, \quad \alpha > 0. \quad (10)$$

GSK5 $H_2\left(\frac{1}{2}, \frac{1}{2}\right) = \text{Log}_q(1)$.

As shown in [9], the unique function H_n, which satisfies [GSK1]-[GSK5], is Sharma–Mittal entropy [6].

In the following paragraphs we will assume that X and Y are discrete jointly distributed random variables taking values from sample spaces $\{x_1, \ldots, x_n\}$ and $\{y_1, \ldots, y_m\}$, and distributed in accordance to $P_X \in \Delta_n$ and $P_Y \in \Delta_m$, respectively. In addition, the joint distribution of X and Y will be denoted in $P_{X,Y} \in \Delta_{nm}$ and the conditional distribution of X given Y will be denoted in $P_{X|Y} = \frac{P_{X,Y}(x,y)}{P_Y(y)} \in \Delta_m$, provided that $P_Y(y) > 0$. We will identify the entropy of a random variable X with the entropy of its distribution P_X and the Sharma–Mittal entropy will be denoted with $H_{\alpha,q}(X) \equiv H_n(P_X)$.

Thus, for a random variable which is distributed to X, Sharma–Mittal entropy can be expressed in

$$H_{\alpha,q}(X) = \frac{1}{1-q} \left(\left(\sum_x P_X(x)^\alpha \right)^{\frac{1-q}{1-\alpha}} - 1 \right), \tag{11}$$

and it can equivalently be expressed as the η_q transformation of Rényi entropy as in

$$H_{\alpha,q}(X) \equiv \eta_q(R_\alpha(X)). \tag{12}$$

Sharma–Mittal entropy, for $\alpha, q \in \mathbb{R}_0^+ \setminus 1$, being a continuous function of the parameters and the sums goes over the support of P_X. Thus, in the case of $q = 1, \alpha \neq 1$, Sharma–Mittal reduces to Rényi entropy of order α [2]

$$R_\alpha(X) \equiv H_{\alpha,1}(X) = \frac{1}{1-\alpha} \log \left(\sum_x P_X(x)^\alpha \right), \tag{13}$$

which further reduces to Shannon entropy for $\alpha = 1, q = 1$, [34]

$$S(X) \equiv H_{1,1}(X) = \sum_x P_X(x) \log P_X(x), \tag{14}$$

while in the case of $q \neq 1, \alpha = 1$ it reduces to Gaussian entropy [5]

$$G_q(X) \equiv H_{1,q}(X) = \frac{1}{(1-q) \ln 2} \left(\prod_{i=1}^n P_X(x)^{P_X(x)} - 1 \right). \tag{15}$$

In addition, Tsallis entropy [3] is obtained for $\alpha = q \neq 1$,

$$T_q(X) \equiv \frac{1}{(1-q) \ln 2} \left(\sum_x P_X(x)^q - 1 \right), \tag{16}$$

while in the case of for $q = 2 - \alpha$ it reduces to the Landsberg–Vedral entropy [4]

$$L_\alpha(X) \equiv H_{\alpha,2-\alpha}(X) = \frac{1}{(\alpha-1) \ln 2} \left(\frac{1}{\sum_x P_X(x)^\alpha} - 1 \right). \tag{17}$$

3. Sharma–Mittal Information Transfer Axioms

One of the main goals of information and communication theories is characterization and analysis of the information transfer between sender X and receiver Y, which communicate through a channel. The sender and receiver are described by probability distributions P_X and P_Y while the communication channel with the input X and the output Y is described by the transition matrix $P_{Y|X}$:

$$P_{Y|X}^{(i,j)} \equiv P_{Y|X}(y_j|x_i). \tag{18}$$

We assume that maximum likelihood detection is performed at the receiver, which is defined by the mapping $d : \{y_1, \ldots, y_m\} \to \{x_1, \ldots, x_n\}$ as follows:

$$d(y_j) = x_i \quad \Leftrightarrow \quad P_{Y|X}(y_j|x_i) > P_{Y|X}(y_j|x_k); \quad \text{for all } k \neq i, \tag{19}$$

assuming that the inequality in (19) is uniquely satisfied. Thus, if the input symbol x_i is sent and the output symbol y_j is received, the x_i will be detected if $x_i = d(y_j)$ and a detection error will be made otherwise, and we define the error function functions $\phi : \{x_1, \ldots, x_m\} \times \{y_1, \ldots, y_m\} \to \{0, 1\}$ as in

$$\phi(x_i, y_j) = \begin{cases} 1, & \text{if } x_i = d(y_j) \\ 0, & \text{otherwise,} \end{cases} \tag{20}$$

the detection error if a symbol x_i is sent

$$P_{err}(x_i) = \sum_{y_j} P_{Y|X}(y_j|x_i)\phi(x_i, y_j); \quad \text{for all} \quad x_i, \tag{21}$$

as well as the average detection error

$$\bar{P}_{err} = \sum_{x_i} P_X(x_i) P_{err}(x_i) = \sum_{x_i, y_j} P_{X,Y}(x, y)\phi(x_i, y_j). \tag{22}$$

Totally destructive channel: A channel is said to be totally destructive if

$$P_{Y|X}^{(i,j)} = P_{Y|X}(y_j|x_i) = P_Y(y_j) = \frac{1}{m}; \quad \text{for all} \quad x_i, \tag{23}$$

i.e., if the sender X and receiver Y are described by independent random variables,

$$X \perp\!\!\!\perp Y \quad \Leftrightarrow \quad P_{X,Y}(x, y) = P_X(x) P_Y(y), \tag{24}$$

where the relationship of independence is denoted in $\perp\!\!\!\perp$. In this case, $\phi_i(y_j) = 1$ for all y_j and the probability of error is $P_{err}(x_i) = 1$; for all x_i, as well as the average probability of error $\bar{P}_{err} = 1$, which means that a correct maximum likelihood detection is not possible.

Perfect communication channel: A channel is said to be perfect if for every x_i,

$$P_{Y|X}(y_j|x_i) > 0, \quad \text{for at least one } y_j \tag{25}$$

and for every y_j

$$P_{Y|X}(y_j|x_i) > 0, \quad \text{for exactly one } x_i. \tag{26}$$

Note that in this case $P_{Y|X}(y_j|x_i)$ can still take a zero value for some y_j and that $\phi_i(y_j) = 0$ for any non-zero $P_{Y|X}(y_j|x_i)$. Thus, the error probability is equal to zero $P_{err}(x_i) = 0$; for all x_i, as well as the average probability of error $\bar{P}_{err} = 0$, which means that perfect detection is possible by means of a maximum likelihood detector.

Noisy channel with non-overlapping outputs: A simple example of a perfect transmission channel is the noisy channel with non-overlapping outputs (NOC), which is schematically described in Figure 1. It is a 2-input $m = 2k$-output channel ($k \in \mathbb{N}$) defined by the transition matrix:

$$P_{Y|X} = \begin{bmatrix} P_{Y|X}(\cdot|x_1) \\ P_{Y|X}(\cdot|x_2) \end{bmatrix} = \begin{bmatrix} \frac{1}{k} & \cdots & \frac{1}{k} & 0 & \cdots & 0 \\ 0 & \cdots & 0 & \frac{1}{k} & \cdots & \frac{1}{k} \end{bmatrix} \tag{27}$$

(in this and in the following matrices, the symbol "$\cdots$" stands for the k-time repletion). In the case of $k = 1$ and $m = 2k = 2$, the channel reduces to the noiseless channel. Although the channel is noisy, the input can always be recovered from the output (if y_j is received and $j \leq k$, the input symbol x_1 is sent, otherwise x_2 is sent). Thus, it is expected that the information which is passed through the channel is equal to the information that can be generated by the input. Note that for a channel input distributed in accordance with

$$P_X = \begin{bmatrix} P_X(x_1) \\ P_X(x_2) \end{bmatrix} = \begin{bmatrix} a \\ 1-a \end{bmatrix}; \quad 0 \leq a \leq 1, \tag{28}$$

the joint probability distribution $P_{X,Y}$ can be expressed as in:

$$P_{X,Y} = \begin{bmatrix} \frac{a}{k} & \cdots & \frac{a}{k} & 0 & \cdots & 0 \\ 0 & \cdots & 0 & \frac{1-a}{k} & \cdots & \frac{1-a}{k} \end{bmatrix} \tag{29}$$

and the output distribution P_Y, which can be obtained by the summations over columns, is

$$P_Y = [P_Y(y_1), \ldots, P_Y(y_m)]^T = \begin{bmatrix} \frac{a}{k}, \ldots, \frac{a}{k}, \frac{1-a}{k}, \ldots, \frac{1-a}{k} \end{bmatrix}^T. \tag{30}$$

Binary symmetric channels: The binary symmetric channel (BSC) is a two input two output channel described by the transition matrix

$$P_{Y|X} = \begin{bmatrix} P_{Y|X}(\cdot|x_1)^T \\ P_{Y|X}(\cdot|x_2)^T \end{bmatrix} = \begin{bmatrix} 1-p & p \\ p & 1-p \end{bmatrix}, \tag{31}$$

which is schematically described in Figure 2. Note that for $p = \frac{1}{2}$ BSC reduces to a totally destructive channel, while in the case of $p = 0$ it reduces to a perfect channel.

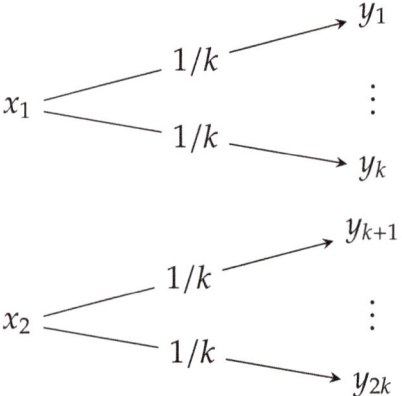

Figure 1. Noisy channel with non-overlapping outputs.

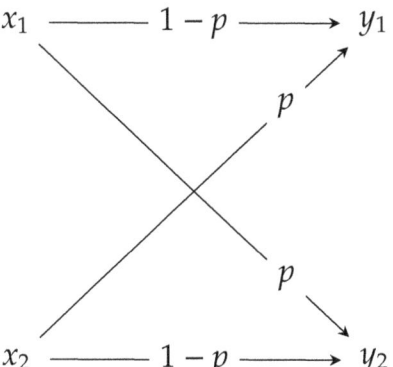

Figure 2. Binary symmetric channel.

Sharma–Mittal Information Transfer Axioms

In this paper, we search for information theoretical measures of information transfer between sender X and receiver Y, which communicate through a channel if the information is measured with Sharma–Mittal entropy. Thus, we are interested in the information transfer measure, $I_{\alpha,q}(X,Y)$, which is called the α-q-mutual information and its maximum,

$$C = \max_{P_X} I_{\alpha,q}(X,Y), \tag{32}$$

which is called the α-q-capacity and which requires the following set of axioms to be satisfied.

($\overline{A}_1$)   The channel cannot convey negative information, i.e.,

$$C_{\alpha,q}(P_{Y|X}) \geq I_{\alpha,q}(X,Y) \geq 0. \tag{33}$$

($\overline{A}_2$)   The information transfer is zero in the case of a totally destructive channel, i.e.,

$$P_{Y|X}(y|x) = \frac{1}{m}, \text{ for all } x,y \quad \Rightarrow \quad I_{\alpha,q}(X,Y) = C_{\alpha,q}(P_{Y|X}) = 0, \tag{34}$$

which is consistent with the conclusion that the average probability of error is one, $\bar{P}_{err} = 1$, in the case of a totally destructive channel.

($\overline{A}_3$)   In the case of perfect transmission, the information transfer is equal to the input information, i.e.,

$$X = Y \quad \Rightarrow \quad I_{\alpha,q}(X,Y) = H_{\alpha,q}(X), \quad C_{\alpha,q}(P_{Y|X}) = \text{Log}_q n, \tag{35}$$

which is consistent with the conclusion that the average probability of error is zero, $\bar{P}_{err} = 0$, in the case of a perfect transmission channel, so that all the information from the input is conveyed.

($\overline{A}_4$)   The channel cannot transfer more information than it is possible to be sent, i.e.,

$$I_{\alpha,q}(X,Y) \leq C_{\alpha,q}(P_{Y|X}) \leq \text{Log}_q n, \tag{36}$$

which means that a channel cannot add additional information.

($\overline{A}_5$)   The channel cannot transfer more information than it is possible to be received, i.e.,

$$I_{\alpha,q}(X,Y) \leq C_{\alpha,q}(P_{Y|X}) \leq \text{Log}_q m, \tag{37}$$

which means that a channel cannot add additional information.

($\overline{A}_6$)   Consistency with the Shannon case:

$$\lim_{q \to 1, \alpha \to 1} I_{\alpha,q}(X,Y) = I(X,Y), \quad \text{and} \quad \lim_{q \to 1, \alpha \to 1} C_{\alpha,q}(P_{Y|X}) = C(P_{Y|X}) \quad (38)$$

Thus, the axioms ($\overline{A}_2$) and ($\overline{A}_3$) ensure that the information measures are consistent with the maximum likelihood detection (19)–(21). On the other hand, the axioms ($\overline{A}_1$), ($\overline{A}_4$) and ($\overline{A}_5$), prevent a situation in which a physical system conveys information in spite of going through a completely destructive channel, or in which the negative information transfer is observed, indicating that the channel adds or removes information by itself, which could be treated as nonphysical behavior without an intuitive explanation. Finally, the property ($\overline{A}_6$) ensure that the information transfer measures can be considered as generalizations of corresponding Shannon measures. For these reasons, we assume that the satisfaction of the properties ($\overline{A}_1$)–($\overline{A}_5$) is mandatory for any reasonable definition of Sharma–Mittal information transfer measures.

4. The α-Mutual Information and the α-Capacity

One of the first proposals for the Rényi mutual information goes back to Arimoto [24], who considered the following definition of mutual information:

$$I_\alpha(X,Y) = \frac{\alpha}{1-\alpha} \log \left(\sum_y \left(\sum_x P_X^{(\alpha)}(x) P_{Y|X}^\alpha(y|x) \right)^{\frac{1}{\alpha}} \right), \quad (39)$$

where the escort distribution $P_{X^{(\alpha)}}$ is defined as in (10), and he also invented an iterative algorithm for the computation of the α-capacity [35], which is defined from the α-mutual information:

$$C_\alpha(P_{Y|X}) = \max_{P_X} I_\alpha(X,Y). \quad (40)$$

Notably, Arimoto's mutual information can equivalently be represented using the conditional Rényi entropy

$$R_\alpha(X|Y) = \frac{\alpha}{\alpha - 1} \log_2 \sum_y P_Y(y) \left(\sum_x P_{X|Y=y}(x)^\alpha \right)^{\frac{1}{\alpha}}, \quad (41)$$

as in

$$I_\alpha(X,Y) \equiv R_\alpha(X) - R_\alpha(X|Y), \quad (42)$$

which can be interpreted as the input uncertainty reduction after the output symbols are received and, in the case of $\alpha \to 1$, the previous definition reduces to the Shannon case. In addition, this measure is directly related to the famous Gallager exponent

$$E_0(\rho, P_X) = -\log \left(\sum_y \left(\sum_x P_X(x) P_{Y|X}^{\frac{1}{1+\rho}}(y|x) \right)^{1+\rho} \right), \quad (43)$$

which has been widely used to establish the upper bound of error probability in channel coded communication systems [36] via the relationship [29]

$$I_\alpha(X,Y) = \frac{\alpha}{1-\alpha} E_0 \left(\frac{1}{\alpha} - 1, P_X^{(\alpha)} \right). \quad (44)$$

In addition, in the case of $\alpha \to 1$, it reduces to

$$I_1(X,Y) = \lim_{\alpha \to 1} I_\alpha(X,Y) = I(X,Y), \quad (45)$$

where

$$I(X,Y) = \sum_{x,y} P_{X,Y}(x,y) \log \frac{P_{X,Y}(x,y)}{P_X(x)P_Y(y)} \qquad (46)$$

stands for Shannon's mutual information [37].

The α-mutual information $I_\alpha(X,Y)$ and the α-capacity $C_\alpha(P_{Y_X})$ satisfy the axioms $(\overline{A_1})$–$(\overline{A_6})$ for $q = 1$ and $\alpha > 0$, as stated by the following theorem, which further justifies their usage as the measures of (maximal) information transfer.

Theorem 1. *The mutual information measures I_α and C_α satisfy the following set of properties:*

(A_1)   *The channel cannot convey negative information, i.e.,*

$$C_\alpha(P_{Y|X}) \geq I_\alpha(X,Y) \geq 0. \qquad (47)$$

(A_2)   *The (maximal) information transfer is zero in the case of a totally destructive channel, i.e.,*

$$P_{Y|X}(y|x) = \frac{1}{m}, \text{ for all } x,y \quad \Rightarrow \quad I_\alpha(X,Y) = C_\alpha(P_{Y|X}) = 0. \qquad (48)$$

(A_3)   *In the case of perfect transmission, the (maximal) information transfer is equal to the (maximal) input information, i.e.,*

$$X = Y \quad \Rightarrow \quad I_\alpha(X,Y) = R_\alpha(X), \quad C_\alpha(P_{Y|X}) = \log n. \qquad (49)$$

(A_4)   *The channel cannot transfer more information than it is possible to be sent, i.e.,*

$$I_\alpha(X,Y) \leq C_\alpha(P_{Y|X}) \leq \log n; \qquad (50)$$

(A_5)   *The channel cannot transfer more information than it is possible to be received, i.e.,*

$$I_\alpha(X,Y) \leq C_\alpha(P_{Y|X}) \leq \log m. \qquad (51)$$

(A_6)   *Consistency with the Shannon case:*

$$\lim_{\alpha \to 1} I_\alpha(X,Y) = I(X,Y), \quad \text{and} \quad \lim_{\alpha \to 1} C_\alpha(P_{Y|X}) = C(P_{Y|X}) \qquad (52)$$

Proof. As shown in [38], $R_\alpha(X|Y) \leq R_\alpha(X)$, and the nonnegativity property (A_1) follows from the definition of Arimoto's mutual information (42). In addition, if $X \perp\!\!\!\perp Y$, then $P_{Y|X}(y|x) = P_Y(y)$ so that the definition (61) implies the property (A_2). Furthermore, in the case of a perfect transmission channel, the mutual information (61) can be represented in

$$I_\alpha(X,Y) = \frac{\alpha}{\alpha-1} \log \frac{\sum_y \left(\sum_x P_X(x)^\alpha P_{Y|X}^\alpha(y|x)\right)^{\frac{1}{\alpha}}}{\left(\sum_x P_X^{(\alpha)}(x)\right)^{\frac{1}{\alpha}}} = \frac{\alpha}{\alpha-1} \log \frac{\sum_y \left(P_X(d(y))^\alpha P_{Y|X}^\alpha(y \mid d(y))\right)^{\frac{1}{\alpha}}}{\left(\sum_x P_X^{(\alpha)}(x)\right)^{\frac{1}{\alpha}}}, \qquad (53)$$

and since

$$\sum_y \left(P_X(d(y))^\alpha P_{Y|X}^\alpha(y \mid d(y))\right)^{\frac{1}{\alpha}} = \sum_y P_X(d(y)) P_{Y|X}(y \mid d(y)) =$$
$$\sum_x \sum_{y:d(y)=x} P_X(d(y)) P_{Y|X}(y \mid d(y)) = \sum_x P_X(x) \sum_{y:d(y)=x} P_{Y|X}(y|x) = 1, \qquad (54)$$

we obtain $I_\alpha(X,Y) = R_\alpha(X)$, which proves the property (A_3). Moreover, from the definition as shown in [38], Arimoto's conditional entropy is positive and satisfies the weak chain rule $R_\alpha(X|Y) \geq R_\alpha(X) - \log m$, so that the properties (A_4) and (A_5) follow from the definition of Arimoto's mutual information (42). Finally, the property (A_6) follows directly from the equation (45) and can be approved using L'Hôpital's rule, which completes the proof of the theorem. □

5. Alternative Definitions of the α-Mutual Information and the α-Channel Capacity

Since Rényi's proposal, there have been several lines of research to find an appropriate definition and characterization of information transfer measures related to Rényi entropy, which are established by the substitution of the Rényi divergence measure

$$D_\alpha(P||Q) = \frac{1}{\alpha - 1} \log\left(\sum_x P(x)^\alpha Q(x)^{1-\alpha}\right), \qquad (55)$$

instead of the Kullback–Leibler one,

$$D(P||Q) = D_1(P||Q) = \sum_x P(x) \log \frac{P(x)}{Q(x)}, \qquad (56)$$

in some of the various definitions which are equivalent in the case of Shannon information measures (46) [29]:

$$\begin{aligned}I(X,Y) &= \min_{Q_Y} \mathbb{E}\left[D_\alpha\left(P_{Y|X}||Q_Y\right)\right] = \min_{Q_Y} \mathbb{E}\left[D_\alpha\left(P_{Y|X}||Q_Y\right)\right] \\ &= \min_{Q_X} \min_{Q_Y} D_\alpha(P_{X,Y}||Q_X Q_Y) = D_\alpha(P_{X,Y}||P_X P_Y) = S(X) - S(X|Y)\end{aligned} \qquad (57)$$

where $S(X|Y)$ stands for the Shannon conditional entropy,

$$S(X|Y) = \sum_{x,y} P_{X,Y}(x,y) \log P_{X|Y}(x|y). \qquad (58)$$

All of these measures are consistent with the Shannon case in view of the property (A_6), but their direct usage as measures of Rényi information transfer leads to a breaking of some the properties (A_1)–(A_5), which justifies the usage of Arimoto's measures from the previous section as appropriate ones in the context of this research. In the following section, we review the alternative definitions.

5.1. Information Transfer Measures by Sibson

Alternative approaches based on Rényi divergence were proposed by Sibson [23] and considered later by several authors in the context of quantum secure communications [39–44], who introduced

$$J_\alpha^1(X;Y) = \min_{Q_Y} D_\alpha\left(P_{Y|X} P_X || Q_Y P_X\right), \qquad (59)$$

which can be represented as in [26]

$$J_\alpha^1(X,Y) = \frac{\alpha}{\alpha-1} \log\left(\sum_y \left(\sum_x P_X(x) P_{Y|X}^\alpha(y|x)\right)^{\frac{1}{\alpha}}\right) \qquad (60)$$

and, in the discrete setting, can be related to the Gallager exponent as in [29]:

$$J_\alpha^1(X,Y) = \frac{\alpha}{1-\alpha} E_0\left(\frac{1}{\alpha} - 1, P_X\right), \qquad (61)$$

which differs from Arimoto's definition (61) since in this case the escort distribution does not participate in the error exponent, but an ordinary one does. However, in the case of a perfect channel for which $X = Y$, the conditional distribution $P_{Y|X}^\alpha(y|x) = 1$ for $x = y$ and zero otherwise, so Sibson's measure (60) reduces to $R_{1/\alpha}(X)$, thus breaking the axiom (A_3). This disadvantage can be overcome by the reparametrization $\alpha \leftrightarrow 1/\alpha$ so that $J_{1/\alpha}^1(X,Y)$

is used as a measure of Rényi information transfer, and the properties of the resulting measure can be considered in a manner similar to the case of Arimoto.

5.2. Information Transfer Measures by Augustin and Csiszar

An alternative definition of Rényi mutual information was also presented by Augustin [25], and later Csiszar [26], who defined

$$J_\alpha^2(X;Y) = \min_{Q_Y} \mathbb{E}\left[D_\alpha\left(P_{Y|X}\|Q_Y\right)\right], \tag{62}$$

However, in the case of perfect transmission, for which $X = Y$, the measure reduces to Shannon entropy

$$J_\alpha^2(X;Y) = S(X), \tag{63}$$

which breaks the axiom (A_3).

5.3. Information Transfer Measures by Lapidoth, Pfister, Tomamichel and Hayashi

A similar obstacle to the case of the Augustin–Csiszar measure can be observed in the case of mutual information which was considered by Lapidoth and Pfister [27] and Tomamichel and Hayashi [28], who proposed

$$J_\alpha^3(X;Y) = \min_{Q_X} \min_{Q_Y} D_\alpha(P_{X,Y}\|Q_X Q_Y). \tag{64}$$

As shown in [27] (Lemma 11), if $X = Y$, then

$$J_\alpha^3(X;Y) = \begin{cases} \frac{\alpha}{1-\alpha} \lim_{\alpha \to \infty} R_\alpha(X) & \text{if } \alpha \in \left[0, \frac{1}{2}\right], \\ R_{\frac{\alpha}{2\alpha-1}}(X) & \text{if } \alpha > \frac{1}{2} \end{cases} \tag{65}$$

so the axiom (A_3) is broken in this case, as well.

Remark 1. *Despite the difference between the definitions of information transfer, in the discrete setting, the alternative definitions discussed above reach the same maximum over the set of input probability distributions, P_X, [26,29,45].*

5.4. Information Transfer Measures by Chapeau-Blondeau, Delahaies, Rousseau, Tridenski, Zamir, Ingber and Harremoes

Chapeau-Blondeau, Delahaies and Rousseau [31], and independently Tridenski, Zamir and Ingber [46] and Harremoes [47], defined the Rényi mutual information using the Rényi divergence (55), so that the mutual information defined using the Rényi divergence

$$J_\alpha^4(X,Y) = D_\alpha(P_{X,Y}\|P_X P_Y) \tag{66}$$

for $\alpha > 0$ and $\alpha \neq 1$, while in the case of $\alpha = 1$ it reduces to Shannon mutual information. However, the ordinal definition can correspond only to a Rényi entropy of order $2 - \alpha$ since in the case of $X = Y$ it reduces to $J_\alpha^4(X,Y) = R_{2-\alpha}(X)$ (see also [47]), which can be overcome by the reparametrization $\alpha = 2 - q$, similar to the case of Sibson's measure. This measure has been discussed in the past with various operational characterizations, and could also be considered as a measure of information transfer, although the satisfaction of all of the axioms (A_1)–(A_6) is not self-evident for general channels.

5.5. Information Transfer Measures by Jizba, Kleinert and Shefaat

Finally, we will mention the definition by Jizba, Kleinert and Shefaat [48],

$$J_\alpha^4(X,Y) \equiv R_\alpha(X) - \hat{R}_\alpha(X|Y), \tag{67}$$

which is defined in the same manner as in Arimoto's case (42), but with another choice of conditional Rényi entropy

$$\hat{R}_\alpha(X|Y) = \frac{1}{1-\alpha} \log \sum_x P_X^{(\alpha)}(x) 2^{(1-\alpha)R_\alpha(X|Y=y)}, \tag{68}$$

which arises from the Generalized Shannon–Khinchin axiom [GSK4] if the pseudo-additivity in the equation (9) is restricted to an ordinary addition, in which case the GSK axioms uniquely determine Rényi entropy [49]. However, despite its wide applicability in the modeling of causality and financial time series, this mutual information can take negative values which breaks the axiom (A_1), which is assumed to be mandatory in this paper. For further discussion of the physicalism of negative mutual information in the domain of financial time series analysis, the reader is referred to [48].

6. The α-q Mutual Information and the α-q-Capacity

In the past several attempts have been done to define an appropriate channel capacity measure which corresponds to instances of the Sharma–Mittal entropy class. All of them follow a similar recipe by which the channel capacity is defined as in (32), as a maximum of appropriately defined mutual information $I_{\alpha,q}$. However, all of the classes consider only special cases of Sharma–Mittal entropy and all of them fail to satisfy at least one of the properties $(\overline{A}_1)$–$(\overline{A}_6)$ which an information transfer has to satisfy, as will be discussed Section 7.

In this section we propose a general measures of the α-q mutual information and the α-q-capacity by the requirement that the axioms $(\overline{A}_1)$–$(\overline{A}_6)$ are satisfied, which could qualify them as appropriate measures of information transfer, without nonphysical properties. The special instances of the α-q (maximal) information transfer measures are also discussed and the analytic expressions for a binary symmetric channel are provided.

6.1. The α-q Information Transfer Measures and Its Instances

The α-q-mutual information (42) is defined using the q-subtraction defined in (6), as follows:

$$I_{\alpha,q}(X,Y) = H_{\alpha,q}(X) \ominus_q H_{\alpha,q}(X|Y), \tag{69}$$

where we introduced the conditional Sharma–Mittal entropy $H_{\alpha,q}(Y|X)$ as in

$$H_{\alpha,q}(X|Y) = \eta_q(R_\alpha(X|Y)) = \frac{1}{(1-q)\ln 2} \left(\left(\sum_y P_Y(y) \left(\sum_x P_{X|Y=y}(x)^\alpha \right)^{\frac{1}{\alpha}} \right)^{\frac{\alpha(1-q)}{\alpha-1}} - 1 \right), \tag{70}$$

$R_\alpha(X|Y)$ stands for Arimoto's definition of the conditional Rényi entropy (41). The expression (69) can also be obtained if the mapping η_q is applied to both sides of the equality (42), by which Arimoto's mutual information is defined, so we may establish the relationship

$$I_{\alpha,q}(X,Y) = \eta_q(I_\alpha(X,Y)) = \eta_q \left(\frac{\alpha}{1-\alpha} \log \left(\sum_y \left(\sum_x P_X^{(\alpha)}(x) P_{Y|X}^\alpha(y|x) \right)^{\frac{1}{\alpha}} \right) \right), \tag{71}$$

which can be represented using the Gallager error exponent (43) as in

$$I_{\alpha,q}(X,Y) = \eta_q \left(\frac{\alpha}{1-\alpha} E_0 \left(\frac{1}{\alpha} - 1, P_X^{(\alpha)} \right) \right) = \frac{1}{(1-q)\ln 2} \left(2^{\frac{\alpha(1-q)}{1-\alpha} E_0 \left(\frac{1}{\alpha} - 1, P_X^{(\alpha)} \right)} - 1 \right). \tag{72}$$

Arimoto's α-q-capacity is now defined in

$$C_{\alpha,q} = \max_{P_X} I_{\alpha,q}(X,Y), \tag{73}$$

and using the fact that η_q is increasing, it can be related with the corresponding α-capacity as in

$$C_{\alpha,q} = \max_{P_X} I_{\alpha,q}(X,Y) = \max_{P_X} \eta_q(I_\alpha(X,Y)) = \eta_q\left(\max_{P_X} I_\alpha(X,Y)\right) = \eta_q\left(C_\alpha(P_{Y|X})\right). \quad (74)$$

Using the expressions (45) and (71), in the case of $\alpha = 1$, the α-q mutual information reduces to

$$I_{1,q} = \frac{1}{(1-q)\ln 2}\left(\prod_{x,y} 2^{P_{X,Y}(x,y)\log\frac{P_{X,Y}(x,y)}{P_X(x)P_Y(y)}} - 1\right)$$

$$= \frac{1}{(1-q)\ln 2}\left(\prod_{x,y}\left(\frac{P_{X,Y}(x,y)}{P_X(x)P_Y(y)}\right)^{P_{X,Y}(x,y)} - 1\right). \quad (75)$$

The α-q-capacity is given in

$$C_{1,q} = \max_{P_X}\left(\frac{1}{(1-q)\ln 2}\left(\prod_{x,y}\left(\frac{P_{X,Y}(x,y)}{P_X(x)P_Y(y)}\right)^{P_{X,Y}(x,y)} - 1\right)\right) \quad (76)$$

and these measures can serve as (maximal) information transfer measures corresponding to Gaussian entropy, which was not considered before in the context of information transmission. Naturally, if in addition $q \to 1$, the measures reduce to Shannon's mutual information and Shannon capacity [37].

Additional special cases of the α-q (maximal) information transfer include the α-mutual information (42) and the α-capacity (40), which are obtained for $q = 1$; the measures which correspond to Tsallis entropy can be obtained for $q = \alpha$ and the ones which correspond to Landsberg–Vedral entropy for $q = 2 - \alpha$. These special instances are listed in Table 1.

As discussed in Section 7, previously considered information measures cover only particular special cases and break at least one of the axioms $(\overline{A}_1)$–$(\overline{A}_5)$, which leads to unexpected and counterintuitive conclusions about the channels, such as negative information transfer and achieving super-capacitance or sub-capacitance [4], which could be treated as a nonphysical behavior. On the other hand, apart from the generality, the α-q information transfer measures proposed in this paper overcame the disadvantages which could qualify them as appropriate measures, as stated in the following theorem.

Theorem 2. *The α-q information transfer measures $I_{\alpha,q}$ and $C_{\alpha,q}$ satisfy the set of the axioms $(\overline{A}_1)$–$(\overline{A}_6)$.*

Proof. The proof is the straightforward application of the mapping η_q to the equations in the α-mutual information properties (A_1)–(A_5), while the $(\overline{A}_6)$ follows from the above discussion. □

Remark 2. *Note that the symmetry $I_{\alpha,q}(X,Y) = I_{\alpha,q}(Y,X)$ does not hold in general in the case of the α-q mutual information nor in the case of the α mutual information [50,51] and if the mutual information is defined so that the symmetry is preserved, some of the axioms $(\overline{A}_1)$–$(\overline{A}_6)$ might be broken. In addition, the alternative definition of the mutual information, $I_{\alpha,q}(Y,X) = H_{\alpha,q}(Y) - H_{\alpha,q}(Y|X)$, which uses an ordinary substraction operator instead of $\ominus_q$ operation, can also be introduced, but in this case the property $(\overline{A}_5)$ might not hold in general, as discussed in Section 7.*

6.2. The α-q-Capacity of Binary Symmetric Channels

As shown by Cai and Verdú [45], the α-mutual information of Arimoto's type I_α is maximized for the uniform distribution $P_X = (1/2, 1/2)$, and Arimoto's α-capacity has the value

$$C_\alpha(BSC) = 1 - r_\alpha(p), \quad (77)$$

where the binary entropy function r_α is defined as

$$r_\alpha(p) = R_\alpha(p, 1-p) = \frac{1}{1-\alpha} \log(p^\alpha + (1-p)^\alpha), \tag{78}$$

for $\alpha > 0$, $\alpha \neq 1$, while in the limit of $\alpha \to 1$, the expression (78) reduces to the well-known result for the Shannon capacity (see Fano [52])

$$C_1(BSC) = \lim_{\alpha \to 1} C_\alpha(BSC) = 1 + p \log p + (1-p) \log(1-p). \tag{79}$$

The analytic expressions for the α-q-capacities of binary symmetric channel's can be obtained from the expressions (74) and (77), so that

$$C_{\alpha,q}(BSC) = \eta_q(C_\alpha(BSC)) = \frac{1}{(1-q)\ln 2} \left(2^{1-q}(p^\alpha + (1-p)^\alpha)^{-\frac{1-q}{1-\alpha}} - 1 \right); \tag{80}$$

in the case of $q = 1$, it reduces to the case of Rényi entropy while, in the case of $\alpha = 1$, to the case of Gaussian entropy (77)

$$C_{1,q}(BSC) = \frac{1}{(1-q)\ln 2} \left(2p^p(1-p)^{1-p} - 1 \right). \tag{81}$$

The analytic expressions for BSC α-q capacities for other instances can straightforwardly be obtained by specifying the values of the parameters, whose instances are listed in Table 1, while the plots of the BSC α-q-capacities, which correspond to the Gaussian and the Tsallis entropies, are shown in Figures 3 and 4.

The α-q-capacity (80) can equivalently be expressed in

$$C_{\alpha,q}(BSC) = \text{Log}_q 2 \ominus_q h_{\alpha,q}(p), \tag{82}$$

where the Sharma–Mittal binary entropy function is defined in

$$h_{\alpha,q}(p) = H_{\alpha,q}(p, 1-p) = \frac{1}{1-q} \left((p^\alpha + (1-p)^\alpha)^{\frac{1-q}{1-\alpha}} - 1 \right), \tag{83}$$

which reduces to the Rényi binary entropy function, in the case of $q = 1$,

$$h_{\alpha,1}(p) = \lim_{q \to 1} h_{\alpha,q}(p) = R_\alpha(p, 1-p) = \frac{1}{1-\alpha} \log(p^\alpha + (1-p)^\alpha)), \tag{84}$$

to the Tsallis binary entropy function, in the case of $\alpha = 1$,

$$h_{q,q}(p) = h_{q,q}(p) = T_q(p, 1-p) = \frac{1}{1-q}(p^q + (1-p)^q - 1), \tag{85}$$

to the Gaussian binary entropy function, in the case of $\alpha = 1$,

$$h_{1,q}(p) = \lim_{\alpha \to 1} h_{\alpha,q}(p) = G_q(p, 1-p) = \frac{1}{(1-q)\ln 2} \left(p^{-(1-q)p}(1-p)^{-(1-q)(1-p)} - 1 \right), \tag{86}$$

and to the Shannon binary entropy function, in the case of $\alpha = q = 1$,

$$h_{1,1}(p) = \lim_{q,\alpha \to 1} h_{\alpha,q}(p) = S(p, 1-p) = -p \log p - (1-p) \log(1-p). \tag{87}$$

The expression (82) can be interpreted similarly as in the Shannon case. Thus, a BSC channel with input X and output Y can be modeled with an input–output relation $Y = X \oplus Z$ where $\oplus$ stands for modulo 2 sum and Z is channel noise taking values from $\{1, 0\}$, distributed in accordance with $(p, 1-p)$. If we measure the information which is lost per bit during transmission with the Sharma–Mittal entropy $H_{\alpha,q}(Z) = h_\alpha(p)$, then $C_{\alpha,q}$ stands for useful information left over for every bit of information received.

Table 1. Instances of the α-q-mutual information for different values of the parameters and corresponding expressions for the BSC α-q-capacities.

$H_{\alpha,q}$	$I_{\alpha,q}$	$C_{\alpha,q}$
S $\alpha = q = 1$	$\sum_{x,y} P_{X,Y}(x,y) \log \frac{P_{X,Y}(x,y)}{P_X(x)P_Y(y)}$	$1 + p \log p + (1-p) \log(1-p)$
R_α $q = 1$	$\frac{\alpha}{1-\alpha} E_0\left(\frac{1}{\alpha} - 1, P_X^{(\alpha)}\right)$	$1 - \frac{\log(p^\alpha + (1-p)^\alpha)}{1-\alpha}$
T_q $q = \alpha$	$\frac{1}{(1-q)\ln 2}\left(2^{qE_0\left(\frac{1}{q}-1, P_X^{(q)}\right)} - 1\right)$	$\frac{1}{(1-q)\ln 2}\left(2^{1-q}(p^q + (1-p)^q)^{-1} - 1\right)$
L_α $q = 2 - \alpha$	$\frac{1}{(\alpha-1)\ln 2}\left(2^{-\alpha E_0\left(\frac{1}{\alpha}-1, P_X^{(\alpha)}\right)} - 1\right)$	$\frac{1}{(1-\alpha)\ln 2}\left(2^{\alpha-1}(p^\alpha + (1-p)^\alpha) - 1\right)$
G_q $\alpha = 1$	$\frac{1}{(1-q)\ln 2}\left(\prod_{x,y}\left(\frac{P_{X,Y}(x,y)}{P_X(x)P_Y(y)}\right)^{P_{X,Y}(x,y)} - 1\right)$	$\frac{1}{(1-q)\ln 2}\left(2^{1-q} p^{(1-q)p}(1-p)^{(1-q)(1-p)} - 1\right)$

$$E_0(\rho, P_X) = -\log\left(\sum_y \left(\sum_x P_X(x) P_{Y|X}^{\frac{1}{1+\rho}}(y|x)\right)^{1+\rho}\right)$$

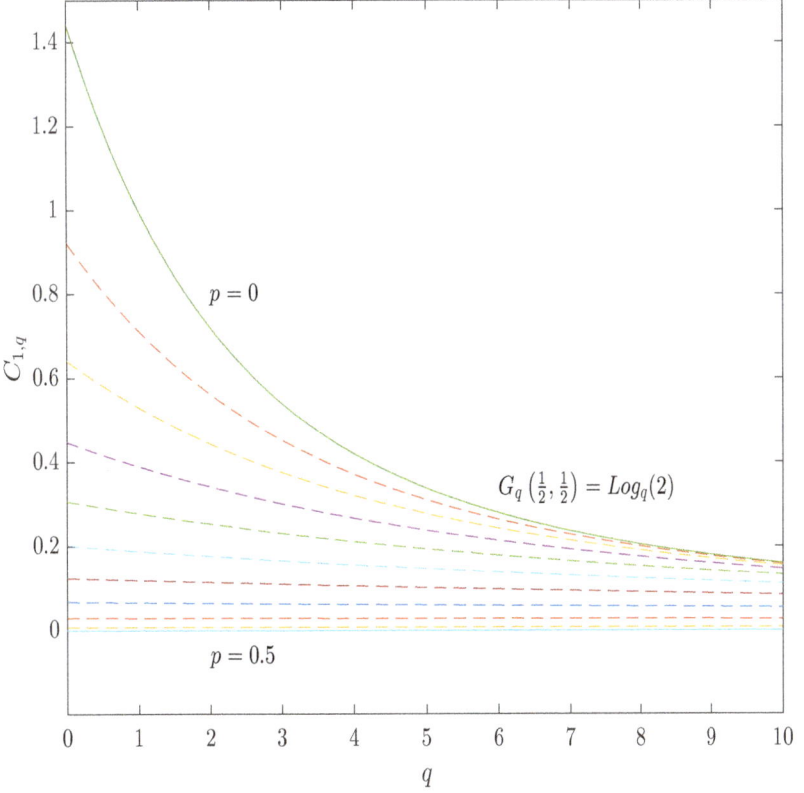

Figure 3. The α-q-capacity of BSC for the Gaussian entropy (the case of $\alpha = 1$) as a function of q for various values of the channel parameter p from 0.5 (totally destructive channel) to 0 (perfect transmission). All of the curves lies between 0 and $\text{Log}_q 2$, which is the maximum value of the Gaussian entropy.

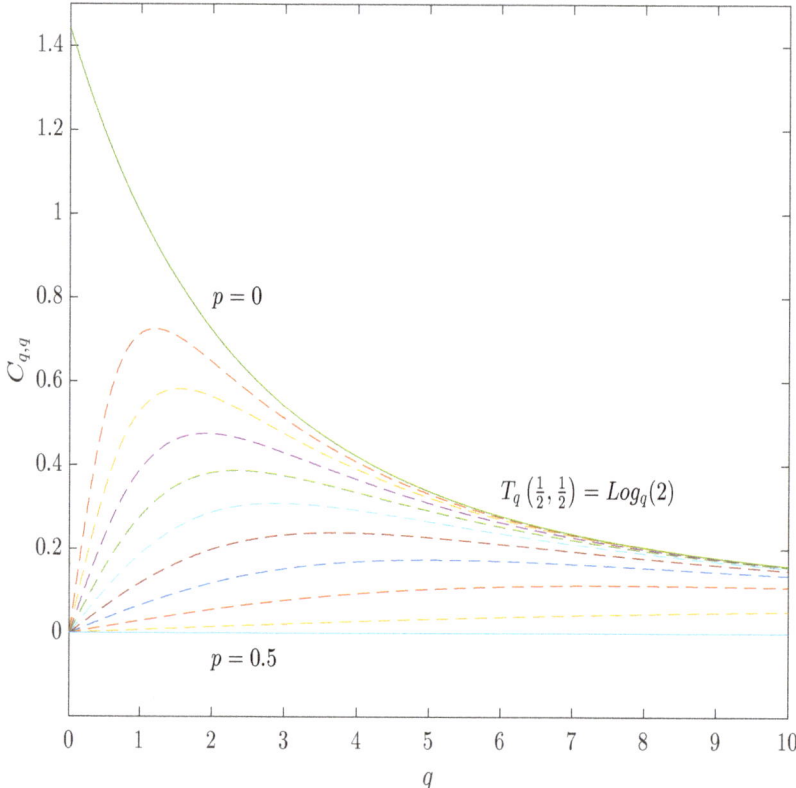

Figure 4. The α-q-capacity of BSC for the Tsallis entropy (the case of $\alpha = q$) as a function of q for various values of the channel parameter p from 0.5 (totally destructive channel) to 0 (perfect transmission). All of the curves lies between 0 and $\text{Log}_q 2$, which is the maximum value of the Tsallis entropy.

7. An Overview of the Previous Approaches to Sharma–Mittal Information Transfer Measures

In this section, we review the previous attempts at a definition of Sharma–Mittal information transfer measures, which are defined from the basic requirement of consistency with the Shannon measure as given by the axiom $(\overline{A}_6)$. However, as we show in the following paragraphs, all of them break at least one of the axioms $(\overline{A}_1)$–$(\overline{A}_5)$, which are satisfied in the case of the α-q (maximal) information transfer measures (69) and (73), in accordance with the discussion in Section 6.

7.1. Daróczy's Capacity

The first considerations of generalized channel capacities and generalized mutual information for the q-entropy go back to Daróczy [30], who introduced conditional Tsallis entropy

$$\bar{T}_q(Y|X) = \sum_x P_X^q(x) T_q(Y|X=x), \tag{88}$$

where the row entropies are defined as in

$$T_q(Y|X=x) = \frac{1}{(1-q)\log(2)} \left(\sum_x P_{Y|X}(y|x)^q - 1 \right) \tag{89}$$

and the mutual information is defined as in

$$J^5_{\alpha,q}(X,Y) = T_q(Y) - \bar{T}_q(Y|X). \tag{90}$$

However, in the case of a totally destructive channel, $X \perp\!\!\!\perp Y$, $P_{Y|X}(y|x) = P_Y(y)$, $T_q(Y|X=x) = T_q(Y)$ and

$$T_q(Y|X) = T_q(Y) \sum_x P_X(x)^q \tag{91}$$

so that

$$J^5_{\alpha,q}(X,Y) = T_q(Y)\left(1 - \sum_x P_X(x)^q\right) = \left(1 - \sum_x P_X(x)^q\right) \operatorname{Log}_q m. \tag{92}$$

This expression is zero for an input probability distribution $P_X = (1,0,\ldots,0)$ and its permutations, but, in general, it is negative for $q < 1$, positive for $q > 1$ and 0 only for $q = 1$, so the axiom $(\bar{A}_2)$ is broken (see Figure 5). As a result, the channel capacity, which is defined in accordance to (32), is zero for $q \leq 1$ and positive for $q > 1$, as illustrated in Figure 6 by the example of BSC for which the Daroczy's channel capacity can be computed as in [30,53]

$$C^5_q(BSC) = \frac{1 - 2^{1-q}}{q-1} - \frac{2^{-q}}{q-1}[1 - (1-p)^q - p^q]. \tag{93}$$

In the same figure, we plotted the graph for the α-q channel capacities proposed in this paper, and all of them remain zero in the case of a totally destructive BSC, as expected.

7.2. Yamano Capacities

Similar problems to the ones mentioned above arise in the case of mutual information and corresponding capacity measures considered by Yamano [33], who addressed the information transmission characterized by Landsberg–Vedral entropy L_q, given in (17).

Thus, the first proposal is based on the mutual information of the form

$$J^6_q(X,Y) = L_q(X) + L_q(Y) - L_q(X,Y), \tag{94}$$

where the joint entropy is defined in

$$L_q(X,Y) = \frac{1}{q-1}\left(\frac{1}{\sum_{x,y} P_{X,Y}(x,y)^q} - 1\right). \tag{95}$$

However, in the case of a fully destructive channel, $P_Y(y) = 1/m$ and $P_{X,Y}(x,y) = P_X(x)/m$, so that

$$J^6_q(X,Y) = \frac{1}{q-1}\left(\frac{1}{\sum_x P_X(x)^q} - 1\right) + \frac{1}{q-1}\left(m^{q-1} - 1\right) - \frac{1}{q-1}\left(m^{q-1}\frac{1}{\sum_x P_X(x)^q} - 1\right), \tag{96}$$

which can be simplified to

$$J^6_q(X,Y) = \frac{1 - m^{q-1}}{q-1}\left(\frac{1}{\sum_x P_X(x)^q} - 1\right). \tag{97}$$

Similarly to the case of Daroczy's capacity, this expression is zero for an input probability distribution $P_X = (1,0,\ldots,0)$ and its permutations but, in general, it is negative for $q > 1$, positive for $q < 1$ and 0 only for $q = 1$, so the axiom $(\bar{A}_2)$ is broken (see Figure 5). In Figure 6 we illustrated the Yamano channel capacity as a function of the parameter q, in the case of two input channels with $P_X = [a, 1-a]$, the channel capacity is zero for $q > 1$ (which is obtained for $P_X = [1,0]$), and

$$C^6_q(BSC) = \frac{1}{q-1}\left(2^q - 1 - 2^{2q-2}\right), \tag{98}$$

for $q > 1$ (which is obtained for $P_X = [1/2, 1/2]$). In the same Figure, we plotted the graph for the α-q channel capacities proposed in this paper, and, as before, all of them remain zero in the case of a totally destructive BSC, as expected.

Further attempts were made in [33], where the mutual information is defined in an analogous manner to (66) and (66), with the generalized divergence measure introduced in [54]. Thus, the alternative measure for mutual information is defined in

$$J_q^7(X,Y) = \frac{1}{(1-q)\ln 2} \frac{1}{\sum_{x,y} P_{X,Y}^q(x,y)} \left[1 - \sum_{x,y} P_{X,Y}(x,y) \left(\frac{P_X(x)P_Y(y)}{P_{X,Y}(x,y)}\right)^{1-q}\right]. \quad (99)$$

However, in the case of the simplest perfect communication channel for which $X = Y$, the mutual information reduces to

$$J_q^7(X,Y) = \frac{1}{(1-q)\ln 2} \frac{1 - \sum_x P_X(x)^{2-q}}{\sum_x P_X(x)^q} \neq L_q(X), \quad (100)$$

which breaks the axiom $(\overline{A}_3)$.

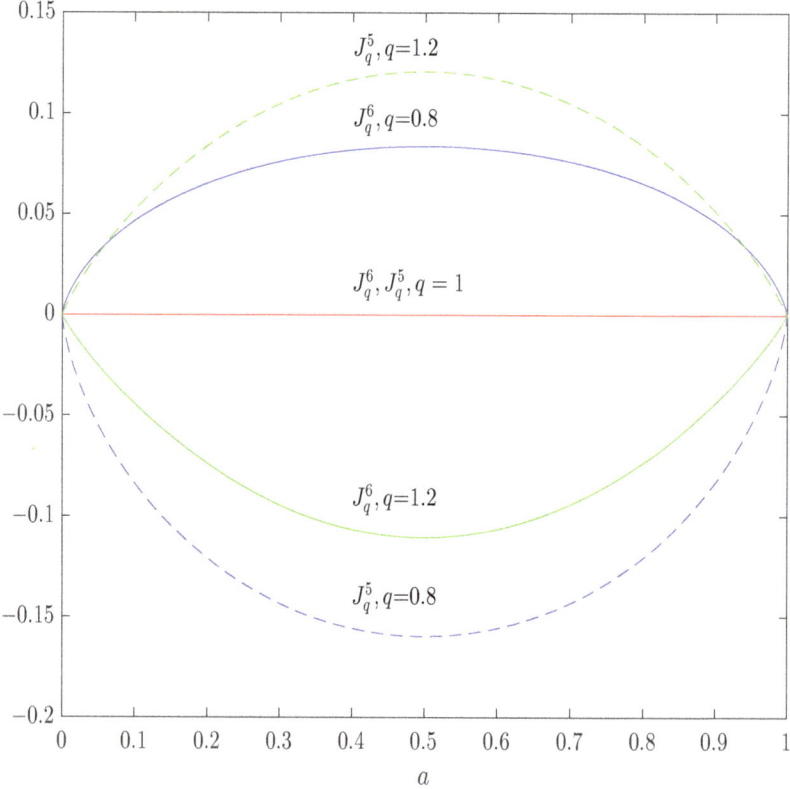

Figure 5. Daróczy's (solid lines) and Yamano's (dashed lines) mutual information in the case of a totally destructive BSC as functions of the input distribution parameter a, $P_X = [a, 1-a]^T$ for different values of q, obtaining negative values for $q < 1$ and $q > 1$, respectively, breaking the axioms $(\overline{A}_1)$ and $(\overline{A}_2)$. The α-q-mutual information is zero; for all q, and satisfies $(\overline{A}_1)$ and $(\overline{A}_2)$.

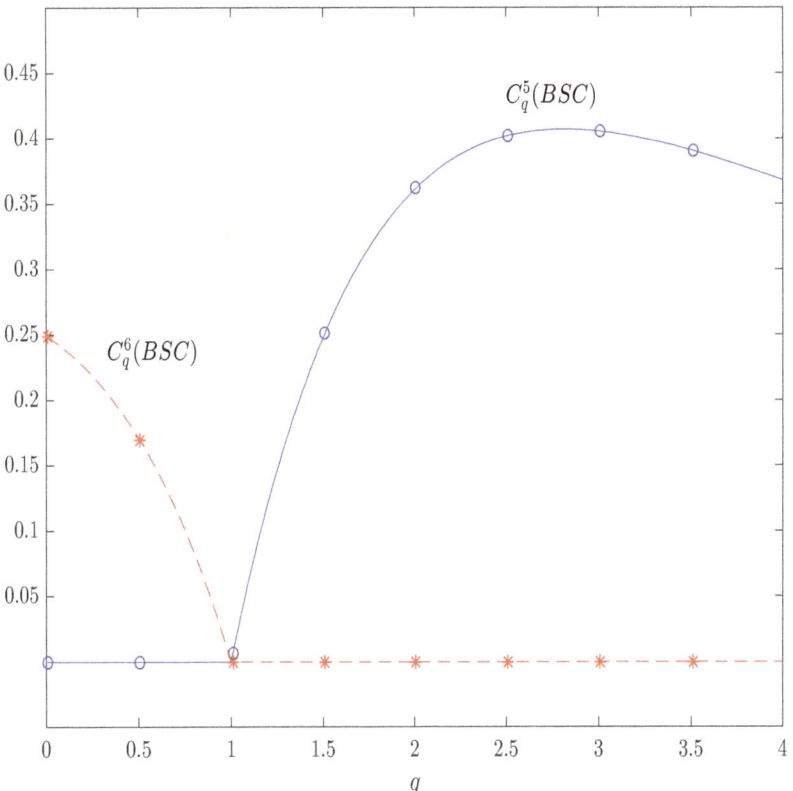

Figure 6. Daróczy's (solid lines) and Yamano's (dashed lines) capacities in the case of totally destructive BSC as functions of the parameter q. In the regions of $q < 1$ and $q > 1$, respectively, the corresponding negative mutual information is maximized for $P_X = [1,0]^T$ (zero capacity) having the positive values outside the regions and breaking the axiom $(\overline{A}_2)$. The α-q-capacity is zero; for all q, and satisfies $(\overline{A}_2)$.

7.3. Landsber–Vedral capacities

To avoid these problems, Landsberg and Vedral [4] proposed the mutual information measure and related channel capacities for the Sharma–Mittal entropy class $H_{\alpha,q}$, particularly considering the choice of $q = \alpha$, which corresponds to Tsallis entropy, $q = 2 - \alpha$, and the case of $q = 1$, which corresponds to the Rényi entropy

$$J^8_{\alpha,q}(X,Y) = H_{\alpha,q}(Y) - \tilde{H}_{\alpha,q}(Y|X), \tag{101}$$

where the conditional entropy $\tilde{H}_{\alpha,q}{}^{LV}(Y|X)$ is defined as in

$$\tilde{H}_{\alpha,q}(Y|X) = \sum_x P_X(x) H_{\alpha,q}(Y|X = x) \tag{102}$$

and

$$H_{\alpha,q}(Y|X = x) = \frac{1}{1-q}\left(\left(\sum_y P_{Y|X}(y|x)^\alpha\right)^{\frac{1-q}{1-\alpha}} - 1\right). \tag{103}$$

Although this definition bears some similarities to the α-q mutual information proposed in formula (69), several key differences can be observed. First of all, it characterizes the information transfer as the output uncertainty reduction after the input symbols are known, instead of input uncertainty reduction, after the output symbols are known (42). In addition, it uses the ordinary—operation instead of the $\ominus_q$ one. In addition, note that the definition of conditional entropy (102) generally differs from the definition proposed in (70).

The definition (101) resolves the issue of the axiom $(\overline{A}_2)$ which appears in the case of the Daroczy capacity, since in the case of a totally destructive channel $(X \perp\!\!\!\perp Y)$, $P_{Y|X}(y|x) = P_Y(y)$ and $L_q(Y|X=x) = L_q(Y)$ and $L_q(Y|X) = L_q(Y)$, so that $I_{\alpha,q}^{lv}(X,Y) = 0$. However, the problems remain with the axiom $(\overline{A}_5)$, which can be observed in the case of a noisy channel with non-overlapping outputs if the number of channel inputs is lower than the number of channel outputs $n < m$. Indeed, in the case of a noisy channel with non-overlapping outputs given by the transition matrix (27), both of the row entropies $L_q(Y|X=x)$ have the same value, which is independent of x

$$H_{\alpha,q}(Y|X=x) = \frac{k^{1-q}-1}{(q-1)\ln 2} = \mathrm{Log}_q k; \quad \text{for} \quad x = x_1, x_2, \tag{104}$$

and the maximal value of Landsberg–Vedral mutual information (101) is obtained only by maximizing $H_{\alpha,q}(Y)$ over P_X, which is achieved if X is uniformly distributed, since in this case Y is uniformly distributed, as well as ($a = \frac{1}{2}$ in (28)), so the maximal value of the output entropy is $H_{\alpha,q}(Y) = \mathrm{Log}_q(2k)$ and the mutual information is maximized for

$$C_{\alpha,q}^8(NOC) = \mathrm{Log}_q(2k) - \mathrm{Log}_q(k), \tag{105}$$

which is greater than $\mathrm{Log}_q(2)$ for $k \geq 2$, i.e., for $m \geq 4$ outputs, so the axiom $(\overline{A}_5)$ is broken, which is illustrated in Figure 7.

7.4. Chapeau-Blondeau–Delahaies–Rousseau Capacities

Following a similar approach to the one in Section 5.4, Chapeau-Blondeau, Delahaies and Rousseau considered the definition of mutual information which corresponds to the Tsallis entropy using Tsallis divergence,

$$D_{q,q}(P||Q) = \frac{1}{q-1}\left(\sum_x P(x)^q Q(x)^{1-q} - 1\right), \tag{106}$$

can be written in

$$\begin{aligned}J_q^9(X,Y) &= D_{q,q}(P_{X,Y}||P_X P_Y) = \eta_q\left(D_q(P_{X,Y}||P_X P_Y)\right) \\ &= \frac{1}{1-q}\left(1 - \sum_{x,y} P_{X,Y}(x,y)^q P_X(x)^{1-q} P_Y(y)^{1-q}\right).\end{aligned} \tag{107}$$

However, this definition is not directly applicable as a measure of information transfer to the Tsallis entropy with index q, since in the case of $X = Y$ it reduces to $J_q^9(X,Y) = T_{2-q}(X)$, and requires the reparametrization $q \leftrightarrow 2-q$, similar to Section 5.4, while the satisfaction of the axioms $(\overline{A}_4)$ and $(\overline{A}_5)$ is not self evident.

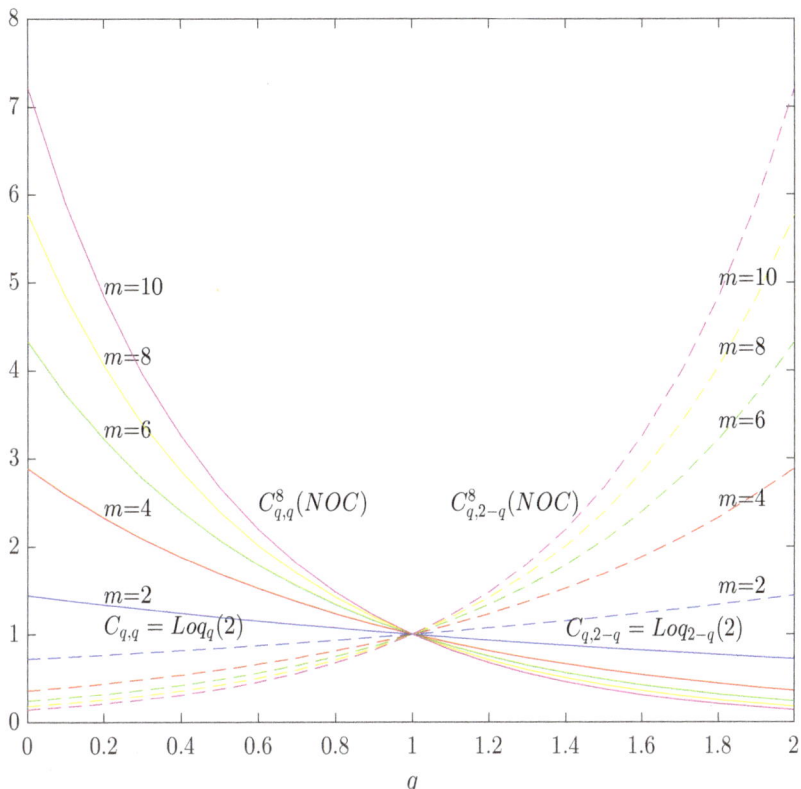

Figure 7. Landsberg–Vedral capacities for the Tsallis (solid lines) and the Landsberg–Vedral (dashed lines) entropies in the case of a (perfect) noisy channel with non-overlapping outputs with m outputs as functions of q, for different values of m. The axiom $(\overline{A_4})$ is broken for all $m > 2$ and satisfied in the case of corresponding α-q-capacities, $C_{q,q}$ and $C_{q,2-q}$.

8. Conclusions and Future Work

A general treatment of the Sharma–Mittal entropy transfer was provided together with the analyses of existing information transfer measures for the non-additive Sharma–Mittal information transfer. It was shown that the existing definitions fail to satisfy at least one of the axioms common to the Shannon case, by which the information transfer has to be non-negative, less than the input and output uncertainty, equal to the input uncertainty in the case of perfect transmission and equal to zero in the case of a totally destructive channel. Thus, breaking some of these axioms implies unexpected and counterintuitive conclusions about the channels, such as achieving super-capacitance or sub-capacitance [4], which could be treated as nonphysical behavior. In this paper, alternative measures of the α-q mutual information and the α-q channel capacity were proposed so that all of the axioms which are broken in the case of the Sharma–Mittal information transfer measures considered before are satisfied, which could qualify them as physically consistent measures of information transfer.

Taking into account the previous research of non-extensive statistical mechanics [3], where the linear growth of the physical quantities has been recognized as a critical property in non-extensive [55] and non-exponentially growing systems [56], and taking into account the previous research from the field of information theory, where the Sharma–Mittal entropy has been considered an appropriate scaling measure which provides extensive information rates [21], the α-q mutual information and the α-q channel capacity seem to be promising measures for the characterization of information transmission in the systems where the Shannon entropy rate diverges or disappears in an infinite time limit. In addition, as was shown in this paper, the proposed information transfer measures are compatible with the maximum likelihood detection, which indicates their potential for operational characterization of coding theory and hypothesis testing problems [26].

Author Contributions: Conceptualization, V.M.I. and I.B.D.; validation, V.M.I. and I.B.D.; formal analysis, V.M.I.; funding acquisition, I.B.D.; project administration, I.B.D.; writing—original draft preparation, V.M.I.; writing—review and editing, V.M.I. and I.B.D. All authors have read and agreed to the published version of the manuscript.

Funding: This research was supported in part by NSF under grants 1907918 and 1828132 and by Ministry of Science and Technological Development, Republic of Serbia, Grants Nos. ON 174026 and III 044006. The APC was funded by NSF under grants 1907918 and 1828132.

Data Availability Statement: Not applicable.

Conflicts of Interest: The authors declare no conflict of interest.

References

1. Ilić, V.M.; Stanković, M.S. A unified characterization of generalized information and certainty measures. *Phys. A Stat. Mech. Appl.* **2014**, *415*, 229–239. [CrossRef]
2. Renyi, A. *Probability Theory*; North-Holland Series in applied mathematics and mechanics; North-Holland Publishing Company: Amsterdam, The Netherlands, 1970.
3. Tsallis, C. Possible generalization of Boltzmann-Gibbs statistics. *J. Stat. Phys.* **1988**, *52*, 479–487. [CrossRef]
4. Landsberg, P.T.; Vedral, V. Distributions and channel capacities in generalized statistical mechanics. *Phys. Lett. A* **1998**, *247*, 211–217. [CrossRef]
5. Frank, T.; Daffertshofer, A. Exact time-dependent solutions of the Renyi Fokker-Planck equation and the Fokker-Planck equations related to the entropies proposed by Sharma and Mittal. *Phys. A Stat. Mech. Appl.* **2000**, *285*, 351–366. [CrossRef]
6. Sharma, B.; Mittal, D. New non-additive measures of entropy for discrete probability distributions. *J. Math. Sci.* **1975**, *10*, 28–40.
7. Tsallis, C. What are the numbers that experiments provide. *Quim. Nova* **1994**, *17*, 468–471.
8. Nivanen, L.; Le Méhauté, A.; Wang, Q.A. Generalized algebra within a nonextensive statistics. *Rep. Math. Phys.* **2003**, *52*, 437–444. [CrossRef]
9. Ilić, V.M.; Stanković, M.S. Generalized Shannon-Khinchin axioms and uniqueness theorem for pseudo-additive entropies. *Phys. A Stat. Mech. Appl.* **2014**, *411*, 138–145. [CrossRef]
10. Jizba, P.; Korbel, J. When Shannon and Khinchin meet Shore and Johnson: Equivalence of information theory and statistical inference axiomatics. *Phys. Rev. E* **2020**, *101*, 042126. [CrossRef]
11. Esteban, M.D.; Morales, D. A summary on entropy statistics. *Kybernetika* **1995**, *31*, 337–346.
12. Lenzi, E.; Scarfone, A. Extensive-like and intensive-like thermodynamical variables in generalized thermostatistics. *Phys. A Stat. Mech. Appl.* **2012**, *391*, 2543–2555. [CrossRef]
13. Frank, T.; Plastino, A. Generalized thermostatistics based on the Sharma-Mittal entropy and escort mean values. *Eur. Phys. J. B Condens. Matter Complex Syst.* **2002**, *30*, 543–549. [CrossRef]
14. Aktürk, O.Ü.; Aktürk, E.; Tomak, M. Can Sobolev inequality be written for Sharma-Mittal entropy? *Int. J. Theor. Phys.* **2008**, *47*, 3310–3320. [CrossRef]
15. Mazumdar, S.; Dutta, S.; Guha, P. Sharma–Mittal quantum discord. *Quantum Inf. Process.* **2019**, *18*, 1–26. [CrossRef]
16. Elhoseiny, M.; Elgammal, A. Generalized Twin Gaussian processes using Sharma–Mittal divergence. *Mach. Learn.* **2015**, *100*, 399–424. [CrossRef]
17. Koltcov, S.; Ignatenko, V.; Koltsova, O. Estimating Topic Modeling Performance with Sharma–Mittal Entropy. *Entropy* **2019**, *21*, 660. [CrossRef]
18. Jawad, A.; Bamba, K.; Younas, M.; Qummer, S.; Rani, S. Tsallis, Rényi and Sharma-Mittal holographic dark energy models in loop quantum cosmology. *Symmetry* **2018**, *10*, 635. [CrossRef]
19. Ghaffari, S.; Ziaie, A.; Moradpour, H.; Asghariyan, F.; Feleppa, F.; Tavayef, M. Black hole thermodynamics in Sharma–Mittal generalized entropy formalism. *Gen. Relativ. Gravit.* **2019**, *51*, 1–11. [CrossRef]
20. Américo, A.; Khouzani, M.; Malacaria, P. Conditional Entropy and Data Processing: An Axiomatic Approach Based on Core-Concavity. *IEEE Trans. Inf. Theory* **2020**, *66*, 5537–5547. [CrossRef]
21. Girardin, V.; Lhote, L. Rescaling entropy and divergence rates. *IEEE Trans. Inf. Theory* **2015**, *61*, 5868–5882. [CrossRef]
22. Ciuperca, G.; Girardin, V.; Lhote, L. Computation and estimation of generalized entropy rates for denumerable Markov chains. *IEEE Trans. Inf. Theory* **2011**, *57*, 4026–4034. [CrossRef]
23. Sibson, R. Information radius. *Z. Wahrscheinlichkeitstheorie Verwandte Geb.* **1969**, *14*, 149–160. [CrossRef]

24. Arimoto, S. Information Mesures and Capacity of Order α for Discrete Memoryless Channels. In *Topics in Information Theory*; Colloquia Mathematica Societatis János Bolyai; Csiszár, I., Elias, P., Eds.; North-Holland Pub. Co.: Amsterdam, The Netherlands, 1977; Volume 16, pp. 41–52.
25. Augustin, U. Noisy Channels. Ph.D. Thesis, Universität Erlangen-Nürnberg, Erlangen, Germany, 1978.
26. Csiszár, I. Generalized cutoff rates and Rényi's information measures. *IEEE Trans. Inf. Theory* **1995**, *41*, 26–34. [CrossRef]
27. Lapidoth, A.; Pfister, C. Two measures of dependence. *Entropy* **2019**, *21*, 778. [CrossRef]
28. Tomamichel, M.; Hayashi, M. Operational interpretation of Rényi information measures via composite hypothesis testing against product and Markov distributions. *IEEE Trans. Inf. Theory* **2017**, *64*, 1064–1082. [CrossRef]
29. Verdú, S. α-mutual information. In Proceedings of the 2015 Information Theory and Applications Workshop (ITA), San Diego, CA, USA, 1–6 February 2015; pp. 1–6.
30. Daróczy, Z. Generalized information functions. *Inf. Control* **1970**, *16*, 36–51. [CrossRef]
31. Chapeau-Blondeau, F.; Rousseau, D.; Delahaies, A. Renyi entropy measure of noise-aided information transmission in a binary channel. *Phys. Rev. E* **2010**, *81*, 051112. [CrossRef]
32. Chapeau-Blondeau, F.; Delahaies, A.; Rousseau, D. Tsallis entropy measure of noise-aided information transmission in a binary channel. *Phys. Lett. A* **2011**, *375*, 2211–2219. [CrossRef]
33. Yamano, T. A possible extension of Shannon's information theory. *Entropy* **2001**, *3*, 280–292. [CrossRef]
34. Shannon, C.E. A mathematical theory of communication. *Bell Syst. Tech. J.* **1948**, *27*, 379–423. [CrossRef]
35. Arimoto, S. Computation of random coding exponent functions. *Inf. Theory IEEE Trans.* **1976**, *22*, 665–671. [CrossRef]
36. Gallager, R. A simple derivation of the coding theorem and some applications. *IEEE Trans. Inf. Theory* **1965**, *11*, 3–18. [CrossRef]
37. Cover, T.M.; Thomas, J.A. *Elements of Information Theory (Wiley Series in Telecommunications and Signal Processing)*; John Wiley & Sons, Inc: Hoboken, NJ, USA, 2006.
38. Fehr, S.; Berens, S. On the conditional Rényi entropy. *Inf. Theory IEEE Trans.* **2014**, *60*, 6801–6810. [CrossRef]
39. Wilde, M.M.; Winter, A.; Yang, D. Strong converse for the classical capacity of entanglement-breaking and Hadamard channels via a sandwiched Rényi relative entropy. *Commun. Math. Phys.* **2014**, *331*, 593–622. [CrossRef]
40. Gupta, M.K.; Wilde, M.M. Multiplicativity of completely bounded p-norms implies a strong converse for entanglement-assisted capacity. *Commun. Math. Phys.* **2015**, *334*, 867–887. [CrossRef]
41. Beigi, S. Sandwiched Rényi divergence satisfies data processing inequality. *J. Math. Phys.* **2013**, *54*, 122202. [CrossRef]
42. Hayashi, M.; Tomamichel, M. Correlation detection and an operational interpretation of the Rényi mutual information. *J. Math. Phys.* **2016**, *57*, 102201. [CrossRef]
43. Hayashi, M.; Tajima, H. Measurement-based formulation of quantum heat engines. *Phys. Rev. A* **2017**, *95*, 032132. [CrossRef]
44. Hayashi, M. Quantum Wiretap Channel With Non-Uniform Random Number and Its Exponent and Equivocation Rate of Leaked Information. *IEEE Trans. Inf. Theory* **2015**, *61*, 5595–5622. [CrossRef]
45. Cai, C.; Verdú, S. Conditional Rényi Divergence Saddlepoint and the Maximization of α-Mutual Information. *Entropy* **2019**, *21*, 969. [CrossRef]
46. Tridenski, S.; Zamir, R.; Ingber, A. The Ziv–Zakai–Rényi bound for joint source-channel coding. *IEEE Trans. Inf. Theory* **2015**, *61*, 4293–4315. [CrossRef]
47. Harremoës, P. Interpretations of Rényi entropies and divergences. *Phys. A Stat. Mech. Its Appl.* **2006**, *365*, 57–62. [CrossRef]
48. Jizba, P.; Kleinert, H.; Shefaat, M. Rényi's information transfer between financial time series. *Phys. A Stat. Mech. Appl.* **2012**, *391*, 2971–2989. [CrossRef]
49. Jizba, P.; Arimitsu, T. The world according to Rényi: Thermodynamics of multifractal systems. *Ann. Phys.* **2004**, *312*, 17–59. [CrossRef]
50. Iwamoto, M.; Shikata, J. Information theoretic security for encryption based on conditional Rényi entropies. In Proceedings of the International Conference on Information Theoretic Security, Singapore, 28–30 November 2013; pp. 103–121.
51. Ilić, V.; Djordjević, I.; Stanković, M. On a general definition of conditional Rényi entropies. *Proceedings* **2018**, *2*, 166. [CrossRef]
52. Fano, R.M. *Transmission of Information*; M.I.T. Press: Cambridge, MA, USA, 1961.
53. Ilic, V.M.; Djordjevic, I.B.; Küeppers, F. On the Daróczy-Tsallis capacities of discrete channels. *Entropy* **2015**, *20*, 2.
54. Yamano, T. Information theory based on nonadditive information content. *Phys. Rev. E* **2001**, *63*, 046105. [CrossRef]
55. Tsallis, C.; Gell-Mann, M.; Sato, Y. Asymptotically scale-invariant occupancy of phase space makes the entropy Sq extensive. *Proc. Natl. Acad. Sci. USA* **2005**, *102*, 15377–15382. [CrossRef]
56. Korbel, J.; Hanel, R.; Thurner, S. Classification of complex systems by their sample-space scaling exponents. *New J. Phys.* **2018**, *20*, 093007. [CrossRef]

Article

Classical and Quantum H-Theorem Revisited: Variational Entropy and Relaxation Processes

Carlos Medel-Portugal [1], Juan Manuel Solano-Altamirano [2] and José Luis E. Carrillo-Estrada [1,*]

[1] Instituto de Física, Benemérita Universidad Autónoma de Puebla, Apdo. Postal. J-48, Puebla 72570, Mexico; cmedel@ifuap.buap.mx
[2] Facultad de Ciencias Químicas, Benemérita Universidad Autónoma de Puebla, 14 Sur y Av. San Claudio, Col. San Manuel, Puebla 72520, Mexico; jmanuel.solano@correo.buap.mx
* Correspondence: carrillo@ifuap.buap.mx

Abstract: We propose a novel framework to describe the time-evolution of dilute classical and quantum gases, initially out of equilibrium and with spatial inhomogeneities, towards equilibrium. Briefly, we divide the system into small cells and consider the local equilibrium hypothesis. We subsequently define a global functional that is the sum of cell H-functionals. Each cell functional recovers the corresponding Maxwell–Boltzmann, Fermi–Dirac, or Bose–Einstein distribution function, depending on the classical or quantum nature of the gas. The time-evolution of the system is described by the relationship $dH/dt \leq 0$, and the equality condition occurs if the system is in the equilibrium state. Via the variational method, proof of the previous relationship, which might be an extension of the H-theorem for inhomogeneous systems, is presented for both classical and quantum gases. Furthermore, the H-functionals are in agreement with the correspondence principle. We discuss how the H-functionals can be identified with the system's entropy and analyze the relaxation processes of out-of-equilibrium systems.

Keywords: non-equilibrium thermodynamics; entropy; variational entropy

Citation: Medel-Portugal, C.; Solano-Altamirano, J.M.; Carrillo-Estrada, J.L.E. Classical and Quantum H-Theorem Revisited: Variational Entropy and Relaxation Processes. *Entropy* **2021**, *23*, 366. https://doi.org/10.3390/e23030366

Academic Editor: Petr Jizba

Received: 27 January 2021
Accepted: 16 March 2021
Published: 19 March 2021

Publisher's Note: MDPI stays neutral with regard to jurisdictional clai-ms in published maps and institutio-nal affiliations.

Copyright: © 2021 by the authors. Licensee MDPI, Basel, Switzerland. This article is an open access article distributed under the terms and conditions of the Creative Commons Attribution (CC BY) license (https://creativecommons.org/licenses/by/4.0/).

1. Introduction

The theoretical bases and the procedures that allow us to describe equilibrium systems are well-established. These procedures can be applied to a wide range of natural systems, including both the macroscopic phenomenological methods (thermodynamics) and the microscopic description (statistical mechanics) (Out-of-equilibrium systems, of course, are still a challenge). For instance, in the kinetic theory of gases, the behavior of a dilute classical gas is described through the Boltzmann transport equation [1], and the time-evolution of a system towards equilibrium is finely accounted for through the Boltzmann H-theorem.

However, for quantum out-of-equilibrium systems, the construction of a kinetic framework with the same level of success and universality as the classical version still presents some fundamental challenges. For instance, to obtain a complete correspondence principle between classical mechanics and quantum mechanics, the form of the quantum analogues of both the Boltzmann H-theorem and the Boltzmann transport equation is inadequate. In this context, Tolman was one of the earliest physicists to propose a quantum H-theorem [2], using a probability transition relationship, the random phases hypothesis, and an H-functional defined in terms of a spatially homogeneous distribution function. Tolman also proposed a potential quantum analogue of the transport equation, in terms of the occupation numbers, by applying time perturbation theory. Additional attempts, under quantum operator formalism, have addressed the description of quantum transport phenomena through the Hamiltonian of the system and the master equation (which is, in these works, the analogue of the Boltzmann transport equation) [3–7]. However, these approaches are not consistent with the classical-quantum correspondence principle.

Similarly, some authors have proposed H-functionals and attempted to proof a quantum H-theorem [8–16]. However, whether or not the homogeneous distribution function hypothesis is assumed or if its framework fulfills the correspondence principle is unclear or not discussed. Since the pioneering work of Tolman, at several stages, there has been some discussion regarding the general validity of the quantum H-theorem, some possible violations of the second law of thermodynamics, and the interpretation of the quantum entropy [10,12–14,16–21].

Nonetheless, the framework to describe spatially non-homogeneous systems is still under construction, although several approaches have been developed. For instance, the celebrated Onsager formulation (linear thermodynamics) [22,23] has been successful in describing irreversible chemical and physical phenomena. However, some descriptions, such as those the internal behavior of gases [24] and the entropy measurement [25,26], cannot be completely addressed with linear thermodynamics.

In addition, some aspects regarding the classical H-theorem and the Boltzmann H-functional require revision to improve their mutual consistency. One example is the modification of the H-theorem to include phenomena stemming from stochastic trajectories, violations of the second law of thermodynamics, the relationship between Shannon's measure of information and the Boltzmann's entropy, and the calculation of thermodynamical quantities and thermalization of specific systems [8,26–29].

To contribute to the construction of a consistent classical and quantum H-theorem, within a formalism that describes out-of-equilibrium non-homogeneous systems, we propose a new theoretical framework. Specifically, for both classical and quantum systems, we include non-homogeneous distribution functions in the H-functionals, and consider non-homogeneous systems in the proofs of the resulting H-theorems. Our proposed H-functionals satisfy the correspondence principle, but more importantly, these functionals describe the time-evolution of spatially non-homogeneous systems towards equilibrium.

The organization of this article is as follows. In Section 2, we highlight, for our purposes, the most fundamental assumptions required to proof the Boltzmann H-theorem, we provide an alternative method to obtain the Maxwell–Boltzmann distribution using the variational method, propose an alternative H-functional for classical systems, and demonstrate the respective H-theorem. In Section 3, we review the Tolman proposal for the quantum version of the H-theorem (quantum H-theorem) and how the Bose–Einstein and Fermi–Dirac distributions are treated within this framework. Subsequently, we present our proposal for a quantum H-functional and the proof of the corresponding quantum H-theorem. In Section 4, we analyze the classical-quantum correspondence between the quantum and classical H-functionals. In Section 5, we explore how relaxation processes occur in a quantum ideal gas and, based on what we call variational entropy, propose a time-evolution equation for the distribution function. Finally, we discuss some key ideas resulting from our approach and close with a summary in Section 6.

2. Classical Scheme

The Boltzmann kinetic theory of gases represents a fundamental connection between the microscopic nature of matter and the phenomenological macroscopic laws of classical thermodynamics. The stochasticity introduced by the molecular chaos hypothesis in the otherwise deterministic kinetics of the particles allows for the demonstration of the celebrated Boltzmann H-theorem. In contrast, in this article, we propose an alternative approach developed using a variational procedure applied to an H-functional. We start this section by briefly accounting for the important elements of the standard derivation of the Boltzmann transport equation and demonstrating the H-theorem, such as they are presented in classical textbooks [1].

2.1. The Boltzmann Transport Equation

The first step in the Boltzmann kinetic theory of gases is defining the distribution function, $f(\vec{r},\vec{v},t)$, as the average number of molecules that, at time t, have position $\vec{r}$ and velocity $\vec{v}$, and are contained in a μ-space volume element $d^3r d^3v$. Assuming a deterministic Newtonian description of molecular motion, as well as the invariance of the μ-space volume measure, one arrives at the Boltzmann rate equation:

$$f(\vec{r}+\vec{v}\delta t, \vec{v}+\vec{F}\delta t, t+\delta t) = f(\vec{r},\vec{v},t) + \left(\frac{\partial f}{\partial t}\right)_{\text{coll}} \delta t. \quad (1)$$

Here, the term $(\partial f/\partial t)_{\text{coll}}$ describes the in and out fluxes from and towards the volume element, due to the collisions. Subsequently, from the previous equation, the integro-differential Boltzmann transport equation is obtained:

$$\left(\frac{\partial}{\partial t} + \vec{v}_1 \cdot \nabla_{\vec{r}} + \frac{\vec{F}}{m} \cdot \nabla_{\vec{v}_1}\right) f_1 = \int d\Omega \int d^3v_2 \sigma(\Omega) |\vec{v}_1 - \vec{v}_2| (\tilde{f}_2 \tilde{f}_1 - f_2 f_1). \quad (2)$$

In Equation (2), Ω is the solid angle, σ is the scattering cross section, $\vec{F}$ the external force applied to the system, and f_1 and f_2 ($\tilde{f}_1$ and $\tilde{f}_2$) are the distribution functions of particles 1 and 2, respectively, before (and after) the collision.

Particle dynamics and the effects of external forces are described by the left-hand side of Equation (2). The right-hand side is derived by considering binary collisions between particles and accepting the *molecular chaos hypothesis*, i.e., it is assumed that the positions and velocities of the particles are not time-correlated.

2.2. A Summary of the H-Theorem and the Maxwell—Boltzmann Distribution

The evolution of a dilute gas towards thermodynamic equilibrium is frequently addressed by first defining the H-functional [1,2]:

$$H_B = \int f(\vec{v},t) \ln f(\vec{v},t) d^3v. \quad (3)$$

Notice that $f_B(\vec{v},t)$ is a spatially homogeneous distribution function. The functional H_B, originally introduced by Boltzmann in 1872, describes a dilute gas occupying a volume V, at temperature T, with total energy E, and total number of free classical particles N. To clearly distinguish the Boltzmann functional H_B, we denote hereafter the Maxwell–Boltzmann distribution function as f_B.

The physically correct spontaneous time-evolution of an out of equilibrium dilute gas is corroborated by the H-theorem. This theorem establishes that if (a) the homogeneous function $f(\vec{v},t)$ satisfies the Boltzmann transport equation and (b) the molecular chaos hypothesis is valid, then the system evolves in such a manner that $dH_B/dt \leq 0$, and if $dH_B/dt = 0$, then the system is in the equilibrium state. The H-theorem is straightforward to prove using Equation (2) [1], and it assures the consistency between our microscopic approach to describe the system's spontaneous time-evolution and the phenomenological observations established by the second law of classical thermodynamics; in fact, H_B can be associated with an entropy density.

On the other hand, considering a dilute gas in equilibrium with no applied external forces, i.e., $(\partial f/\partial t) = 0$ and f is independent of $\vec{r}$, we can directly prove that the equilibrium distribution function obtained from Equation (2) is precisely the Maxwell–Boltzmann distribution function. The proof of the above first requires identification of the sufficient condition for f to render a null r.h.s. of Equation (2). Such an f, which we denote here as f_0, must satisfy

$$f_0(\vec{v}_2')f_0(\vec{v}_1') - f_0(\vec{v}_2)f_0(\vec{v}_1) = 0. \quad (4)$$

Subsequently, the Maxwell–Boltzmann distribution function can be obtained by taking the logarithm of Equation (4) and conserved mechanical quantities (see ([1], ch. 4.2)).

Before introducing our proposed H-functional, we must state that in defining H_B, Equation (3), it is assumed that the distribution function f is spatially homogeneous. This assumption simplifies the demonstration of the H theorem. However, it also introduces a limited conception of the out-of-equilibrium condition of the gas. Given the relatively simple nature of a dilute gas, one of the salient features of an out-of-equilibrium condition is the existence of inhomogeneities in the system, which is not considered in the above.

2.3. Non-Homogeneous Classical H-Functional

As we saw in the previous section, the validity of the H-theorem relies significantly on assuming that the distribution function is homogeneous and the molecular chaos hypothesis is fulfilled. To extend the previous procedure to systems with non-homogeneous distribution functions, which might allow for the study of systems in a more general out-of-equilibrium condition, we introduce a modified H-functional. For the sake of clarity and simplicity, we use primed functions and quantities to denote the classical case to differentiate them from the quantum analogues.

Our proposed classical H-functional, denoted as $\mathcal{H}'$, describes a dilute classical gas occupying a volume V. In our theoretical treatment, we divide this volume into K cells, which, without loss of generality, have identical volumes, $\delta V_M = V/K$, $M = 1, \ldots, K$. Each cell of index M has the following local functions, properties, and variables: an H-functional, $\mathcal{H}'_M$, a homogeneous distribution function, $f'_M(\vec{v}, t)$, number of particles, $\mathcal{N}'_M$, temperature, T'_M, and energy, $\mathcal{E}'_M$. Taken as a whole, the system has an energy E, and a global number of free classical particles N. We also assume that the system is perfectly isolated, and that the number of particles in each cell is sufficiently large, so as to obtain accurate averages. We start our analysis by proposing the following inhomogeneous H-functional:

$$\mathcal{H}'(t) = \sum_{M=1}^{K} \int_{\delta V_M} f'_M(\vec{v}, t) \ln f'_M(\vec{v}, t) \mathrm{d}^3 v. \tag{5}$$

The distribution functions, $\{f'_M(\vec{v}, t)\}$, depend implicitly on the position of the cells, relative to the global system, and on the velocity $\vec{v}$ and time t. Notice that each f'_M can be formally extended to the complete coordinate space by defining each f'_M to be zero outside the M-th cell, in such a manner that the distribution function of the complete system is a piece-wise sum of $\{f'_M(\vec{v}, t)\}$:

$$f'(\vec{r}, \vec{v}, t) = \sum_{M=1}^{K} f'(\vec{r}_M, \vec{v}, t) = \sum_{M=1}^{K} f'_M(\vec{v}, t). \tag{6}$$

Here, $\vec{r}_M$ is the center of the cell of index M, and $f'_M(\vec{v}, t) \neq f'_N(\vec{v}, t)$ for $M \neq N$. This extended definition allows us to omit the symbol δV_M in all integrals performed over the cell volume. In terms of $f'_M(\vec{v}, t)$ and a local variable of energy, $\epsilon(\vec{v})$, we have

$$\mathcal{H}'_M = \int f'_M(\vec{v}, t) \ln f'_M(\vec{v}, t) \mathrm{d}^3 v, \tag{7}$$

$$\mathcal{N}'_M = \int f'_M(\vec{v}, t) \mathrm{d}^3 v, \tag{8}$$

$$\mathcal{E}'_M = \int f'_M(\vec{v}, t) \epsilon(\vec{v}) \mathrm{d}^3 v. \tag{9}$$

Notice that assuming every f'_M to be homogeneous implies that we are accepting the validity of the local equilibrium hypothesis. In addition, the set $\{f'_M(\vec{v}, t)\}$ must satisfy the following restrictions:

$$\sum_{M=1}^{K} \int f'_M(\vec{v}, t) \mathrm{d}^3 v = N \tag{10}$$

and

$$\sum_{M=1}^{K} \int f'_M(\vec{v}, t) \epsilon(\vec{v}) \mathrm{d}^3 v = E. \tag{11}$$

We now use the variational method to find the extremal of $\mathcal{H}'$, consistent with restrictions (7)–(11) together with the corresponding Lagrange multipliers $\{\alpha_M\}$ and $\{\beta_M\}$. This yields

$$\begin{aligned}\frac{\delta \mathcal{H}'}{\delta f'_J(\vec{v}')} &= \sum_{M=1}^{K}\int \frac{\delta}{\delta f'_J(\vec{v}')}[f'_M(\vec{v})\ln f'_M(\vec{v})]\mathrm{d}^3 v - \sum_{M=1}^{K}\alpha_M\int \frac{\delta f'_M(\vec{v})}{\delta f'_J(\vec{v}')}\mathrm{d}^3 v \\ &\quad - \sum_{M=1}^{K}\beta_M\int \epsilon(\vec{v})\frac{\delta f'_M(\vec{v})}{\delta f'_J(\vec{v}')}\mathrm{d}^3 v \\ &= \ln f'_J(\vec{v}') + 1 - \alpha_J - \beta_J \epsilon(\vec{v}') = 0.\end{aligned} \qquad (12)$$

Solving the last line for $f'_J(\vec{v}')$ renders

$$f'_J(\vec{v}') = C\exp(\alpha_J + \beta_J \epsilon(\vec{v}')) \qquad (13)$$

where C is a constant. We notice that by applying the variational procedure on $\mathcal{H}'$, we predict that when equilibrium is reached, the distribution function of each cell has the form of the Maxwell–Boltzmann distribution function, which is consistent with the local equilibrium assumption.

2.3.1. Properties of $\mathcal{H}'$ for Systems in Equilibrium

If the complete system is in equilibrium without external forces applied to the gas, from classical thermodynamics of systems in equilibrium, we ascertain that the local number of particles and the local energy do not depend on the cell number. In a statistical sense, this is

$$\mathcal{N}'_M = \mathcal{N}' \equiv \bar{\mathcal{N}}' \qquad (14)$$

and

$$\mathcal{E}'_M = \mathcal{E}' \equiv \bar{\mathcal{E}}'. \qquad (15)$$

In Equations (14) and (15) the bar implies averaged properties over the complete system. Moreover, the global distribution function is homogeneous, hence f'_M does not depend on the cell number M (i.e., $f'_M(\vec{v},t) = f'(\vec{v},t)$, $\forall M$). Several properties arise directly from this, e.g., from Equations (14) and (15) $E = \sum_M \mathcal{E}'_M = K\mathcal{E}'$, $N = \sum_M \mathcal{N}'_M = K\mathcal{N}'$. Here we have used Equations (8) and (9). Equation (5), in terms of Equation (3), can be rewritten as:

$$\mathcal{H}'(t) = \int \sum_{M=1}^{K} [f'(\vec{v},t)\ln f'(\vec{v},t)]\mathrm{d}^3 v = K\int f'(\vec{v},t)\ln f'(\vec{v},t)\mathrm{d}^3 v = KH_B(t). \qquad (16)$$

To identify $\mathcal{H}'$ with the entropy, we need to show that $\mathcal{H}'(t)$ is extensive, with respect to $Kf'(\vec{v},t)$. This is shown by analyzing the following expression:

$$\begin{aligned}\int [Kf'(\vec{v},t)]\ln[Kf'(\vec{v},t)]\mathrm{d}^3 v &= \int [(K\ln K)f'(\vec{v},t) + Kf'(\vec{v},t)\ln f'(\vec{v},t)]\mathrm{d}^3 v \\ &= K\int \left\{f'(\vec{v},t)\left[\ln K + \ln f'(\vec{v},t)\right]\right\}\mathrm{d}^3 v.\end{aligned} \qquad (17)$$

We observe that if the number of particles in the μ-space, $f'(\vec{v},t)$, is much larger than the number of cells, K, then the first term of Equation (17) is negligible, and consequently:

$$\int [Kf'(\vec{v},t)]\ln[Kf'(\vec{v},t)]\mathrm{d}^3 v \approx K\int f'(\vec{v},t)\ln f'(\vec{v},t)\mathrm{d}^3 v, \qquad (18)$$

i.e., $\mathcal{H}'_M$ is extensive, and the sum $\sum_M \mathcal{H}'_M$ is the H-functional of the complete system, which reduces to the Boltzmann H-functional. Therefore, $\mathcal{H}'$ can be identified with the entropy density of the system.

Furthermore, in equilibrium, the Lagrange multipliers are position- and time-independent thus $f'_M(\vec{v})$ reduces to

$$f'_M(\vec{v}) = C\exp(\alpha + \beta\epsilon(\vec{v})) \equiv \tilde{f}'(\vec{v}), \quad M = 1, \ldots, K. \tag{19}$$

The constant C can be omitted, which is shown by defining the following $\mathcal{H}''$ functional:

$$\mathcal{H}''(t) = \sum_{M=1}^{K} \int [f'_M(\vec{v},t) \ln f'_M(\vec{v},t) - f'_M(\vec{v},t)] \mathrm{d}^3 v. \tag{20}$$

Since $\sum_{M=1}^{K} \int f'_M(\vec{v},t) \mathrm{d}^3 v = N$ (a constant), and because we are mainly interested in the time-derivative of $\mathcal{H}''$, C can be conveniently omitted. In other words, $\mathcal{H}'$ leads to the Maxwell–Boltzmann distribution function of systems in equilibrium.

2.3.2. Proof of the H-Theorem for Non-Homogeneous Distributions

Throughout this section, we consider a classical gas with an initial condition close to the equilibrium, which ensures that the local equilibrium hypothesis remains valid during the time-evolution of the system. Also, we use the following definitions for the deviations of concentration and energy, relative to the equilibrium values:

$$\mathcal{N}'_M(t) = \int f'_M(\vec{v},t) \mathrm{d}^3 v = \tilde{\mathcal{N}}' + \Delta'_M(t) \tag{21}$$

and

$$\mathcal{E}'_M(t) = \int f'_M(\vec{v},t)\epsilon(\vec{v}) \mathrm{d}^3 v = \tilde{\mathcal{E}}' + \delta'_M(t). \tag{22}$$

Here $\tilde{\mathcal{N}}' = N/K$ and $\tilde{\mathcal{E}}' = E/K$ are the cell particle number and the cell energy in equilibrium, respectively, which are given by

$$\tilde{\mathcal{N}}' = \int \tilde{f}'(\vec{v}) \mathrm{d}^3 v \quad \text{and} \quad \tilde{\mathcal{E}}' = \int \tilde{f}'(\vec{v})\epsilon(\vec{v}) \mathrm{d}^3 v, \tag{23}$$

where we have used $\tilde{f}'(\vec{v})$ as defined in Equation (19). In Equations (21) and (22), Δ'_M and δ'_M are considered deviations relative to $\tilde{\mathcal{N}}'$ and $\tilde{\mathcal{E}}'$, respectively. For systems that are sufficiently close to equilibrium, it is reasonable to expect first that $\Delta'_M(t) \ll \tilde{\mathcal{N}}'$ and $\delta'_M(t) \ll \tilde{\mathcal{E}}'$, and second that Δ'_M and δ'_M are sufficiently large compared to the fluctuations of $\tilde{\mathcal{N}}'$ and $\tilde{\mathcal{E}}'$. Similarly, we can assume that every local distribution function can be written as

$$f'_M(\vec{v},t) = \tilde{f}'(\vec{v})(1 + g'_M(\vec{v},t)), \quad 1 \gg |g'_M(\vec{v},t)|. \tag{24}$$

With the previous considerations, in the following, we proof an alternative H-theorem, considering the H-functional, $\mathcal{H}'$, defined by Equation (5).

We commence by differentiating Equation (5) with respect to time:

$$\frac{\mathrm{d}\mathcal{H}'}{\mathrm{d}t} = \sum_{M=1}^{K} \int [1 + \ln f'_M(\vec{v},t)] \dot{f}'_M(\vec{v},t) \mathrm{d}^3 v. \tag{25}$$

(Starting here, we use the standard notation $\dot{h} \equiv (\mathrm{d}h/\mathrm{d}t)$). Substituting Equation (24) into Equation (25) yields

$$\frac{\mathrm{d}\mathcal{H}'}{\mathrm{d}t} = \sum_{M=1}^{K} \int \tilde{f}'(\vec{v})[1 + \ln\{\tilde{f}'(\vec{v}) + \tilde{f}'(\vec{v})g'_M(\vec{v},t)\}] \dot{g}'_M(\vec{v},t) \mathrm{d}^3 v. \tag{26}$$

The logarithmic term of Equation (26) expanded up to the first-order term of its Taylor series, around $g'_M(\vec{v},t) = 0$, is

$$\ln[\tilde{f}'(\vec{v}) + \tilde{f}'(\vec{v})g'_M(\vec{v},t)] \approx \ln[\tilde{f}'(\vec{v})] + g'_M(\vec{v},t), \tag{27}$$

and substituting this into Equation (26) gives

$$\frac{d\mathcal{H}'}{dt} = \sum_{M=1}^{K} \int \tilde{f}'(\vec{v})[1 + \ln \tilde{f}'(\vec{v}) + g'_M(\vec{v},t)]\dot{g}'_M(\vec{v},t)\mathrm{d}^3v. \tag{28}$$

Substituting $\ln \tilde{f}'(\vec{v}) = \exp(\alpha + \beta\epsilon(\vec{v}))$, see Equation (19) and subsequent text, and omitting C we obtain

$$\frac{d\mathcal{H}'}{dt} = \sum_{M=1}^{K} \int \tilde{f}(\vec{v})[\alpha + \beta\epsilon(\vec{v})]\dot{g}_M(\vec{v},t)\mathrm{d}^3v + \sum_{M=1}^{K} \int \tilde{f}(\vec{v})g_M(\vec{v},t)\dot{g}_M(\vec{v},t)\mathrm{d}^3v. \tag{29}$$

From the definitions of $\mathcal{N}'_M$ and $\mathcal{E}'_M$ —Equations (21) and (22)—and $f'_M(\vec{v},t)$ —Equation (24)— it is straightforward to show that

$$\int \tilde{f}(\vec{v})g_M(\vec{v},t)\mathrm{d}^3v = \Delta_M(t) \quad \Rightarrow \quad \int \tilde{f}(\vec{v})\dot{g}_M(\vec{v},t)\mathrm{d}^3v = \dot{\Delta}_M(t), \tag{30}$$

$$\int \tilde{f}(\vec{v})g_M(\vec{v},t)\epsilon(\vec{v})\mathrm{d}^3v = \delta_M(t) \quad \Rightarrow \quad \int \tilde{f}(\vec{v})\dot{g}_M(\vec{v},t)\epsilon(\vec{v})\mathrm{d}^3v = \dot{\delta}_M(t), \tag{31}$$

and as a consequence of $\sum_{M=1}^{K} \Delta_M(t) = \sum_{M=1}^{K} \delta_M(t) = 0$, we find

$$\sum_{M=1}^{K} \dot{\Delta}_M(t) = \sum_{M=1}^{K} \dot{\delta}_M(t) = 0. \tag{32}$$

Therefore, due to Equations (30)–(32), Equation (29) simplifies to

$$\frac{d\mathcal{H}'}{dt} = \sum_{M=1}^{K} \int \tilde{f}(\vec{v})g_M(\vec{v},t)\dot{g}_M(\vec{v},t)\mathrm{d}^3v. \tag{33}$$

To clearly determine the time-evolution of Equation (33), we split the summation over M into two terms:

$$\frac{d\mathcal{H}'}{dt} = \sum_{J}^{L} \int \tilde{f}'(\vec{v}) g_J'^{+}(\vec{v},t) \dot{g}_J'^{+}(\vec{v},t) \mathrm{d}^3v + \sum_{J}^{P} \int \tilde{f}'(\vec{v}) g_J'^{-}(\vec{v},t) \dot{g}_J'^{-}(\vec{v},t) \mathrm{d}^3v \tag{34}$$

where $L + P = K$. The above split is made based on the assumption that for any given initial state of the system, at t_0, some cells will have either a $g'_J(\vec{v},t_0) \geq 0$ or a $g'_J(\vec{v},t_0) < 0$, which we denote as $\dot{g}_J'^{+}(\vec{v},t)$ or $\dot{g}_J'^{-}(\vec{v},t)$, respectively.

If the system's initial state is sufficiently close to equilibrium, it is physically appropriate to assume that $|g'_M(\vec{v},t_0)| \to 0$ as $t \to \infty$ in a monotonous manner, thus $\dot{g}_J'^{+}(\vec{v},t) \leq 0$ and $\dot{g}_J'^{-}(\vec{v},t) > 0$, for $t \geq t_0$. Consequently, Equation (34) can be re-written as

$$\frac{d\mathcal{H}'}{dt} = -\left[\sum_{J}^{L} \int \tilde{f}(\vec{v})|g_J^{+}(\vec{v},t)||\dot{g}_J^{+}(\vec{v},t)|\mathrm{d}^3v + \sum_{J}^{P} \int \tilde{f}(\vec{v})|g_J^{-}(\vec{v},t)||\dot{g}_J^{-}(\vec{v},t)|\mathrm{d}^3v\right]. \tag{35}$$

Since every integrand in Equation (35) is positive, for all t and $\vec{v}$, and $|g'_M(\vec{v},t_0)| \to 0$ as $t \to \infty$, it follows that

$$\frac{d\mathcal{H}'}{dt} \leq 0. \qquad \text{QED.} \tag{36}$$

In summary, considering a gas occupying a volume V (which is divided into K small cells), with a total energy E and N classical free particles, whose initial state is not in equilibrium, but sufficiently close to equilibrium, then the functional

$$\mathcal{H}'(t) = \sum_{M=1}^{K} \int f'_M(\vec{v},t) \ln f'_M(\vec{v},t) \mathrm{d}^3v \tag{37}$$

where $f'_M(\vec{v}, t)$ is the cell distribution function, which satisfies $d\mathcal{H}'/dt \leq 0$, and the equality relation is attained at $t \to \infty$. In Equation (37), $f'_M(\vec{v}, t)$ is the Maxwell–Boltzmann distribution function, which in general is different for different cells—i.e., the complete system can be non-homogeneous—and each $f'_M(\vec{v}, t)$ is compatible with the cell properties, such as number of particles, $\mathcal{N}'_M$, energy, $\mathcal{E}'_M$, temperature, T_M, and Legendre multipliers α_M and β_M.

3. Quantum Scheme

The classical H-theorem is still considered one pillar on which classical statistical physics is founded. Unfortunately, despite multiple attempts [3,10,11,14–16], the generality of the classical H-theorem has no equally robust quantum match. In this section, we propose and analyze an alternative quantum H-functional using the variational method. We start by briefly outlining a typical textbook demonstration of the quantum H-theorem [2], and subsequently present the analysis of our proposed H-functional.

3.1. H-Theorem and the Fermi–Dirac and Bose–Einstein Distribution Functions

Consider a dilute gas of N non-interacting quantum particles (either bosons or fermions), contained by a vessel of volume V, temperature T, and total energy E. Starting from the Boltzmann definition of entropy, the quantum H functional is

$$H_T = -\ln G, \qquad (38)$$

where G describes the total number of accessible quantum states of the gas that satisfy the above conditions [2]. The quantum H-theorem can be demonstrated as follows. G can be divided into groups of neighboring states, g_k, and certain occupation numbers, n_k, can be associated with each of these groups. Thus, the above functional takes the form

$$H_T = \sum_i n_i \ln n_i - (n_i \pm g_i) \ln(g_i \pm n_i) \pm g_i \ln g_i, \qquad (39)$$

where the upper and lower signs are for bosons and fermions, respectively. Thus the time derivative of Equation (39) is

$$\frac{dH_T}{dt} = \sum_\kappa [\ln n_\kappa - \ln(g_\kappa \pm n_\kappa)] \frac{dn_\kappa}{dt}. \qquad (40)$$

Assuming that the energy exchange between particles is produced by interparticle collisions, and using perturbation theory, the rate of change in the number of particles in a group κ is

$$\begin{aligned}\frac{dn_\kappa}{dt} &= -\sum_{\lambda,(\mu\nu)} A_{\kappa\lambda,\mu\nu} n_\kappa n_\lambda (g_\mu \pm n_\mu)(g_\nu \pm n_\nu) \\ &+ \sum_{\lambda,(\mu\nu)} A_{\mu\nu,\kappa\lambda} n_\mu n_\nu (g_\kappa \pm n_\kappa)(g_\lambda \pm n_\lambda).\end{aligned} \qquad (41)$$

Here $A_{\kappa\lambda,\mu\nu} n_\kappa n_\lambda (g_\mu \pm n_\mu)(g_\nu \pm n_\nu)$ is the expected number of collisions per unit time, in which two particles will be moved from groups (κ, λ) to (μ, ν), and the tensor $A_{\kappa\lambda,\mu\nu}$ is given by

$$A_{\kappa\lambda,\mu\nu} = \frac{4\pi^2}{h} \frac{|I_1 \pm I_2|^2}{\Delta\epsilon}. \qquad (42)$$

In Equation (42), $\Delta\epsilon$ is the net energy change occurring during the collision and $|I_1 - I_2|^2 = |V_{mn,kl}|^2$, where $V_{mn,kl}$ is the element of the transition matrix of a binary collision. It is important to remark that in deriving Equation (41), the equal a priori probabilities and the random a priori phase hypotheses were assumed valid. The random a priori phase hypothesis can be considered an analogue of the molecular chaos hypothesis [15], as it is the mechanism by which stochasticity is introduced into the system.

Substituting Equation (41) into Equation (40), it is straightforward to prove that

$$\frac{dH_T}{dt} \leq 0. \tag{43}$$

At equilibrium (at $t \to \infty$), $dn_\kappa/dt = 0$, hence from Equation (41)

$$\ln\frac{n_\kappa}{g_\kappa \pm n_\kappa} + \ln\frac{n_\lambda}{g_\lambda \pm n_\lambda} = \ln\frac{n_\mu}{g_\mu \pm n_\mu} + \ln\frac{n_\nu}{g_\nu \pm n_\nu}. \tag{44}$$

Considering that energy is conserved during the collision, the Bose–Einstein or Fermi–Dirac distribution functions can be recovered from Equation (44):

$$n_\kappa = \frac{g_\kappa}{\exp(\alpha + \beta\epsilon_\kappa) \mp 1}. \tag{45}$$

In other words, at equilibrium $dH_T/dt = 0$, the distribution function obtained from Equation (39) is the expected distribution function.

3.2. Out-of-Equilibrium, Non-Homogeneous Quantum Systems

Consider a dilute gas enclosed by a perfectly isolated vessel of volume V, with total energy E, and total number of quantum particles N, which can be free fermions or bosons. For our purposes, the volume V is divided into K small cells, each of which has constant volume $\delta V_M = V/K$ ($M = 1, \ldots, K$), temperature T_M, energy ϵ_M, number of particles $\mathcal{N}_M$, and distribution function, $\{f_{Mn}(t)\}$. Hereafter we use the following short-hand notation:

$$f_{Mn}(t) \equiv f(\vec{r}_M, \epsilon_n, t), \tag{46}$$

where $\vec{r}_M$ is the radius vector pointing at the center of the M-th cell. $f_{Mn}(t)$ represents the number of particles contained in the M-th cell that occupies the energy level ϵ_n at time t. Since the particles are considered to be free, the energy levels should not depend on the cell properties, i.e., the energy spectrum, $\{\epsilon_n\}$, is the same for all cells; thus, there is no need to label ϵ_n with an index M.

We propose the following functional as an alternative H-functional for quantum non-homogeneous dilute gases:

$$\mathcal{H}(t) = \sum_{M=1}^{K}\sum_n \left[f_{Mn}(t)\ln f_{Mn}(t) \right.$$
$$\left. \pm \left(1 \mp f_{Mn}(t)\right)\ln(1 \mp f_{Mn}(t)) \right]\delta V_M. \tag{47}$$

Here, the upper and lower signs refer to fermions and bosons, respectively. In addition, when needed, each cell has an associated local chemical potential, α_M, and a local H-functional, which is defined by

$$\mathcal{H}_M(t) = \sum_n \left[f_{Mn}(t)\ln f_{Mn}(t) \pm \left(1 \mp f_{Mn}(t)\right)\ln\left(1 \mp f_{Mn}(t)\right) \right]\delta V_M. \tag{48}$$

Therefore, $\mathcal{N}_M$ and $\mathcal{E}_M$ as functions of time are given by

$$\mathcal{N}_M(t) = \sum_n f_{Mn}(t)\delta V_M \tag{49}$$

and

$$\mathcal{E}_M(t) = \sum_n f_{Mn}(t)\epsilon_n \delta V_M, \tag{50}$$

which are, for the whole system, constrained by the micro-canonical restrictions

$$\sum_{M=1}^{K}\left[\sum_{n}f_{Mn}(t)\right]\delta V_M = \sum_{M=1}^{K}\mathcal{N}_M(t)\delta V_M = N, \tag{51}$$

and

$$\sum_{M=1}^{K}\left[\sum_{n}f_{Mn}(t)\epsilon_n\right]\delta V_M = \sum_{M=1}^{K}\mathcal{E}_M(t)\delta V_M = E. \tag{52}$$

Applying the variational method to $\mathcal{H}$, and using the Lagrange multipliers $\{\alpha_M\}$ and $\{\beta_M\}$, we readily obtain (see also the discussion related to Equation (12)):

$$\ln\left(\frac{1 \mp f_{Mn}(t)}{f_{Mn}(t)}\right) = -\alpha_M(t) - \beta_M(t)\epsilon_n, \tag{53}$$

and solving for $f_{Mn}(t)$ yields

$$f_{Mn}(t) = \frac{1}{\exp\left(-\alpha_M(t) - \beta_M(t)\epsilon_n\right) \pm 1}. \tag{54}$$

Thus, in this zero-order approximation, the form of equilibrium distribution functions is conserved.

3.2.1. Properties of $\mathcal{H}$ for Systems in Equilibrium

If the system is in equilibrium, the temperature becomes homogeneous throughout the complete system. Also, the local number of particles, the local energy, and the Lagrange multipliers do not depend on the cell number, and they should be homogeneous. This is represented by

$$\mathcal{N}_M(t \to \infty) \equiv \bar{\mathcal{N}} = N/K, \tag{55}$$

$$\mathcal{E}_M(t \to \infty) \equiv \bar{\mathcal{E}} = E/K, \tag{56}$$

$$\alpha_M = \bar{\alpha}, \quad \forall M, \tag{57}$$

and

$$\beta_M = \bar{\beta}, \quad \forall M. \tag{58}$$

Substituting Equations (57) and (58) into Equation (54) yields the distribution function of each cell in equilibrium:

$$\tilde{f}_{Mn} = \tilde{f}_n = \frac{1}{\exp\left(-\bar{\alpha} - \bar{\beta}\epsilon_n\right) \pm 1}, \quad \forall M. \tag{59}$$

Using the above equation, we can recover the distribution function and the entropy of a dilute quantum gas in equilibrium as follows. Setting $\bar{\alpha} = \mu/kT$ and $\bar{\beta} = -1/kT$, and substituting them into Equation (59), it renders the Fermi–Dirac and Bose–Einstein distribution functions:

$$\tilde{f}_n = \frac{1}{\exp\left(\frac{\epsilon_n - \mu}{kT}\right) \pm 1}, \tag{60}$$

and substituting Equation (59) into the negative of Equation (47), the entropy of a quantum ideal gas is

$$S = \sum_{M=1}^{K}\sum_{n}\left[\left(\frac{1}{\exp\left(-\bar{\alpha} - \bar{\beta}\epsilon_n\right) \pm 1}\right)\ln\left(\frac{1}{\exp\left(-\bar{\alpha} - \bar{\beta}\epsilon_n\right) \pm 1}\right)\right]$$
$$\pm \ln\left[\prod_{M=1}^{K}\prod_{n}\left(1 \mp \frac{1}{\exp\left(-\bar{\alpha} - \bar{\beta}\epsilon_n\right) \pm 1}\right)\right]\delta V_M. \tag{61}$$

This quantity is what we refer to as "variational entropy," and this name reflects the fact that it was obtained via the variational method.

3.2.2. Proof of the Quantum H-Theorem for Non-Homogeneous Systems

For quantum systems, we also accept the validity of the local equilibrium hypothesis for every cell in the system. This allows us to define non-homogeneous systems, wherein thermodynamic quantities are well-defined on a per-cell basis. In terms of the equilibrium properties, we have

$$\mathcal{N}_M(t) = \sum_n f_{Mn}(t) = \bar{\mathcal{N}} + \Delta_M(t) \tag{62}$$

and

$$\mathcal{E}_M(t) = \sum_n \epsilon_n f_{Mn}(t) = \bar{\mathcal{E}} + \delta_M(t). \tag{63}$$

In Equations (62) and (63) $\bar{\mathcal{N}}$ and $\bar{\mathcal{E}}$ are the cell particle number and the cell energy in equilibrium, which are given by Equations (55) and (56), and Δ_M and δ_M are deviations from $\bar{\mathcal{N}}$ and $\bar{\mathcal{E}}$, respectively, with $\Delta_M(t) \ll \bar{\mathcal{N}}$ and $\delta_M(t) \ll \bar{\mathcal{E}}$.

In the present context, $|\Delta_M|$ and $|\delta_M|$ are sufficiently large to not be fluctuations of the system, and sufficiently small so that the local equilibrium hypothesis is valid for $t > 0$ (we set $t_0 = 0$, and t_0 is the initial time at which the system is prepared). Therefore, the distribution functions can be rewritten as

$$f_{Mn}(t) = \bar{f}_n(1 + g_{Mn}(t)), \quad 1 \gg |g_{Mn}(t)|, \tag{64}$$

from which it follows, by substituting Equation (64) into Equations (62) and (63), that Δ_M and δ_M satisfy

$$\Delta_M(t) = \sum_n \bar{f}_n g_{nM} \tag{65}$$

and

$$\delta_M(t) = \sum_n \bar{f}_n g_{nM} \epsilon_n. \tag{66}$$

An additional consideration is necessary for treating Fermi gases. Since, for these systems, $f_{Mn}(t) \leq 1$, we have

$$1 - \bar{f}_n - \bar{f}_n g_{nM} \geq 0 \implies \frac{1}{\bar{f}_n} \geq 1 + g_{nM}. \tag{67}$$

$\bar{f}_n = 1$ is certainly satisfied if the system temperature is zero. In this state, all energy levels below and including the Fermi energy are occupied, thus the system will necessarily be homogeneous, and consequently, $g_{nM} = 0$. In this article, we will omit this scenario and will only discuss Fermi gases with non-zero temperatures.

To proof the quantum H-theorem, we start by taking the time-derivative of Equation (47):

$$\frac{d\mathcal{H}(t)}{dt} = \sum_n \sum_{M=1}^{K} \dot{f}_{nM}(t) \ln\left[\frac{f_{nM}(t)}{1 \mp f_{nM}(t)}\right] \delta V_M. \tag{68}$$

Subsequently, we substitute Equation (64) into the above equation to obtain

$$\begin{aligned}\frac{d\mathcal{H}(t)}{dt} &= \sum_n \sum_{M=1}^{K} \bar{f}_n \ln\left[\frac{\bar{f}_n(1 + g_{nM})}{1 \mp \bar{f}_n(1 + g_{nM})}\right] \dot{g}_{nM} \delta V_M \\ &= \sum_n \sum_{M=1}^{K} \bar{f}_n \{\ln[\bar{f}_n + \bar{f}_n g_{nM}] \dot{g}_{nM} - \ln[1 \mp \bar{f}_n \mp \bar{f}_n g_{nM}] \dot{g}_{nM} \} \delta V_M. \end{aligned} \tag{69}$$

The logarithmic terms, corresponding to Fermi and Bose gases, are approximated through a Taylor series around $\bar{f}_n g_{nM} = 0$ as

$$\ln[1 \mp \bar{f}_n \mp \bar{f}_n g_{nM}] \approx \ln[1 \mp \bar{f}_n] \mp \frac{\bar{f}_n}{1 \mp \bar{f}_n} g_{nM} \tag{70}$$

and

$$\ln[\bar{f}_n + \bar{f}_n g_{nM}] \approx \ln[\bar{f}_n] + g_{nM}, \tag{71}$$

respectively. Equation (70) is valid because, for non-extremely degenerated Fermi gases, $1 - \bar{f}_n \gg \bar{f}_n |g_{nM}|$ and Equation (71) is fulfilled because, for Boson gases, $1 + \bar{f}_n \gg \bar{f}_n |g_{nM}|$ when $\bar{f}_n \gg \bar{f}_n |g_{nM}|$.

Combining Equations (69)–(71),

$$\frac{d\mathcal{H}}{dt} = \sum_n \sum_{M=1}^{K} \bar{f}_n \{(\ln \bar{f}_n + g_{nM}) \dot{g}_{nM}\} \delta V_M \delta \epsilon_n$$
$$- \sum_n \sum_{M=1}^{K} \bar{f}_n \left\{ \left(\ln[1 \mp \bar{f}_n] \mp \left[\frac{\bar{f}_n}{1 \mp \bar{f}_n} \right] g_{nM} \right) \dot{g}_{nM} \right\} \delta V_M \tag{72}$$

and substituting Equation (59) into Equation (72):

$$\frac{d\mathcal{H}}{dt} = \sum_n \sum_{M=1}^{K} \bar{f}_n \left\{ (\bar{\alpha} + \bar{\beta} \epsilon_n) \dot{g}_{nM} + g_{nM} \left(1 \pm e^{\bar{\alpha} + \bar{\beta} \epsilon_n} \right) \dot{g}_{nM} \right\} \delta V_M. \tag{73}$$

Since both the total number of particles and the total energy of the system are constant, it follows from Equations (51), (52), (62) and (63) that

$$\frac{dN}{dt} = \sum_{M=1}^{K} \dot{\mathcal{N}}_M \delta V_M = \sum_{M=1}^{K} \sum_n \bar{f}_n \dot{g}_{nM} \delta V_M = \sum_{M=1}^{K} \dot{\Delta}_M(t) \delta V_M = 0 \tag{74}$$

and

$$\frac{dE}{dt} = \sum_{M=1}^{K} \dot{\mathcal{E}}_M \delta V_M = \sum_{M=1}^{K} \sum_n \bar{f}_n \dot{g}_{nM} \epsilon_n \delta V_M = \sum_{M=1}^{K} \dot{\delta}_M(t) \delta V_M = 0. \tag{75}$$

Substitute the previous expression in Equation (73) to obtain

$$\frac{d\mathcal{H}}{dt} = \sum_n e^{\bar{\alpha} + \bar{\beta} \epsilon_n} \sum_{M=1}^{K} g_{nM} \dot{g}_{nM} \delta V_M \leq 0. \qquad \text{QED.} \tag{76}$$

To obtain the far right side of Equation (76), we have used the relationship $g_{nM} \dot{g}_{nM} \leq 0$ for $t > 0$. This can be proven by simply arguing that, in the initial state, if a cell is described by $g_{nM}(t_0) > 0$ then $g_{nM}(t) \geq 0$ and $\dot{g}_{nM}(t) \leq 0$, and if $g_{nM}(t_0) < 0$ then $g_{nM}(t) \leq 0$ and $\dot{g}_{nM}(t) \geq 0$. Here we have exploited the fact that the system in equilibrium is homogeneous, and that, by accepting the local equilibrium hypothesis, $g_{Mn}(t)$ is a monotonic function and $g_{nM} \to 0$ as $t \to \infty$ as the system approaches the equilibrium state. Another approach to prove Equation (76) consists of splitting the cells into two subsets, just as we did in the classical scenario.

Briefly, considering a dilute quantum gas contained in a vessel of volume V (divided into K small cells), with total energy E and N quantum free particles, which initially is out of equilibrium—but in such a manner that the local equilibrium hypothesis is valid— the functional

$$\mathcal{H}(t) = \sum_{M=1}^{K} \sum_n \left[f_{Mn}(t) \ln f_{Mn}(t) \pm \left(1 \mp f_{Mn}(t) \right) \ln \left(1 \mp f_{Mn}(t) \right) \right] \delta V_M, \tag{77}$$

where f_{Mn} is the M-th cell distribution function, evolves in time such that $d\mathcal{H}/dt \leq 0$, and the equality condition is attained when the system reaches the equilibrium state. In Equation (77), and for a Fermi (Bose) gas, f_{Mn} corresponds to the Fermi–Dirac (Bose–Einstein) distribution function for each cell. Locally, each cell is in equilibrium, although the complete system may be non-homogeneous, and is characterized by the respective f_{Mn}, number of particles $\mathcal{N}_M$, energy $\mathcal{E}_M$, temperature T_M, and Legendre multipliers α_M and β_M.

4. Quantum—Classical Correspondence

In Sections 2 and 3, we saw that the variational method can be applied to H-functionals, which correctly describes the behavior of classical and quantum dilute gases, with regard to their respective time-evolution. Both H-functionals defined in Equations (5) and (47) also recover the well-known distribution functions, either Maxwell–Boltzmann for a classical gas, Fermi–Dirac for a Fermi gas, or Bose–Einstein for a Bose gas. Nevertheless, the functionals (5) and (47) are seemingly different, and in this section, we show they are related by the correspondence principle.

We start by arguing that, in equilibrium, it is straightforward to proof that Equation (60) collapses into Equation (19) by taking the limit wherein the degeneration parameter $\xi \equiv \exp\left(-(\epsilon - \mu)/(k_B T)\right) \gg 1$. Alternatively, a more general approach to show the quantum–classical correspondence consists of analyzing the collapse from the quantum to the classical H-functionals within the appropriated limit. For the case treated here, this limit is $f_{nM} \approx 0$ for several reasons. Systems at very low temperatures, in which the quantum effects cannot be ignored, are obviously excluded from the current analysis. In systems at sufficiently high temperatures, the particles occupy almost exclusively high-energy levels. Furthermore, the energy spectrum approaches a continuum, as is expected by taking the limit $\hbar \to 0$, and the number of particles per level is very close to zero.

Subsequently, we substitute $f_{nM} \approx 0$ into Equation (47) to obtain

$$\mathcal{H} = \sum_{M=1}^{K} \sum_n [f_{nM}(t) \ln(f_{nM}(t))] \delta V_M = \sum_{M=1}^{K} \sum_n [f_M(\epsilon_n, t) \ln(f_M(\epsilon_n, t))] \delta V_M. \tag{78}$$

Finally, the sum over the quantum energy levels can be replaced by an integral over the velocities by invoking both the uncertainty principle and the fact that, for free particles, the continuum energy spectrum can be written as a function of the velocity. Hence the quantum H-functional transforms, in the classical limit, to

$$\mathcal{H} = \sum_{M=1}^{K} \int C' f_M(\vec{v}, t) \ln\left[C' f_M(\vec{v}, t)\right] d\vec{v}, \tag{79}$$

where C' collects the appropriate constants stemming from writing the energy spectrum as a function of $\vec{v}$.

5. Relaxation Processes in Degenerated Quantum Gases

To obtain a time-evolution equation for an out-of-equilibrium quantum gas, we propose the following approach. We start by evaluating $\Delta \mathcal{H} = \mathcal{H}(t_2) - \mathcal{H}(t_1)$, where our quantum H-functional—Equation (47)—is evaluated at different times t_1 and t_2, with $t_2 > t_1$. This yields

$$\Delta \mathcal{H} = \sum_{M=1}^{K} \sum_n [f''_{nM} \ln f''_{nM} - f'_{nM} \ln f'_{nM} \\ \pm (1 \mp f''_{nM}) \ln(1 \mp f''_{nM}) \mp (1 \mp f'_{nM}) \ln(1 \mp f'_{nM})] \delta V_M. \tag{80}$$

In the above equation, and for the rest of this section, we use the short-hand notation $f_{nM}(t_2) \equiv f''_{nM}$ and $f_{nM}(t_1) \equiv f'_{nM}$. Subsequently, in Equation (80), we replace the distribution functions with their expressions in terms of deviations from equilibrium—Equation (64)—which renders:

$$\Delta \mathcal{H} = \sum_{M=1}^{K} \sum_{n} \left[\tilde{f}_n(1 + g''_{nM}) \ln \tilde{f}_n(1 + g''_{nM}) - \tilde{f}_n(1 + g'_{nM}) \ln \tilde{f}_n(1 + g'_{nM}) \right.$$
$$\pm (1 \mp \tilde{f}_n\{1 + g''_{nM}\}) \ln \left(1 \mp \tilde{f}_n\{1 + g''_{nM}\}\right)$$
$$\left. \mp (1 \mp \tilde{f}_n\{1 + g'_{nM}\}) \ln \left(1 \mp \tilde{f}_n\{1 + g'_{nM}\}\right) \right] \delta V_M. \tag{81}$$

Subsequently, we expand the logarithmic terms up to the first-order in g'_{nM} and g''_{nM} and rearrange the result, which gives

$$\Delta \mathcal{H} = \sum_{M=1}^{K} \sum_{n} [\tilde{f}_n(1 + \ln \tilde{f}_n) - \tilde{f}_n\{\ln(1 \mp \tilde{f}_n) \mp 1\}](g''_{nM} - g'_{nM}) \delta V_M. \tag{82}$$

Finally we divide Equation (82) by $\Delta t \equiv t_2 - t_1$, and take the limit $\Delta t \to 0$ to obtain

$$\frac{d\mathcal{H}}{dt} = \sum_{M=1}^{K} \sum_{n} [\tilde{f}_n(1 + \ln \tilde{f}_n) - \tilde{f}_n\{\ln(1 \mp \tilde{f}_n) \mp 1\}] \left(\frac{g_{nM}}{dt} \right) \delta V_M. \tag{83}$$

Equation (83) is, within our framework, the time-evolution equation for g_{nM}. Clearly, to describe a realistic situation, providing a specific approximation for the deviation function g_{nM} is required. This subject will be explored in future work.

6. Comments and Remarks

The demonstration of the classical H-theorem usually begins by assuming that the gas, despite being initially out of equilibrium, can be described by a spatially homogeneous distribution function. Subsequently, the time-evolution of the system occurs in such a manner that $dH/dt \leq 0$. Therefore, this approach does not describe the evolution to equilibrium of systems with spatial inhomogeneities. To address this issue, in this article, we proposed a framework that may be useful to describe the time-evolution of initially non-homogeneous systems. To this end, we divided the system into small cells to conceive a system wherein the local equilibrium hypothesis is valid in each cell but in such a manner that the total system is not homogeneous. Systems that satisfy the previous conditions will evolve towards equilibrium, and the evolution occurs according to $d\mathcal{H}'/dt \leq 0$, Equation (5) and $d\mathcal{H}/dt \leq 0$, and Equation (47), for classical and quantum gases, respectively. Consequently, this approach can be considered an extension of the H-theorem for more realistic out-of-equilibrium systems.

The classical and quantum H-functionals, $\mathcal{H}'$ and $\mathcal{H}$, respectively, correctly recover the most-probable distribution functions in out-of-equilibrium states (locally) and when the system attains the global equilibrium state. The relaxation process of the system is described by monotonic functions that account for deviations from the global equilibrium.

It is clear that for describing the relaxation process of a concrete system, it is necessary to know, at least to some approximation, the specific forms of the monotonic functions g'_M and g_{nM}, for classical and quantum systems. Whereas the complete analysis of these functions is beyond the scope of the present work, some of their properties can be predicted, e.g., they must be consistent both with the system relaxation times and the mechanisms of energy transfer between cells.

An important aspect of the framework proposed in this work is related to the entropy of systems out-of-equilibrium. Because the functionals $\mathcal{H}'$ and $\mathcal{H}$ can be related to the entropy of dilute gases, either classical or quantum, the fact that these functionals are defined over a system divided into cells enables their use for defining the entropy of out-of-equilibrium systems, other than dilute gases. Specifically, and derived from our previous work (e.g., [30,31]), the $\mathcal{H}'$ and $\mathcal{H}$ functionals may serve to describe the entropy, as well as the entropy generation, occurring during the growth of complex physical systems, such as fractals. Possibly, studying these systems might also shed light on the explicit functional form of g'_M and g_{nM}.

In summary, we proposed a variational procedure to demonstrate the classical and quantum H-theorems, which allowed us to describe, at a mesoscopic local view (cell-scale), the time-evolution of an out-of-equilibrium and spatially non-homogeneous system moving towards the equilibrium condition. In principle, this approach would permit the investigation of the transport phenomena inherent to the equilibration process, occurring in a system with a spatially inhomogeneous out-of-equilibrium initial condition.

Author Contributions: Conceptualization, J.L.E.C.-E.; methodology, C.M.-P., J.M.S.-A. and J.L.E.C.-E.; formal analysis, C.M.-P.; validation, J.M.S.-A. and J.L.E.C.-E.; writing, C.M.-P, J.M.S.-A. and J.L.E.C.-E.; supervision, J.M.S.-A. and J.L.E.C.-E.; project administration J.M.S.-A. and J.L.E.C.-E.; funding acquisition, J.L.E.C.-E. All authors have read and agreed to the published version of the manuscript.

Funding: This research was funded by CONACyT (National Council of Science and Technology), grant No. A1-S-39909, and PRODEP-SEP.

Acknowledgments: CMP acknowledges the Conacyt PhD scholarship.

Conflicts of Interest: The authors declare no conflict of interest.

Abbreviations

The following abbreviations are used in this manuscript:

H_B	The original Boltzmann H-functional
$f_B = f(\vec{v})$	The Maxwell–Boltzmann distribution function
$\mathcal{H}'$	Our H-functional for a classical dilute gas
$f'_M = f'(\vec{r}_M, \vec{v}, t)$	The classical distribution function of a cell centered at $\vec{r}_M$
H_T	The H-functional proposed by Tolman
$\mathcal{H}$	Our H-functional for a quantum dilute gas
$f_{nM} = f_M(\vec{r}_M, \epsilon_n, t)$	The quantum distribution function of a cell centered at $\vec{r}_M$

References

1. Huang, K. *Statistical Mechanics*; John Wiley & Sons, Inc.: Hoboken, NJ, USA, 1963.
2. Tolman, R. *The Principles of Statistical Mechanics*; Dover: Mineola, NY, USA, 1979.
3. Grabert, H.; Weidlich, W. Masterequation, H-theorem and stationary solution for coupled quantum systems. *Z. Für Phys.* **1974**, *268*, 139–143. [CrossRef]
4. Rivas, Á.; Martín-Delgado, M.A. Topological Heat Transport and Symmetry-Protected Boson Currents. *Sci. Rep.* **2017**, *7*, 1–9. [CrossRef] [PubMed]
5. Amato, G.; Breuer, H.; Wimberger, S.; Rodríguez, A.; Buchleitner, A. Noninteracting many-particle quantum transport between finite reservoirs. *Phys. Rev. A.* **2020**, *102*, 022207. [CrossRef]
6. Nicacio, F.; Ferraro, A.; Imparato, A.; Paternostro, M.; ao, F.S. Thermal transport in out-of-equilibrium quantum harmonic chains. *Phys. Rev. E* **2015**, *91*, 042116. [CrossRef] [PubMed]
7. Hussein, R.; Kohler, S. Quantum transport, master equations, and exchange fluctuations. *Phys. Rev. B.* **2014**, *89*, 205424. [CrossRef]
8. Gorban, A. General H-theorem and Entropies that Violate the Second Law. *Entropy* **2014**, *16*, 2408–2432. [CrossRef]
9. Ben-Naim, A. Entropy, Shannon, Measure of Information and Boltzmann, H-Theorem. *Entropy* **2017**, *19*, 48. [CrossRef]
10. Silva, R.; Anselmo, D.H.A.L.; Alcaniz, J.S. Nonextensive quantum H-theorem. *Europhys. Lett.* **2010**, *89*, 10004. [CrossRef]
11. Roeck, W.D.; Maes, C.; K. Netočný, K. Quantum macrostates, equivalence of ensembles, and an H-theorem. *J. Math. Phys.* **2006**, *47*, 073303. [CrossRef]
12. Acharya, J.; Issa, I.; Shende, N.V.; Wagner, A.B. Measuring Quantum Entropy. In Proceedings of the 2019 IEEE International Symposium on Information Theory (ISIT), Paris, France, 7–12 July 2019; pp. 3012–3016. [CrossRef]
13. Kastner, R. On Quantum Collapse as a Basis for the Second Law of Thermodynamics. *Entropy* **2017**, *19*, 106. [CrossRef]
14. Han, X.; Wu, B. Entropy for quantum pure states and quantum H theorem. *Phys. Rev. E.* **2015**, *91*, 062106. [CrossRef] [PubMed]
15. Das, B.; Biswas, S. Proof of quantum mechanical H-theorem beyond binary collisions in quantum gases. *J. Stat. Mech Theory Exp.* **2018**, *2018*, 103101. [CrossRef]
16. Von Neumann, J. Proof of the ergodic theorem and the H-theorem in quantum mechanics. *Eur. Phys. J. H* **2010**, *35*, 201–237. [CrossRef]

17. Brown, H.; Myrvold, W. Boltzmann's H-theorem, its limitations, and the birth of (fully) statistical mechanics. *arXiv* **2008**, arXiv:0809.1304.
18. Dragoljub, A. Paradoxes of Thermodynamics and Statistical Physics. *arXiv* **2009**, arXiv:0912.1756.
19. Syros, C. Quantum thermodynamics - H-theorem and the second law. *Lett. Math. Phys.* **1999**, *50*, 29–43. [CrossRef]
20. Lesovik, G.B.; Ledevev, A.V.; Sadovskyy, I.A.; Suslov, M.V.; Vinokur, V.M. H-theorem in Quantum Physics. *Sci. Rep.* **2016**, *6*. [CrossRef]
21. Lesovik, G.; Sadovskyy, I.; Suslov, M.; Lebedev, A.; Vinokur, V. Arrow of time and its reversal on the IBM quantum computer. *Sci. Rep.* **2019**, *9*, 4396. [CrossRef]
22. Keizer, J. *Statistical Thermodynamics of Nonequilibrium Processes*; Springer: New York, NY, USA, 1987.
23. Onsager, L. Reciprocal Relations in Irreversible Processes. I. *Phys. Rev.* **1931**, *37*, 405–426. [CrossRef]
24. Zanotto, E.; Mauro, J. Comment on Glass Transition, Crystallization of Glass-Forming Melts, and Entropy. *Entropy* **2018**, *20*, 103.
25. Schmelzer, J.; Tropin, T. Glass Transition, Crystallization of Glass-Forming Melts, and Entropy. *Entropy* **2018**, *20*, 103. [CrossRef] [PubMed]
26. Nemilov, S. On the Possibility of Calculating Entropy, Free Energy, and Enthalpy of Vitreous Substances. *Entropy* **2018**, *20*, 187. [CrossRef]
27. Li, S.; Cao, B. On Entropic Framework Based on Standard and Fractional Phonon Boltzmann Transport Equations. *Entropy* **2019**, *21*, 204. [CrossRef] [PubMed]
28. Gring, M.; Kuhnert, M.; Langen, T.; Kitagawa, T.; Rauer, B.; Schreitl, M.; Mazets, I.; Smith, D.; Demler, E.; Schmiedmayer, J. Relaxation and Prethermalization in an Isolated Quantum System. *Science* **2012**, *337*, 1318–1322. [CrossRef]
29. Wang, Q.; Kaabouchiu, A.E. From Random Motion of Hamiltonian Systems to Boltzmann, H Theorem and Second Law of Thermodynamics: a Pathway by Path Probability. *Entropy* **2014**, *16*, 885–894. [CrossRef]
30. Nicolás-Carlock, J.; Solano-Altamirano, J.; Carrillo-Estrada, J. The dynamics of the angular and radial density correlation scaling exponents in fractal to non-fractal morphodynamics. *Chaos Solitons Fractals* **2020**, *133*, 109649. [CrossRef]
31. Nicolás-Carlock, J.R.; Carrillo-Estrada, J.L.; Dossetti, V. Fractality à la carte: a general particle aggregation model. *Sci. Rep.* **2016**, *6*, 19505. [CrossRef]

Article

From Rényi Entropy Power to Information Scan of Quantum States

Petr Jizba [1,†,*], Jacob Dunningham [2,†] and Martin Prokš [1,†]

1. FNSPE, Czech Technical University in Prague, Břehová 7, 115 19 Praha 1, Czech Republic; proksma6@fjfi.cvut.cz
2. Department of Physics and Astronomy, University of Sussex, Brighton BN1 9QH, UK; J.Dunningham@sussex.ac.uk
* Correspondence: p.jizba@fjfi.cvut.cz; Tel.: +420-775-317-307
† These authors contributed equally to this work.

Abstract: In this paper, we generalize the notion of Shannon's entropy power to the Rényi-entropy setting. With this, we propose generalizations of the de Bruijn identity, isoperimetric inequality, or Stam inequality. This framework not only allows for finding new estimation inequalities, but it also provides a convenient technical framework for the derivation of a one-parameter family of Rényi-entropy-power-based quantum-mechanical uncertainty relations. To illustrate the usefulness of the Rényi entropy power obtained, we show how the information probability distribution associated with a quantum state can be reconstructed in a process that is akin to quantum-state tomography. We illustrate the inner workings of this with the so-called "cat states", which are of fundamental interest and practical use in schemes such as quantum metrology. Salient issues, including the extension of the notion of entropy power to Tsallis entropy and ensuing implications in estimation theory, are also briefly discussed.

Keywords: Rényi entropy; Tsallis entropy; entropic uncertainty relations; quantum metrology

Citation: Jizba, P.; Dunningham, J.; Prokš, M. From Rényi Entropy Power to Information Scan of Quantum States. *Entropy* **2021**, *23*, 334. https://doi.org/10.3390/e23030334

Academic Editor: Andrew Barron

Received: 18 February 2021
Accepted: 9 March 2021
Published: 12 March 2021

Publisher's Note: MDPI stays neutral with regard to jurisdictional claims in published maps and institutional affiliations.

Copyright: © 2021 by the authors. Licensee MDPI, Basel, Switzerland. This article is an open access article distributed under the terms and conditions of the Creative Commons Attribution (CC BY) license (https://creativecommons.org/licenses/by/4.0/).

1. Introduction

The notion of entropy is undoubtedly one of the most important concepts in modern science. Very few other concepts can compete with it in respect to the number of attempts to clarify its theoretical and philosophical meaning [1]. Originally, the notion of entropy stemmed from thermodynamics, where it was developed to quantify the annoying inefficiency of steam engines. It then transmuted into a description of the amount of disorder or complexity in physical systems. Though many such attempts were initially closely connected with the statistical interpretation of the phenomenon of heat, in the course of time, they expanded their scope far beyond their original incentives. Along those lines, several approaches have been developed in attempts to quantify and qualify the entropy paradigm. These have been formulated largely independently and with different applications and goals in mind. For instance, in *statistical physics*, entropy counts the number of distinct microstates compatible with a given macrostate [2], in *mathematical statistics*, it corresponds to the inference functional for an updating procedure [3], and in *information theory*, it determines a limit on the shortest attainable encoding scheme [2,4].

Particularly distinct among these are the information-theoretic entropies (ITEs). This is not only because they discern themselves through their firm operational prescriptions in terms of coding theorems and communication protocols [5–9], but because they also offer an intuitive measure of disorder phrased in terms of missing information about a system. Apart from innate issues in communication theory, ITEs have also proved to be indispensable tools in other branches of science. Typical examples are provided by chaotic dynamical systems and multifractals (see, e.g., [10] and citations therein). Fully developed turbulence, earthquake analysis, and generalized dimensions of strange attractors provide further examples [11]. An especially important arena for ITEs in the past two decades has been quantum mechanics (QM) with applications ranging from quantum estimation and coding theory to quantum entanglement. The catalyst has been an infusion of new ideas from (quantum) information theory [12–15], functional analysis [16,17], condensed matter theory [18,19], and cosmology [20,21]. On the experimental front, the use of ITEs has been stimulated not only by new high-precision instrumentation [22,23] but also by, e.g., recent advances in stochastic thermodynamics [24,25] or observed violations of Heisenberg's error-disturbance uncertainty relations [26–30].

In his seminal 1948 paper, Shannon laid down the foundations of modern information theory [5]. He was also instrumental in pointing out that, in contrast with discrete signals or messages where information is quantified by (Shannon's) entropy, the cases with continuous variables are less satisfactory. The continuous version of Shannon's entropy (SE)— the so-called differential entropy, may take negative values [5,31], and so it does not have the same status as its discrete-variable counterpart. To solve a number of information-theoretic problems related to continuous cases Shannon shifted the emphasis from the differential entropy to yet another object—entropy power (EP). The EP describes the variance of a would-be Gaussian random variable with the same differential entropy as the random variable under investigation. EP was used by Shannon [5,6] to bound the capacity of non-Gaussian additive noise channels. Since then, the EP has proved to be essential in a number of applications ranging from interference channels to secrecy capacity [32–36]. It has also led to new advances in information parametric statistics [37,38] and network information theory [39]. Apart from its significant role in information theory, the EP has found wide use in pure mathematics, namely in the theory of inequalities [39] and mathematical statistics and estimation theory [40].

Recent developments in information theory [41], quantum theory [42,43], and complex dynamical systems in particular [10,44,45] have brought about the need for a further extension of the concept of ITE beyond Shannon's conventional type. Consequently, numerous generalizations have started to proliferate in the literature ranging from additive entropies [31,46] through a rich class of non-additive entropies [47–52] to more exotic types of entropies [53]. Particularly prominent among such generalizations are ITEs of Rényi and Tsallis, which both belong to a broader class of so-called Uffink entropic functionals [54,55]. Both Rényi entropy (RE) and Tsalli entropy (TE) represent one-parameter families of deformations of Shannon's entropy. An important point related to the RE is that the RE is not just a theoretical construct, but it has a firm operational meaning in terms of various coding theorems [8,9]. Consequently, REs, along with their associated Rényi entropy powers (REPs), are, in principle, experimentally accessible [8,56,57]. That is indeed the case in specific quantum protocols [58–60]. In addition, REPs of various orders are often used as convenient measures of entanglement—e.g., REP of order 2, i.e., N_2 represents *tangle* τ (with $\sqrt{\tau}$ being *concurrence*) [61], $N_{1/2}$ is related to both *fidelity F* and *robustness R* of a pure state [62], N_∞ quantifies the Bures distance to the closest separable pure state [63], etc. Even though our main focus here will be on REs and REPs since they are more pertinent in information theory, we will include some discussion related to Tsallis entropy powers at the end of this paper.

The aim of this paper is twofold. First, we wish to appropriately extend the notion of SE-based EP to the RE setting. In contrast to our earlier works on the topic [13,64], we will do it now by framing REP in the context of RE-based estimation theory. This will be done by judiciously generalizing such key notions as the De Bruijn identity, isoperimetric inequality (and ensuing Cramér–Rao inequality), and Stam inequality. In contrast to other similar works on the subject [65–68], our approach is distinct in three key respects: (a) we consistently use the notion of escort distribution and escort score vector in setting up the generalized De Bruijn identity and Fisher information matrix, (b) we generalize Stam's uncertainty principle, and (c) Rényi EP is related to variance of the reference Gaussian distribution rather than the Rényi maximizing distribution. As a byproduct, we derive within such a generalized estimation theory framework the Rényi-EP-based quantum uncertainty relations (REPUR) of Schrödinger–Roberston type. The REPUR obtained coincides with our earlier result [13] that was obtained in a very different context by means of the Beckner–Babenko theorem. This in turn serves as a consistency check of the proposed generalized estimation theory. Second, we identify interesting new playgrounds for the Rényi EPs obtained. In particular, we asked ourselves a question: assuming one is able in specific quantum protocols to measure Rényi EPs of various orders, how does this constrain the underlying quantum state distribution? To answer this question, we invoke the concept of the *information distribution* associated with a given quantum state. The latter contains a complete "information scan" of the underlying state distribution. We set up a reconstruction method based on Hausdorff's moment problem [69] to show explicitly how the information probability distribution associated with a given quantum state can be numerically reconstructed from EPs. This is a process that is analogous to quantum-state tomography. However, whereas tomography extracts the full density matrix from an ensemble using many measurements on a tomographically complete basis, the EP reconstruction method extracts the probability density on a given basis. This is an alternative approach that may be advantageous, for example, in quantum metrology schemes, where only knowledge of the local probability density rather than the full quantum state is needed [70].

The paper is structured as follows. In Section 2, we introduce the concept of Rényi's EP. With quantum metrology applications in mind, we discuss this in the framework of estimation theory. First, we duly generalize the notion of Fisher information (FI) by using a Rényi entropy version of De Bruijn's identity. In this connection, we emphasize the role of the so-called *escort distribution*, which appears naturally in the definition of higher-order *score functions*. Second, we prove the RE-based isoperimetric inequality and ensuing Cramér–Rao inequality and find how the knowledge of Fisher information matrix restricts possible values of Rényi's EP. Finally, we further illuminate the role of Rényi's EP by deriving (through the Stam inequality) Rényi's EP-based quantum uncertainty relations for conjugate observables. To flesh this out, the second part of the paper is devoted to the development of the use of Rényi EPs to extract the quantum state from incomplete data. This is of particular interest in various quantum metrology protocols. To this end, we introduce in Section 3 the concepts of information distribution, and, in Section 4, we show how cumulants of the information distribution can be obtained from knowledge of the EPs. With the cumulants at hand, one can reconstruct the underlying information distribution in a process which we call an *information scan*. Details of how one could explicitly realize such an information scan for quantum state PDFs are provided in Section 5. There we employ generalized versions of Gram–Charlier A and the Edgeworth expansion. In Section 6, we illustrate the inner workings of the information scan using the example of a so-called *cat state*. This state is of interest in applications of quantum physics such as quantum-enhanced metrology, which is concerned with the optimal extraction of information from measurements subject to quantum mechanical effects. The cat state we consider is a superposition of the vacuum state and a coherent state of the electromagnetic field; two cases are studied comprising different probabilistic weightings of the superposition state corresponding to *balanced* and *unbalanced* cat states. Section 7 is dedicated to EPs based on Tsallis entropy. In particular, we show that Rényi and Tsallis EPs coincide with each other. This, in turn, allows us to phrase various estimation theory inequalities in terms of TE. In Section 7, we end with conclusions. For the reader's convenience, we relegate some technical issues concerning the generalized De Bruijn identity and associated isoperimetric and Stam inequalities to three appendices.

2. Rényi Entropy Based Estimation Theory and Rényi Entropy Powers

In this section, we introduce the concept of Rényi's EP. With quantum metrology applications in mind, we discuss this in the framework of estimation theory. This will not only allow us to find new estimation inequalities, such as the Rényi-entropy-based De Bruijn identity, isoperimetric inequality, or Stam inequality, but it will also provide a convenient technical and conceptual frame for deriving a one-parameter family of Rényi-entropy-power-based quantum-mechanical uncertainty relations.

2.1. Fisher Information—Shannon's Entropy Approach

First, we recall that the Fisher information matrix $\mathbb{J}(\mathcal{X})$ of a random vector $\{\mathcal{X}_i\}$ in $\mathbb{R}^D$ with the PDF $\mathcal{F}(x)$ is defined as [38]

$$\mathbb{J}(\mathcal{X}) = \text{cov}(V(\mathcal{X})), \tag{1}$$

where the covariance matrix is associated with the random zero-mean vector—the so-called *score vector*, as

$$V(x) = \nabla \mathcal{F}(x)/\mathcal{F}(x). \tag{2}$$

A corresponding trace of $\mathbb{J}(\mathcal{X})$, i.e.,

$$J(\mathcal{X}) = \text{Tr}(\mathbb{J}(\mathcal{X})) = \text{var}(V(\mathcal{X})) = \mathbb{E}(V^2(\mathcal{X})), \tag{3}$$

is known as the Fisher information. Both the FI and FI matrix can be conveniently related to Shannon's differential entropy via De Bruijn's identity [66,67].

De Bruijn's identity: Let $\{\mathcal{X}_i\}$ be a random vector in $\mathbb{R}^D$ with the PDF $\mathcal{F}(x)$ and let $\{\mathcal{Z}_i^G\}$ be a Gaussian random vector (noise vector) with zero mean and unit-covariance matrix, independent of $\{\mathcal{X}_i\}$. Then,

$$\frac{d}{d\epsilon}\mathcal{H}(\mathcal{X} + \sqrt{\epsilon}\mathcal{Z}^G)|_{\epsilon=0} = \frac{1}{2}J(\mathcal{X}), \tag{4}$$

where

$$\mathcal{H}(\mathcal{X}) = -\int_{\mathbb{R}^D} \mathcal{F}(x) \log \mathcal{F}(x)\, dx, \tag{5}$$

is Shannon's differential entropy (measured in *nats*). In the case when the independent additive noise $\{\mathcal{Z}_i\}$ is non-Gaussian with zero mean and covariance matrix $\Sigma = \text{cov}(\mathcal{Z})$, then the following generalization holds [67]:

$$\frac{d}{d\epsilon}\mathcal{H}(\mathcal{X} + \sqrt{\epsilon}\mathcal{Z})|_{\epsilon=0} = \frac{1}{2}\text{Tr}(\mathbb{J}(\mathcal{X})\Sigma). \tag{6}$$

The key point about De Bruijn's identity is that it provides a very useful intuitive interpretation of FI, namely, FI quantifies the sensitivity of transmitted (Shannon type) information to an arbitrary independent additive noise. An important aspect that should be stressed in this context is that FI as a quantifier of sensitivity depends only on the covariance of the noise vector, and thus it is independent of the shape of the noise distribution. This is because De Bruijn's identity remains unchanged for both Gaussian and non-Gaussian additive noise with the same covariance matrix.

2.2. Fisher Information—Rényi's Entropy Approach

We now extend the notion of the FI matrix to the Rényi entropy setting. A natural way to do it is via an extension of De Bruijn's identity to Rényi entropies. In particular, the following statement holds:

Generalized De Bruijn's identity: Let $\{\mathcal{X}_i\}$ be a random vector in $\mathbb{R}^D$ with the PDF $\mathcal{F}(x)$ and let $\{\mathcal{Z}_i\}$ be an independent (generally non-Gaussian) noise vector with the zero mean and covariance matrix $\Sigma = \text{cov}(\mathcal{Z})$, then, for any $q > 0$

$$\frac{d}{d\epsilon}\mathcal{I}_q(\mathcal{X} + \sqrt{\epsilon}\mathcal{Z})|_{\epsilon=0} = \frac{1}{2q}\text{Tr}(\mathbb{J}_q(\mathcal{X})\Sigma), \tag{7}$$

where

$$\mathcal{I}_q = \frac{1}{1-q} \log \int_{\mathbb{R}^D} \mathcal{F}^q(x) dx, \quad q > 0, \qquad (8)$$

is *Rényi's differential entropy* (measured in *nats*) with $\mathcal{I}_1 = \mathcal{H}$. The ensuing FI matrix of order q has the explicit form

$$\mathbb{J}_q(\mathcal{X}) = \text{cov}_q(V_q(\mathcal{X})), \qquad (9)$$

with the score vector

$$V_q(x) = \nabla \rho_q(x)/\rho_q(x) = q \nabla \mathcal{F}(x)/\mathcal{F}(x) = qV(x). \qquad (10)$$

Here, $\rho_q = \mathcal{F}^q / \int_{\mathbb{R}^D} \mathcal{F}^q dx$ is the so-called *escort distribution* [71]. The "cov$_q$" denotes the covariance matrix computed with respect to ρ_q. Proofs of both the conventional (i.e., Shannon entropy based) and generalized (i.e., Rényi entropy based) De Bruijn's identity are provided in Appendix A. There we also discuss some further useful generalizations of De Bruijn's identity. Finally, as in the Shannon case, we define the FI of order q—denoted as $J_q(\mathcal{X})$, as

$$\text{Tr}(\mathbb{J}_q(\mathcal{X})) \equiv J_q(\mathcal{X}). \qquad (11)$$

2.3. Rényi's Entropy Power and Generalized Isoperimetric Inequality

Similarly as in conventional estimation theory, one can expect that there should exist a close connection between the FI matrix $\mathbb{J}_q(\mathcal{X})$ and the corresponding Rényi entropy power $N_p(\mathcal{X})$. In Shannon's information theory, such a connection is phrased in terms of isoperimetric inequality [67]. Here, we prove that a similar relationship works also in Rényi's information theory.

Let us start by introducing the concept of Rényi's entropy power. This is defined as the solution of the equation [13,64]

$$\mathcal{I}_p(\mathcal{X}) = \mathcal{I}_p\left(\sqrt{N_p(\mathcal{X})} \cdot \mathcal{Z}^G\right), \qquad (12)$$

where $\{\mathcal{Z}_i^G\}$ represents a Gaussian random vector with a zero mean and unit covariance matrix. Thus, $N_p(\mathcal{X})$ denotes the variance of a would be Gaussian distribution that has the same Rényi information content as the random vector $\{\mathcal{X}_i\}$ described by the PDF $\mathcal{F}(x)$. Expression (12) was studied in [13,64,72], where it was shown that the only class of solutions of (12) is

$$N_p(\mathcal{X}) = \frac{1}{2\pi} p^{-p'/p} \exp\left(\frac{2}{D} \mathcal{I}_p(\mathcal{X})\right), \qquad (13)$$

with $1/p + 1/p' = 1$ and $p \in \mathbb{R}^+$. In addition, when $p \to 1_+$, one has $N_p(\mathcal{X}) \to N(\mathcal{X})$, where $N(\mathcal{X})$ is the conventional Shannon entropy power [5]. In this latter case, one can use the *asymptotic equipartition property* [55,73] to identify $N(\mathcal{X})$ with "typical size" of a state set, which in the present context is the effective support set size for a random vector. This, in turn, is equivalent to Einstein's entropic principle [74]. In passing, it should be noted that the form of the Rényi EP expressed in (13) is not universally accepted version. In a number of works, it is defined merely as an exponent of RE, see, e.g., [75,76]. Our motivation for the form (13) is twofold: first, it has a clear interpretation in terms of variances of Gaussian distributions and, second, it leads to simpler formulas, cf. e.g., Equation (22).

Generalized isoperimetric inequality: Let $\{\mathcal{X}_i\}$ be a random vector in $\mathbb{R}^D$ with the PDF $\mathcal{F}(x)$. Then,

$$\frac{1}{D} N_q(\mathcal{X}) J_q(\mathcal{X}) \geq N_q(\mathcal{X}) [\det(\mathbb{J}_q(\mathcal{X}))]^{1/D} \geq 1, \qquad (14)$$

where the Rényi parameter $q \geq 1$. We relegate the proof of the generalized isoperimetric inequality to Appendix B.

It is also worth noting that the relation (14) implies another important inequality. By using the fact that the Shannon entropy is maximized (among all PDF's with identical covariance matrix $\boldsymbol{\Sigma}$) by the Gaussian distribution, we have $N_1(\mathcal{X}) \leq \det(\boldsymbol{\Sigma})^{1/D}$ (see, e.g., [77]). If we further employ that $\mathcal{I}_q$ is a monotonously decreasing function of q, see, e.g., [31,78], we can write (recall that $q \geq 1$)

$$\frac{q^{1/(q-1)}}{e} N_q \leq N_1 = \frac{\exp(\frac{2}{D}\mathcal{I}_1)}{2\pi e} \leq \det(\boldsymbol{\Sigma})^{1/D}. \tag{15}$$

The isoperimetric inequality (14) then implies

$$\det(\boldsymbol{\Sigma}(\mathcal{X})) \geq \frac{\left(q^{1/(q-1)}\right)^D}{e^D \det(\mathbb{J}_q(\mathcal{X}))} \geq \frac{1}{e^D \det(\mathbb{J}_q(\mathcal{X}))}. \tag{16}$$

We can further use the inequality

$$\frac{1}{D}\text{Tr}(\mathbb{A}) \geq [\det(\mathbb{A})]^{1/D}, \tag{17}$$

(valid for any positive semi-definite $D \times D$ matrix $\mathbb{A}$) to write

$$\sigma^2(\mathcal{X}) = \frac{1}{D}\text{Tr}(\boldsymbol{\Sigma}(\mathcal{X})) = \frac{1}{D}\sum_{i=1}^{D}\text{Var}(\mathcal{X}_i) \geq \frac{Dq^{1/(q-1)}}{eJ_q(\mathcal{X})} \geq \frac{D}{eJ_q(\mathcal{X})}, \tag{18}$$

where σ^2 is an average variance per component.

Relations (16)–(18) represent the q-generalizations of the celebrated Cramér–Rao information inequality. In the limit of $q \to 1$, we recover the standard Cramér–Rao inequality that is widely used in statistical inference theory [38,79]. A final logical step needed to complete the proof of REPURs is represented by the so-called generalized Stam inequality. To this end, we first define the concept of *conjugate random variables*. We say that random vectors $\{\mathcal{X}_i\}$ and $\{\mathcal{Y}_i\}$ in $\mathbb{R}^D$ are conjugate if their respective PDF's $\mathcal{F}(x)$ and $\mathcal{G}(y)$ can be written as

$$\mathcal{F}(x) = |\varphi_{\mathcal{F}}(x)|^2 / \|\varphi_{\mathcal{F}}\|_2^2, \qquad \mathcal{G}(y) = |\varphi_{\mathcal{G}}(y)|^2 / \|\varphi_{\mathcal{G}}\|_2^2, \tag{19}$$

where the (generally complex) probability amplitudes $\varphi_{\mathcal{F}}(x) \in L_2(\mathbb{R}^D)$ and $\varphi_{\mathcal{G}}(y) \in L_2(\mathbb{R}^D)$ are mutual Fourier images, i.e.,

$$\varphi_{\mathcal{F}}(x) = \hat{\varphi}_{\mathcal{G}}(x) = \int_{\mathbb{R}^D} e^{2\pi i x \cdot y} \varphi_{\mathcal{G}}(y) \, dy, \tag{20}$$

and analogously for $\varphi_{\mathcal{G}}(y) = \hat{\varphi}_{\mathcal{F}}(y)$. With this, we can state the generalized Stam inequality.

Generalized Stam inequality (Stam's uncertainty principle): Let $\{\mathcal{X}_i\}$ and $\{\mathcal{Y}_i\}$ be conjugate random vectors in $\mathbb{R}^D$. Then,

$$16\pi^2 N_q(\mathcal{Y}) \geq [\det(\mathbb{J}_r(\mathcal{X}))]^{1/D}, \tag{21}$$

is valid for any $r \in [1, \infty)$ and $q \in [1/2, 1]$ that are connected via the relation $1/r + 1/q = 2$. In particular, if we define $r' = 2r$ and $q' = 2q$, then r' and q' are Hölder conjugates. A proof of the generalized Stam inequality is provided in Appendix C.

Let us now consider Hölder conjugate indices p and q with $p \in [2, \infty)$ (so that $q \in [1, 2]$). Combining the isoperimetric inequality (14) together with the generalized Stam inequality (21), we obtain the following one-parameter class of REP-based inequalities

$$
\begin{aligned}
N_{p/2}(\mathcal{X}) N_{q/2}(\mathcal{Y}) &= N_{p/2}(\mathcal{X}) \frac{[\det(\mathbb{J}_{p/2}(\mathcal{X}))]^{1/D}}{[\det(\mathbb{J}_{p/2}(\mathcal{X}))]^{1/D}} N_{q/2}(\mathcal{Y}) \\
&\geq \frac{N_{q/2}(\mathcal{Y})}{[\det(\mathbb{J}_{p/2}(\mathcal{X}))]^{1/D}} \geq \frac{1}{16\pi^2}.
\end{aligned}
\quad (22)
$$

By symmetry, the role of q and p can be reversed. In Refs. [13,64], we presented an alternative derivation of inequalities (22) that was based on the Beckner–Babenko theorem. There it was also proved that the inequality saturates if and only if the distributions involved are Gaussian. The only exception to this rule is for the asymptotic values $p = 1$ and $q = \infty$ (or vice versa) where the saturation happens whenever the peak of $\mathcal{F}(x)$ and tail of $\mathcal{G}(y)$ (or vice versa) are Gaussian.

The passage to quantum mechanics is quite straightforward. First, we realize that, in QM, the Fourier conjugate wave functions are related via two reciprocal relations

$$
\begin{aligned}
\psi_{\mathcal{F}}(x) &= \int_{\mathbb{R}^D} e^{iy \cdot x/\hbar} \psi_{\mathcal{G}}(y) \frac{dy}{(2\pi\hbar)^{D/2}}, \\
\psi_{\mathcal{G}}(y) &= \int_{\mathbb{R}^D} e^{-iy \cdot x/\hbar} \psi_{\mathcal{F}}(x) \frac{dx}{(2\pi\hbar)^{D/2}}.
\end{aligned}
\quad (23)
$$

The Plancherel (or Riesz–Fischer) equality implies that, when $\|\psi_{\mathcal{F}}\|_2 = 1$, then also automatically $\|\psi_{\mathcal{G}}\|_2 = 1$ (and vice versa). Thus, the connection between amplitudes $\varphi_{\mathcal{F}}$ and $\varphi_{\mathcal{G}}$ from (19) and amplitudes $\psi_{\mathcal{F}}$ and $\psi_{\mathcal{G}}$ from (23) is

$$
\begin{aligned}
\varphi_{\mathcal{F}}(x) &= (2\pi\hbar)^{D/4} \psi_{\mathcal{F}}(\sqrt{2\pi\hbar} x), \\
\varphi_{\mathcal{G}}(y) &= (2\pi\hbar)^{D/4} \psi_{\mathcal{G}}(\sqrt{2\pi\hbar} y).
\end{aligned}
\quad (24)
$$

The factor $(2\pi\hbar)^{D/4}$ ensures that also $\varphi_{\mathcal{F}}$ and $\varphi_{\mathcal{G}}$ functions are normalized (in the sense of $\|\dots\|_2$) to unity; however, due to Equation (19), it might be easily omitted. The corresponding Rényi EPs change according to

$$
\begin{aligned}
N_{p/2}(\mathcal{X}) &\equiv N_{p/2}(\mathcal{F}) \mapsto N_{p/2}(|\psi_{\mathcal{F}}|^2) = 2\pi\hbar N_{p/2}(\mathcal{F}), \\
N_{q/2}(\mathcal{Y}) &\equiv N_{q/2}(\mathcal{G}) \mapsto N_{q/2}(|\psi_{\mathcal{G}}|^2) = 2\pi\hbar N_{q/2}(\mathcal{G}),
\end{aligned}
\quad (25)
$$

and hence REP-based inequalities (22) acquire in the QM setting a simple form

$$
N_{p/2}(|\psi_{\mathcal{F}}|^2) N_{q/2}(|\psi_{\mathcal{G}}|^2) \geq \frac{\hbar^2}{4}.
\quad (26)
$$

This represents an infinite tower of mutually distinct (generally irreducible) REPURs [13].

At this point, some comments are in order. First, historically, the most popular quantifier of quantum uncertainty has been *variance* because it is conceptually simple and relatively easily extractable from experimental data. The variance determines the measure of uncertainty in terms of the fluctuation (or spread) around the mean value, which, while useful for many distributions, does not provide a sensible measure of uncertainty in a number of important situations including multimodal [12,13,64] and heavy-tailed distributions [13,14,64]. To deal with this, a multitude of alternative (non-variance based) measures of uncertainty in quantum mechanics (QM) have emerged. Among these, a particularly prominent role is played by information entropies such as the Shannon entropy [63], Rényi entropy [63,64], Tsallis entropy [80], associated differential entropies, and their quantum-information generalizations [13,15,64]. REPURs (26) fit into this framework of entropic QM URs. In connection with (26), one might observe that the conventional URs based on variances—so-called Robertson–Schrödinger URs [81,82]) and Shannon differential entropy based URs (e.g., Hirschman or Białynicki–Birula URs [15,83]) naturally appear as special cases in this hierarchy. Second, the ITEs enter quantum information theory typically in three distinct ways: (a) as a measure of the quantum information content (e.g., how many qubits are needed to encode the message without loss of information), (b) as a measure of the classical information content (e.g., amount of information in bits that can be recovered from the quantum system) and (c) to quantify the entanglement of pure and mixed bipartite quantum states. Logarithms in base 2 are used because, in quantum information, one quantifies entropy in bits and qubits (rather than nats). This in turn also modifies Rényi's EP as

$$\frac{1}{2\pi} p^{-p'/p} e\left(\frac{2}{b}\cdots\right) \ \mapsto \ \frac{1}{2\pi} p^{-p'/p} 2\left(\frac{2}{b}\cdots\right). \tag{27}$$

In the following, we will employ this QM practice.

3. Information Distribution

To put more flesh on the concept of Rényi's EP, we devote the rest of this paper to the development of the methodology and application of Rényi EPs in extracting quantum states from incomplete data. The technique of quantum tomography is widely used for this purpose and involves making many different measurements on an ensemble of identical copies of a quantum state with a tomographically complete measurement basis [84,85]. This process is very measurement-intensive, scaling exponentially with the number of particles and so methods have been developed to approximate it with fewer measurements [86].

However, the method of Rényi EPs provides an efficient alternative approach. Instead of reconstructing the full quantum state, this process extracts the PDF of the quantum state in a given basis. For a broad class of quantum metrology problems, local rather than global approaches are preferred [70] and, for these, the local PDF of the state at each sensor is needed rather than the full density matrix. With this in mind, we first start with the notion of the information distribution.

Let $\mathcal{F}(x)$ be the PDF for the random variable $\mathcal{X}$. We define the *information random variable* $i_\mathcal{X}(\mathcal{X})$ so that $i_\mathcal{X}(x) = \log_2 1/\mathcal{F}(x)$. In other words, $i_\mathcal{X}(x)$ represents the information in x with respect to $\mathcal{F}(x)$. In this connection, it is expedient to introduce the cumulative distribution function for $i_\mathcal{X}(\mathcal{X})$ as

$$\wp(y) = \int_{-\infty}^{y} d\wp(i_\mathcal{X}) = \int_{\mathbb{R}^D} \mathcal{F}(x)\theta(\log_2 \mathcal{F}(x) + y)dx. \tag{28}$$

The function $\wp(y)$ thus represents the probability that the random variable $i_\mathcal{X}(\mathcal{X})$ is less than or equal to y. We have denoted the corresponding probability measure as $d\wp(i_\mathcal{X})$. Taking the Laplace transform of both sides of (28), we get

$$\mathcal{L}\{\wp\}(s) = \int_{\mathbb{R}^D} \mathcal{F}(x) \frac{e^{s\log_2 \mathcal{F}(x)}}{s} dx = \frac{\mathbb{E}\left[e^{s\log_2 \mathcal{F}}\right]}{s}, \tag{29}$$

where $\mathbb{E}[\cdots]$ denotes the mean value with respect to $\mathcal{F}$. By assuming that $\wp(x)$ is smooth, then the PDF associated with $i_\mathcal{X}(\mathcal{X})$—the so-called *information PDF*—is

$$g(y) = \frac{d\wp(y)}{dy} = \mathcal{L}^{-1}\left\{\mathbb{E}\left[e^{s\log_2 \mathcal{F}}\right]\right\}(y). \tag{30}$$

Setting $s = (p-1)\log 2$, we have

$$\mathcal{L}\{g\}(s = (p-1)\log 2) = \mathbb{E}\left[2^{(1-p)i_\mathcal{X}}\right]. \tag{31}$$

The mean here is taken with respect to the PDF g. Equation (31) can also be written explicitly as

$$\int_{\mathbb{R}^D} dx \, \mathcal{F}^p(x) = \int_{\mathbb{R}} g(y) 2^{(1-p)y} dy. \tag{32}$$

Note that, when $\mathcal{F}^p$ is integrable for $p \in [1,2]$, then (32) ensures that the moment-generating function for $g(x)$ PDF exists. Thus, in particular, the moment-generating function exists when $\mathcal{F}(x)$ represents Lévy α-stable distributions, including the heavy-tailed stable distributions (i.e, PDFs with the Lévy stability parameter $\alpha \in (0,2]$). The same holds for $\hat{\mathcal{F}}$ and $p' \in [2, \infty)$ due to the Beckner–Babenko theorem [13,87,88].

4. Reconstruction Theorem

Since $\mathcal{L}\{g\}(s)$ is the *moment-generating function* of the random variable $i_{\mathcal{X}}(\mathcal{X})$, one can generate all moments of the PDF $g(x)$ (if they exist) by taking the derivatives of $\mathcal{L}\{g\}$ with respect to s. From a conceptual standpoint, it is often more useful to work with cumulants rather than moments. Using the fact that the *cumulant generating function* is simply the (natural) logarithm of the moment-generating function, we see from (32) that the differential RE is a reparametrized version of the cumulant generating function of the information random variable $i_{\mathcal{X}}(\mathcal{X})$. In fact, from (31), we have

$$\mathcal{I}_p(\mathcal{X}) = \frac{1}{(1-p)} \log_2 \mathbb{E}\left[2^{(1-p)i_{\mathcal{X}}}\right]. \tag{33}$$

To understand the meaning of REPURs, we begin with the cumulant expansion (33), i.e.,

$$p \mathcal{I}_{1-p}(\mathcal{X}) = \log_2 e \sum_{n=1}^{\infty} \frac{\kappa_n(\mathcal{X})}{n!} \left(\frac{p}{\log_2 e}\right)^n, \tag{34}$$

where $\kappa_n(\mathcal{X}) \equiv \kappa_n(i_{\mathcal{X}})$ denotes the n-th cumulant of the information random variable $i_{\mathcal{X}}(\mathcal{X})$ (in units of $bits^n$). We note that

$$\begin{aligned}\kappa_1(\mathcal{X}) &= \mathbb{E}[i_{\mathcal{X}}(\mathcal{X})] = \mathcal{H}(\mathcal{X}), \\ \kappa_2(\mathcal{X}) &= \mathbb{E}\left[i_{\mathcal{X}}(\mathcal{X})^2\right] - (\mathbb{E}[i_{\mathcal{X}}(\mathcal{X})])^2,\end{aligned} \tag{35}$$

i.e., they represent the Shannon entropy and *varentropy*, respectively. By employing the identity

$$\mathcal{I}_{1-p}(\mathcal{X}) = \frac{D}{2} \log_2\left[2\pi(1-p)^{-1/p} N_{1-p}(\mathcal{X})\right], \tag{36}$$

we can rewrite (34) in the form

$$\log_2[N_{1-p}(\mathcal{X})] = \log_2\left[\frac{(1-p)^{1/p}}{2\pi}\right] + \frac{2}{D} \sum_{n=1}^{\infty} \frac{\kappa_n(\mathcal{X})}{n!} \left(\frac{p}{\log_2 e}\right)^{n-1}. \tag{37}$$

From (37), one can see that

$$\begin{aligned}\kappa_n(\mathcal{X}) &= \frac{nD}{2}(\log_2 e)^{n-1} \frac{d^{n-1} \log_2[N_{1-p}(\mathcal{X})]}{dp^{n-1}}\bigg|_{p=0} \\ &+ \frac{D}{2}(\log_2 e)^n [(n-1)! + \delta_{1n} \log 2\pi],\end{aligned} \tag{38}$$

73

where δ_{1n} is the Kronecker delta function that has a value of one if $n = 1$, or zero otherwise. In terms of the Grünwald–Letnikov derivative formula (GLDF) [89], we can rewrite (38) as

$$\kappa_n(\mathcal{X}) = \lim_{\Delta \to 0} \frac{nD}{2} \frac{(\log_2 e)^n}{\Delta^{n-1}} \sum_{k=0}^{n-1} (-1)^k \binom{n-1}{k} \log[N_{1+k\Delta}(\mathcal{X})]$$

$$+ \frac{D}{2}(\log_2 e)^n [(n-1)! + \delta_{1n} \log 2\pi]. \quad (39)$$

Thus, in order to determine the first m cumulants of $i_\mathcal{X}(\mathcal{X})$, we need to know all $N_1, N_{1+\Delta}, \ldots, N_{1+(m-1)\Delta}$ entropy powers. In practice, Δ corresponds to a characteristic resolution scale for the entropy index which will be chosen appropriately for the task at hand, but is typically of the order 10^{-2}. Note that the last term in (38) and (39) can be also written

$$\frac{D}{2}(\log_2 e)^n[(n-1)! + \delta_{1n} \log 2\pi] = \kappa_n(\mathcal{Z}_G^\mathbf{1}) \equiv \kappa_n(i_\mathcal{Y}), \quad (40)$$

with $\mathcal{Y}$ being the random variable distributed with respect to the *Gaussian* distribution $\mathcal{Z}_G^\mathbf{1}$ with the *unit* covariance matrix.

When all the cumulants exist, then the problem of recovering the underlying PDF for $i_\mathcal{X}(\mathcal{X})$ is equivalent to the *Stieltjes* moment problem [90]. Using this connection, there are a number of ways to proceed; the PDF in question can be reconstructed e.g., in terms of sums involving orthogonal polynomials (e.g., the Gram–Charlier A series or the Edgeworth series [91]), the inverse Mellin transform [92], or via various maximum entropy techniques [93]. Pertaining to this, the theorem of Marcinkiewicz [94] implies that there are no PDFs for which $\kappa_m = \kappa_{m+1} = \ldots = 0$ for $m \geq 3$. In other words, the cumulant generating function cannot be a finite-order polynomial of degree greater than 2. The important exceptions, and indeed the only exceptions to Marcinkiewicz's theorem are the *Gaussian* PDFs that can have the first two cumulants nontrivial and $\kappa_3 = \kappa_4 = \ldots = 0$. Thus, apart from the special case of Gaussian PDFs where only N_1 and $N_{1+\Delta}$ are needed, one needs to work with as many entropy powers $N_{1+k\Delta}, k \in \mathbb{N}$ (or ensuing REPURs) as possible to receive as much information as possible about the structure of the underlying PDF. In theory, the whole infinite tower of REPURs would be required to uniquely specify a system's information PDF. Note that, for *Gaussian* information PDFs, one needs only N_1 and $N_{1+\Delta}$ to reconstruct the PDF uniquely. From (37) and (39), we see that knowledge of N_1 corresponds to $\kappa_1(\mathcal{X}) = \mathcal{H}(\mathcal{X})$ while $N_{1+\Delta}$ further determines κ_2, i.e., the varentropy. Since N_1 is involved (via (39)) in the determination of all cumulants, it is the most important entropy power in the tower. Thus, the entropy powers of a given process have an equivalent meaning to the PDF: they describe the morphology of uncertainty of the observed phenomenon.

We should stress that the focus of the reconstruction theorem we present is on cumulants κ_n which can be directly used for a shape estimation of $g(x)$ but not $\mathcal{F}(x)$. However, by knowing $g(y)$, we have a complete "information scan" of $\mathcal{F}(x)$. Such an information scan is, however, not unique, indeed two PDFs that are rearrangements of each other—i.e., *equimeasurable* PDFs, have identical $\wp(y)$ and $g(y)$. Even though equimeasurable PDFs cannot be distinguished via their entropy powers, they can be, as a rule, distinguished via their respective momentum-space PDFs and associated entropy powers. Thus, the information scan has a tomographic flavor to it. From the multi-peak structure of $g(y)$, one can determine the *number* and *height* of the stationary points. These are invariant characteristics of a given family of equimeasurable PDFs. This will be further illustrated in Section 6.

5. Information Scan of Quantum-State PDF

With knowledge of the entropy powers, the question now is how we can reconstruct the information distribution $g(x)$. The inner workings of this will now be explicitly illustrated with the (generalized) Gram-Charlier A expansion. However, other—often more efficient methods—are also available [91]. Let κ_n be cumulants obtained from entropy powers and let $G(x)$ be some reference PDF whose cumulants are γ_k. The information PDF $g(x)$ can be then written as [91]

$$g(x) = \exp\left[\sum_{k=1}^{\infty}(\kappa_k - \gamma_k)(-1)^k \frac{(d^k/dx^k)}{k!}\right] G(x). \tag{41}$$

With hindsight, we choose the reference PDF $G(x)$ to be a shifted gamma PDF, i.e.,

$$G(x) \equiv \mathcal{G}(x|a,\alpha,\beta) = \frac{e^{-(x-a)/\beta}(x-a)^{\alpha-1}}{\beta^{\alpha}\Gamma[\alpha]}, \tag{42}$$

with $a < x < \infty$, $\beta > 0$, $\alpha > 0$. In doing so, we have implicitly assumed that the $\mathcal{F}(y)$ PDF is in the first approximation equimeasurable with the Gaussian PDF. To reach a corresponding matching, we should choose $a = \log_2(2\pi\sigma^2)/2$, $\alpha = 1/2$ and $\beta = \log_2 e$. Using the fact that [95]

$$(\beta)^{k+1/2} \frac{d^k \mathcal{G}(x|a,1/2,\beta)}{k! dx^k} = \left(\frac{x-a}{\beta}\right)^{-k} L_k^{(-1/2-k)}\left(\frac{x-a}{\beta}\right) \mathcal{G}(x|a,1/2,\beta), \tag{43}$$

(where L_k^{δ} is an associated Laguerre polynomial of order k with parameter δ) and given that $\kappa_1 = \gamma_1 = \alpha\beta + a = \log_2(2\pi\sigma^2 e)/2$, and $\gamma_k = \Gamma(k)\alpha\beta^k = (\log_2 e)^k/2$ for $k > 1$ we can write (41) as

$$g(x) = \mathcal{G}(x|a,1/2,\beta)\left[1 + \frac{(\kappa_2 - \gamma_2)}{\beta^{1/2}(x-a)^2} L_2^{(-5/2)}\left(\frac{x-a}{\beta}\right)\right.$$
$$\left. - \frac{(\kappa_3 - \gamma_3)}{\beta^{1/2}(x-a)^3} L_3^{(-7/2)}\left(\frac{x-a}{\beta}\right) + \cdots \right]. \tag{44}$$

If needed, one can use a relationship between the moments and the cumulants (Faà di Bruno's formula [94]) to recast the expansion (44) into more familiar language. For the Gram–Charlier A expansion, various formal convergence criteria exist (see, e.g., [91]). In particular, the expansion for nearly Gaussian equimeasurable PDFs $\mathcal{F}(y)$ converges quite rapidly and the series can be truncated fairly quickly. Since in this case one needs fewer κ_k's in order to determine the information PDF $g(x)$, only EPs in the small neighborhood of the index 1 will be needed. On the other hand, the further the $\mathcal{F}(y)$ is from Gaussian (e.g., heavy-tailed PDFs), the higher the orders of κ_k are required to determine $g(x)$, and hence a wider neighborhood of the index 1 will be needed for EPs.

6. Example—Reconstruction Theorem and (Un)Balanced Cat State

We now demonstrate an example of the reconstruction in the context of a quantum system. Specifically, we consider cat states that are often considered in the foundations of quantum physics as well as in various applications, including solid state physics [96] and quantum metrology [97]. The form of the state we consider is $|\psi\rangle = \mathcal{N}(|0\rangle + \nu|\alpha/\nu\rangle)$, where $\mathcal{N} = [1 + 2\nu\exp(-\alpha^2/2\nu^2) + \nu^2]^{-1/2}$ is the normalization factor, $|0\rangle$ is the vacuum state, $\nu \in \mathbb{R}$ a weighting factor, and $|\alpha\rangle$ is the coherent state given by

$$|\alpha\rangle = e^{-\alpha^2/2}\sum_{n=0}^{\infty}\frac{\alpha^n}{\sqrt{n!}}|n\rangle, \tag{45}$$

(taking $\alpha \in \mathbb{R}$). For $\nu = 1$, we refer to the state as a *balanced cat state* (BCS) and for $\nu \neq 1$, as an *unbalanced cat state* (UCS). Changing the basis of $|\psi\rangle$ to the eigenstates of the general quadrature operator

$$\hat{Y}_\theta = \frac{1}{\sqrt{2}}\left(\hat{a}e^{-i\theta} + \hat{a}^\dagger e^{i\theta}\right), \tag{46}$$

where $\hat{a}$ and $\hat{a}^\dagger$ are the creation and annihilation operators of the electromagnetic field, we find the PDF for the general quadrature variable y_θ to be

$$\mathcal{F}(y_\theta) = \mathcal{N}^2 \pi^{-\frac{1}{2}} e^{-y_\theta^2} \left|1 + \nu \exp\left[-\frac{\alpha^2}{\nu^2 2}\left(1 + e^{2i\theta} - 2\sqrt{2}e^{i\theta}\frac{\nu}{\alpha}y_\theta\right)\right]\right|^2, \tag{47}$$

where $\mathcal{N}$ is the normalization constant. Setting $\theta = 0$ and $\nu = 1$ returns the PDF of the BCS for the position-like variable y_0. With this, the Rényi EPs $N_{1-p}(\chi)$ are calculated and found to be constant across varying p. This is because $\mathcal{F}(y_0)$ for the BCS is in fact a piecewise rearrangement of a Gaussian PDF (yet has an overall non-Gaussian structure) as depicted in Figure 1, thus $N_{1-p}(\chi) = \sigma^2$ for all p, where σ^2 is the variance of the 'would be Gaussian'. Taking the reference PDF to be $G(x) = \mathcal{G}(x|a, \alpha, \beta)$, with $a = \log_2(2\pi\sigma^2)/2$, $\alpha = 1/2$ and $\beta = \log_2(e)$, it is evident that $(\kappa_k - \gamma_k) = 0$ for all $k \geq 1$, and from the Gram–Charlier A series (41), a perfect matching in the reconstruction is achieved. Furthermore, it can be shown that the variance of (47) increases with α, i.e., the variance increases as the peaks of the PDF diverge, which is in stark contrast to the Rényi EPs which remain constant for increasing α. This reveals the shortcomings of variance as a measure of uncertainty for non-Gaussian PDFs.

The peaks, located at $\mathcal{F}(y_\theta) = 2^{-a_j^\dagger}$, where j is an index labelling the distinct peaks, give rise to sharp singularities in the target $g(x)$. With regard to the BCS position PDF, distributions of the conjugate parameter $\mathcal{F}(y_{\pi/2})$ distinguish $\mathcal{F}(y_0)$ from its equimeasurable Gaussian PDF and hence the Rényi EPs also distinguish the different cases. The number of available cumulants k is computationally limited, but, as this grows, information about the singularities will be recovered in the reconstruction. In the following, we show how the tail convergence and location of a singularity for $g(x)$ can be reconstructed using $k = 5$.

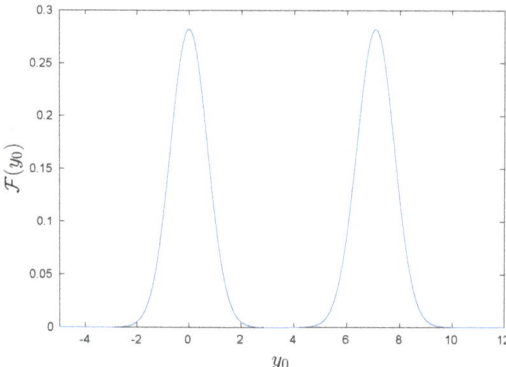

Figure 1. Probability distribution function of a balanced cat state (BCS) for the quantum mechanical state's position-like quadrature variable with $\alpha = 5$. This clearly displays an overall non-Gaussian structure; however, as this is a piecewise rearrangement of a Gaussian PDF for all α, we have that $N_{1-p} = \sigma^2$ for all p and α.

We consider the case of a UCS with $\nu = 0.97$, $\alpha = 10$ and we take $\theta = 0$ in Equation (47) to find the PDF in the y_0 quadrature which is non-Gaussian for all piecewise rearrangements.

As such, all REPs N_{1-p} vary with p and consequently all cumulants κ_k carry information on $g(x)$. Here, we choose to reconstruct the UCS information distribution by means of the Edgeworth series [91] so that

$$g(x) = \exp\left[n \sum_{j=2}^{\infty}(\kappa_j - \gamma_j)\frac{(-1)^j}{j!}\frac{d^j}{dx^j}n^{-j/2}\right]G(x), \qquad (48)$$

where the reference PDF $G(x)$ is again the shifted gamma distribution. Using the Edgeworth series, the information PDF is approximated by expanding in orders of n, which has the advantage over the Gram–Charlier A expansion discussed above of bounding the errors of the approximation. For the particular UCS of interest, expanding to order $n^{-3/2}$ reveals convergence toward the analytic form of the information PDF shown as the target line in Figure 2. This shows that, for a given characteristic resolution, control over the first five Rényi EPs can be enough for a useful information scan of a quantum state with an underlying non-Gaussian PDF. In the example shown in Figure 2, we see that the information scan accurately predicts the tail behavior as well as the location of the singularity, which corresponds to the second (lower) peak of $\mathcal{F}(y_0)$.

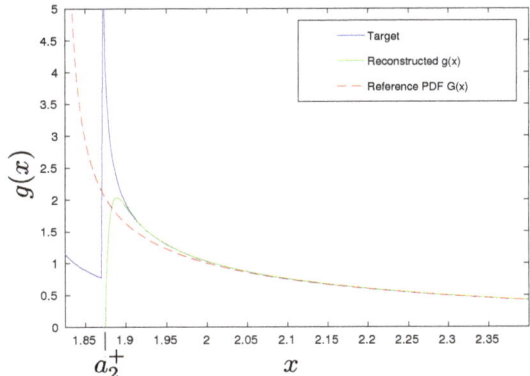

Figure 2. Reconstructed information distribution of an unbalanced cat state with $\nu = 0.97$ and $\alpha = 10$. The Edgeworth expansion has been used here to order $n^{-3/2}$ requiring control of the first five REPs. Good convergence of the tail behavior is evident as well as the location of the singularity corresponding to the second peak; a_2^+ corresponds to the value of x at the point of intersection with the second (lower) peak of $\mathcal{F}(y_0)$.

7. Entropy Powers Based on Tsallis Entropy

Let us now briefly comment on the entropy powers associated with yet another important differential entropy, namely *Tsallis differential entropy*, which is defined as [47]

$$\mathcal{S}_q(\mathcal{F}) = \frac{1}{(1-q)}\left[\int_{\mathbb{R}^D}(\mathcal{F}^q(x) - \mathcal{F}(x))dx\right], \qquad (49)$$

where, as before, the PDF $\mathcal{F}(x)$ is associated with a random vector $\{\mathcal{X}_i\}$ in $\mathbb{R}^D$.

Similarly to the RE case, the Tsallis entropy power $N_p^T(\mathcal{X})$ is defined as the solution of the equation

$$\mathcal{S}_q(\mathcal{X}) = \mathcal{S}_q^T\left(\sqrt{N_q^T(\mathcal{X})}\cdot\mathcal{Z}^G\right). \qquad (50)$$

The ensuing entropy power has not been studied in the literature yet, but it can be easily derived by observing that the following scaling property for differential Tsallis entropy holds, namely

$$S_q(a\mathcal{X}) = S_q(\mathcal{X}) \oplus_q \ln_q |a|^D, \tag{51}$$

where $a \in \mathbb{R}$ and the q-deformed sum and logarithm are defined as [11]: $x \oplus_q y = x + y + (1-q)xy$ and $\ln_q x = (x^{1-q} - 1)/(1-q)$, respectively. Relation (51) results from the following chain of identities:

$$\begin{aligned}
S_q(a\mathcal{X}) &= \frac{1}{1-q}\left[\int_{\mathbb{R}^D} d^D y \left(\int_{\mathbb{R}^D} d^D x \delta(y - ax)\mathcal{F}(x)\right)^q - 1\right] \\
&= \frac{1}{1-q}\left[|a|^{D(1-q)} \int_{\mathbb{R}^D} d^D y \mathcal{F}^q(y) - 1\right] \\
&= |a|^{D(1-q)}\left(S_q(\mathcal{X}) + \frac{1}{1-q}\right) - \frac{1}{1-q} \\
&= |a|^{D(1-q)} S_q(\mathcal{X}) + \ln_q |a|^D \\
&= \left[(1-q)\ln_q |a|^D + 1\right] S_q(\mathcal{X}) + \ln_q |a|^D \\
&= S_q(\mathcal{X}) \oplus_q \ln_q |a|^D.
\end{aligned} \tag{52}$$

We can further use the simple fact that

$$S_q(\mathcal{Z}_G) = \ln_q (2\pi q^{q'/q})^{D/2}. \tag{53}$$

Here, q and q' satisfy $1/q + 1/q' = 1$ with $q \in \mathbb{R}^+$. By combining (50), (51), and (53), we get

$$S_q(\mathcal{X}) = \ln_q (2\pi q^{q'/q})^{D/2} \oplus_q \ln_q (N_q^T)^D/2 = \ln_q (2\pi q^{q'/q} N_q^T)^{D/2}, \tag{54}$$

where we have used the sum rule from the q-deformed calculus: $\ln_q x \oplus_q \ln_q y = \ln_q xy$. Equation (54) can be resolved for N_p^T by employing the q-exponential, i.e., $e_q^x = [1 + (1-q)x]^{1/(1-q)}$, which (among others) satisfies the relation $e_q^{\ln_q x} = \ln_q(e_q^x) = x$. With this, we have

$$N_q^T(\mathcal{X}) = \frac{1}{2\pi} q^{-q'/q} \left[\exp_q(S_q(\mathcal{X}))\right]^{2/D} = \frac{1}{2\pi} q^{-q'/q} \exp_{1-(1-q)D/2}\left(\frac{2}{D} S_q(\mathcal{X})\right). \tag{55}$$

In addition, when $q \to 1_+$, one has

$$\lim_{q \to 1} N_q^T(\mathcal{X}) = \frac{1}{2\pi e} \exp\left(\frac{2}{D} \mathcal{H}(\mathcal{X})\right) = N(\mathcal{X}), \tag{56}$$

where $N(\mathcal{X})$ is the conventional Shannon entropy power and $\mathcal{H}(\mathcal{X})$ is the Shannon entropy [5].

In connection with Tsallis EP, we might notice one interesting fact, namely by starting from Rényi's EP (considering RE in nats), we can write

$$\begin{aligned}
N_q(\mathcal{X}) &= \frac{1}{2\pi} q^{-q'/q} \exp\left(\frac{2}{D} \mathcal{I}_q(\mathcal{X})\right) = \frac{1}{2\pi} q^{-q'/q} \left(\int d^D x \mathcal{F}^q(x)\right)^{2/(D(1-q))} \\
&= \frac{1}{2\pi} q^{-q'/q} \left[e_q^{S_q^T(\mathcal{X})}\right]^{2/D} = N_q^T(\mathcal{X}).
\end{aligned} \tag{57}$$

Here, we have used a simple identity

$$\left(\int d^D x \mathcal{F}^q(x)\right)^{1/(1-q)} = \left[(1-q)\mathcal{S}_q^T(\mathcal{X}) + 1\right]^{1/(1-q)} = e_q^{\mathcal{S}_q^T(\mathcal{X})}. \tag{58}$$

Thus, we have obtained that Rényi and Tsallis EPs coincide with each other. In particular, Rényi's EPI (22) can be equivalently written in the form

$$N_{p/2}^T(\mathcal{X}) N_{q/2}^T(\mathcal{Y}) \geq \frac{1}{16\pi^2}. \tag{59}$$

Similarly, we could rephrase the generalized Stam inequality (21) and generalized isoperimetric inequality (14) in terms of Tsallis EPs. Though such inequalities are quite interesting from a mathematical point of view, it is not yet clear how they could be practically utilized in the estimation theory as there is no obvious operational meaning associated with Tsallis entropy (e.g., there is no coding theorem for Tsallis entropy). On the other hand, Tsallis entropy is an important concept in the description of entanglement [98]. For instance, Tsallis entropy of order 2 (also known as linear entropy) directly quantifies state purity [63].

8. Conclusions

In the first part of this paper, we have introduced the notion of Rényi's EP. With quantum metrology applications in mind, we carried out our discussion in the framework of estimation theory. In doing so, we have generalized the notion of Fisher information (FI) by using a Rényi entropy version of De Bruijn's identity. The key role of the escort distribution in this context was highlighted. With Rényi's EP at hand, we proved the RE-based isoperimetric and Stam inequalities. We have further clarified the role of Rényi's EP by deriving (through the generalized Stam inequality) a one-parameter family of Rényi EP-based quantum mechanical uncertainty relations. Conventional variance-based URs of Robertson-Schrödinger and Shannon differential entropy-based URs of Hirschman or Białynicki-Birula naturally appear as special cases in this hierarchy of URs. Interestingly, we found that the Tsallis entropy-based EP coincided with Rényi's EP provided that the order is the same. This might open quite a new, hitherto unknown role for Tsallis entropy in estimation theory.

The second part of the paper was devoted to developing the application of Rényi's EP for extracting quantum states from incomplete data. This is of particular interest in various quantum metrology protocols. To that end, we introduced the concepts of information distribution and showed how cumulants of the information distribution can be obtained from knowledge of EPs of various orders. With cumulants thus obtained, one can reconstruct the underlying information distribution in a process which we call an information scan. A numerical implementation of this reconstruction procedure was technically realized via Gram-Charlier A and Edgeworth expansion. For an explicit illustration of the information scan, we used the non-Gaussian quantum states—(un)balanced cat states. In this case, it was found that control of the first five significant Rényi EPs gave enough information for a meaningful reconstruction of the information PDF and brought about non-trivial information on the original balanced cat state PDF, such as asymptotic tail behavior or the heights of the peaks.

Finally, let us stress one more point. Rényi EP-based quantum mechanical uncertainty relations (26) basically represent a one-parameter class of inequalities that constrain higher-order cumulants of state distributions for conjugate observables [13]. In connection with this, the following two questions are in order. Assuming one is able to control Rényi EPs of various orders: (i) how do such Rényi EPs constrain the underlying state distribution and (ii) how do the ensuing REPURs restrict the state distributions of conjugate observables? The first question was tackled in this paper in terms of the information distribution and reconstruction theorem. The second question is more intriguing and has not yet been properly addressed. Work along these lines is presently under investigation.

Author Contributions: Conceptualization, P.J. and J.D.; Formal analysis, M.P.; Methodology, P.J. and M.P.; Validation, M.P.; Visualization, J.D.; Writing—original draft, P.J.; Writing—review & editing, J.D. All authors have read and agreed to the published version of the manuscript.

Funding: P.J. and M.P. were supported by the Czech Science Foundation Grant No. 19-16066S. J.D. acknowledges support from DSTL and the UK EPSRC through the NQIT Quantum Technology Hub (EP/M013243/1).

Institutional Review Board Statement: Not applicable.

Informed Consent Statement: Not applicable.

Data Availability Statement: Not applicable.

Conflicts of Interest: The authors declare no conflict of interest.

Abbreviations

The following abbreviations are used in this manuscript:

ITE	Information-theoretic entropy
UR	Uncertainty relation
RE	Rényi entropy
TE	Tsallis entropy
REPUR	Rényi entropy-power-based quantum uncertainty relation
QM	Quantum mechanics
EP	Entropy power
FI	Fisher information
PDF	Probability density function
EPI	Entropy power inequality
REP	Rényi entropy power
BCS	Balanced cat state
UCS	Unbalanced cat state

Appendix A

Here, we provide an intuitive proof of the generalized De Bruijn identity.

Generalized De Bruijn identity I: By denoting the PDF associated with a random vector $\{\mathcal{X}_i\}$ as $\mathcal{F}(x)$ and the noise PDF as $\mathcal{G}(z)$, we might write the LHS of (7) as

$$\frac{d}{d\epsilon}\mathcal{I}_q(\mathcal{X} + \sqrt{\epsilon}\,\mathcal{Z})|_{\epsilon=0}$$

$$= \frac{1}{1-q}\frac{d}{d\epsilon}\log\left[\int_{\mathbb{R}^D} dy \left(\int_{\mathbb{R}^D} dx \int_{\mathbb{R}^D} dz\, \delta^{(D)}(y - (x + \sqrt{\epsilon}z))\mathcal{F}(x)\mathcal{G}(z)\right)^q\right]\Bigg|_{\epsilon=0}$$

$$= \frac{1}{1-q}\frac{d}{d\epsilon}\log\left[\int_{\mathbb{R}^D} dy \left(\int_{\mathbb{R}^D} dz\, \mathcal{F}(y - \sqrt{\epsilon}z)\mathcal{G}(z)\right)^q\right]\Bigg|_{\epsilon=0}$$

$$= \frac{1}{1-q}\frac{d}{d\epsilon}\log\left\{\int_{\mathbb{R}^D} dy \left[\int_{\mathbb{R}^D} dz\, \left(\mathcal{F}(y) - \sqrt{\epsilon}z_i\nabla_i\mathcal{F}(y)\right.\right.\right.$$

$$\left.\left.\left. + \frac{1}{2}\epsilon z_i z_j \nabla_i \nabla_j \mathcal{F}(y) + \mathcal{O}(\epsilon^{3/2})\right)\mathcal{G}(z)\right]^q\right\}\Bigg|_{\epsilon=0}$$

$$= \frac{q}{1-q}\left[\int_{\mathbb{R}^D} dy\, \rho_q(y) \Sigma_{ij} \frac{\nabla_i \nabla_j \mathcal{F}(y)}{2\mathcal{F}(y)}\right] = \frac{q}{2}\left[\int_{\mathbb{R}^D} dy\, \rho_q(y) \Sigma_{ij} V_i(y) V_j(y)\right]$$

$$= \frac{q}{2}\text{Tr}[\text{cov}_q(V)\Sigma] = \frac{1}{2q}\text{Tr}[\text{cov}_q(V_q)\Sigma] = \frac{1}{2q}\text{Tr}(\mathbb{J}_q \Sigma)\,. \tag{A1}$$

It should be noted that our manipulations make sense only for any $q > 0$, as only in these cases are RE and escort distributions well defined. The right-hand-side of (A1) can also be equivalently written as

$$\frac{1}{2q}\mathbb{E}_q\{[(V_q)_i - \mathbb{E}_q((V_q)_i)]\Sigma_{ij}[(V_q)_j - \mathbb{E}_q((V_q)_j)]\},$$

$$= \frac{1}{2q}\mathbb{E}\{[(\mathcal{Z}_i - \mathbb{E}(\mathcal{Z}_i)](\mathbb{J}_q)_{ij}(\mathcal{X})[\mathcal{Z}_j - \mathbb{E}(\mathcal{Z}_j)]\}, \quad (A2)$$

where the mean $\mathbb{E}_q\{\ldots\}$ is performed with respect to the escort distribution ρ_q, while $\mathbb{E}$ with respect to $\mathcal{G}$ distribution.

We note in passing that the conventional De Bruijn's identity (6) emerges as a special case when $q \to 1$. For the Gaussian noise vector, we can generalize the previous derivation in the following way:

Generalized De Bruijn's identity II: Let $\{\mathcal{X}_i\}$ be a random vector in $\mathbb{R}^D$ with the PDF $\mathcal{F}(x)$ and let $\{\mathcal{Z}_i\}$ be an independent Gaussian noise vector with the zero mean and covariance matrix $\Sigma = \text{cov}(\mathcal{Z}^G)$, then,

$$\frac{d}{d\Sigma_{ij}}\mathcal{I}_q(\mathcal{X} + \mathcal{Z}^G)|_{\Sigma=0} = \frac{q}{1-q}\left[\int_{\mathbb{R}^D}dy\,\rho_q(y)\frac{\nabla_i\nabla_j\mathcal{F}(y)}{2\mathcal{F}(y)}\right]$$

$$= \frac{1}{2q}\left[\int_{\mathbb{R}^D}dy\,\rho_q(y)(V_q)_i(V_q)_j\right] = \frac{1}{2q}(\mathbb{J}_q)_{ij}. \quad (A3)$$

The right-hand-side is equivalent to

$$\frac{1}{2q}\mathbb{E}_q\{[(V_q)_i - \mathbb{E}_q((V_q)_i)][(V_q)_j - \mathbb{E}_q((V_q)_j)]\}. \quad (A4)$$

To prove the identity (A3), we might follow the same line of reasoning as in (A1). The only difference is that, while in (A1) we had a small parameter ϵ in which one could expand to all orders of correlation functions and easily perform differentiation and limit $\epsilon \to 0$ for any noise distribution (with zero mean), the same procedure can not be done in the present context for a generic noise distribution. In fact, only the Gaussian distribution has the property that the higher-order correlation functions and their derivatives with respect to Σ_{ij} are small when Σ is small. The latter is a consequence of the Marcinkiewicz theorem [99].

Appendix B

Here, we prove the *Generalized isoperimetric inequality* from Section 2. The starting point is the *entropy-power inequality* (EPI) [64]: Let $\mathcal{X}_1$ and $\mathcal{X}_2$ be two independent continuous random vectors in $\mathbb{R}^D$ with probability densities $\mathcal{F}^{(1)} \in \ell^q(\mathbb{R}^D)$ and $\mathcal{F}^{(2)} \in \ell^p(\mathbb{R}^D)$, respectively. Suppose further that $\lambda \in (0,1)$ and $r > 1$, and let

$$q = \frac{r}{(1-\lambda) + \lambda r}, \quad p = \frac{r}{\lambda + (1-\lambda)r}, \quad (A5)$$

then the following inequality holds:

$$N_r(\mathcal{X}_1 + \mathcal{X}_2) \geq \left(\frac{N_q(\mathcal{X}_1)}{1-\lambda}\right)^{1-\lambda}\left(\frac{N_p(\mathcal{X}_2)}{\lambda}\right)^\lambda. \quad (A6)$$

Let us now consider a Gaussian noise vector $\mathcal{Z}^G$ (independent of $\mathcal{X}$) with zero mean and covariance matrix Σ. Within this setting, we can write the following EPIs:

$$N_r(\mathcal{X} + \mathcal{Z}^G) \geq \epsilon^\lambda \left(\frac{1}{1-\lambda}\right)^{1-\lambda} \left(\frac{1}{\lambda}\right)^\lambda [N_q(\mathcal{X})]^{1-\lambda}, \tag{A7}$$

$$N_r(\mathcal{X} + \mathcal{Z}^G) \geq \epsilon^{1-\lambda} \left(\frac{1}{1-\lambda}\right)^{1-\lambda} \left(\frac{1}{\lambda}\right)^\lambda [N_p(\mathcal{X})]^\lambda, \tag{A8}$$

with $\epsilon \equiv \det(\Sigma)^{1/D}$. Here, we have used the simple fact that $N_r(\mathcal{Z}^G) = \det(\Sigma)^{1/D}$, irrespective of the value of r.

Let us now fix r and maximize the RHS of inequality (A7) with respect to λ and q provided we keep the constraint condition (A5). This yields the condition extremum

$$\lambda = \frac{\epsilon}{N_q(\mathcal{X})} \exp\left[q(1-q)\frac{d \log N_q(\mathcal{X})}{dq}\right] + \mathcal{O}(\epsilon^2). \tag{A9}$$

With this, q turns out to be

$$q = r + \frac{\epsilon(1-r)r}{N_r(\mathcal{X})} \exp\left[(1-r)r\frac{d \log N_r(\mathcal{X})}{dr}\right] + \mathcal{O}(\epsilon^2), \tag{A10}$$

which in the limit $\epsilon \to 0$ reduces to $q = r \geq 1$. The latter implies that $p = 1$. The result (A10) implies that the RHS of (A7) reads

$$N_q(\mathcal{X}) + \epsilon \exp\left[(1-r)r\frac{d \log N_r(\mathcal{X})}{dr}\right]\left[1 - (1-r)r\frac{d \log N_r(\mathcal{X})}{dr}\right] + \mathcal{O}(\epsilon^2). \tag{A11}$$

Should we have started with the p index, we would arrive at an analogous conclusion. To proceed, we stick, without loss of generality, to the inequality (A7). This implies that

$$\begin{aligned}
N_r(\mathcal{X} + \mathcal{Z}^G) &\geq N_q(\mathcal{X}) \\
&+ \epsilon \exp\left[(1-r)r\frac{d \log N_r(\mathcal{X})}{dr}\right]\left[1 - (1-r)r\frac{d \log N_r(\mathcal{X})}{dr}\right] \\
&+ \mathcal{O}(\epsilon^2) \\
&= N_r(\mathcal{X}) + [N_q(\mathcal{X}) - N_r(\mathcal{X})] \\
&+ \epsilon \exp\left[(1-r)r\frac{d \log N_r(\mathcal{X})}{dr}\right]\left[1 - (1-r)r\frac{d \log N_r(\mathcal{X})}{dr}\right] \\
&+ \mathcal{O}(\epsilon^2) \\
&\geq N_r(\mathcal{X}) + \epsilon \exp\left[(1-r)r\frac{d \log N_r(\mathcal{X})}{dr}\right] + \mathcal{O}(\epsilon^2). \tag{A12}
\end{aligned}$$

To proceed, we employ the identity $\log N_r(\mathcal{X}) = 2/D[\mathcal{I}_r(\mathcal{X}) - \mathcal{I}_r(\mathcal{Z}_{\mathbf{I}}^G)]$ with $\mathcal{Z}_{\mathbf{I}}^G$ representing a Gaussian random vector with zero mean and *unit* covariance matrix, and the fact that $\mathcal{I}_r$ is monotonously decreasing function of r, i.e., $d\mathcal{I}_r/dr \leq 0$ (see, e.g., Ref. [78]). With this, we have

$$\exp\left[(1-r)r\frac{d\log N_r(\mathcal{X})}{dr}\right] \geq \exp\left[\frac{2(r-1)r}{D}\frac{d\mathcal{I}_r(\mathcal{Z}_1^G)}{dr}\right]$$

$$= \exp\left[(r-1)r\frac{d}{dr}\left(\frac{1}{r-1}\log r\right)\right]$$

$$= er^{r/(r-1)} \geq \frac{e^2}{r}. \quad (A13)$$

Consequently, Equation (A12) can be rewritten as

$$\frac{N_r(\mathcal{X}+\mathcal{Z}^G) - N_q(\mathcal{X})}{\Sigma_{ij}} \geq \frac{\epsilon}{\Sigma_{ij}}\frac{e^2}{r} + \mathcal{O}(\epsilon^2/\Sigma_{ij}). \quad (A14)$$

At this stage, we are interested in the $\Sigma_{ij} \to 0$ limit. In order to find the ensuing leading order behavior of ϵ/Σ_{ij}, we can use L'Hospital's rule, namely

$$\frac{\epsilon}{\Sigma_{ij}} = \frac{d\epsilon}{d\Sigma_{ij}} = \frac{d}{d\Sigma_{ij}}\exp\left[\frac{1}{D}\text{Tr}(\log \Sigma)\right] = \frac{\epsilon}{D}(\Sigma^{-1})_{ij}. \quad (A15)$$

Now, we neglect the sub-leading term of order $\mathcal{O}(\epsilon^2/\Sigma_{ij})$ in (A14) and take $\det(\ldots)^{1/D}$ on both sides. This gives

$$\det\left(\frac{dN_r(\mathcal{X}+\mathcal{Z}^G)}{d\Sigma_{ij}}\right)^{1/D}\Bigg|_{\Sigma=0} = \frac{1}{rD}N_r(\mathcal{X})[\det(\mathbb{J}_r(\mathcal{X}))]^{1/D} \geq \frac{e^2}{rD} \geq \frac{1}{rD}, \quad (A16)$$

or equivalently

$$N_r(\mathcal{X})[\det(\mathbb{J}_r(\mathcal{X}))]^{1/D} \geq 1. \quad (A17)$$

At this stage, we can use the inequality of arithmetic and geometric means to write (note that $\mathbb{J}_r = \text{cov}_r(V_r)$ is a positive semi-definite matrix)

$$\frac{1}{D}\text{Tr}(\mathbb{J}_r(\mathcal{X})) \geq [\det(\mathbb{J}_r(\mathcal{X}))]^{1/D}. \quad (A18)$$

Consequently, we have

$$\frac{1}{D}N_r(\mathcal{X})\text{Tr}(\mathbb{J}_r(\mathcal{X})) = \frac{1}{D}N_r(\mathcal{X})J_r(\mathcal{X}) \geq N_r(\mathcal{X})[\det(\mathbb{J}_r(\mathcal{X}))]^{1/D} \geq 1, \quad (A19)$$

as stated in Equation (14).

Appendix C

In this appendix, we prove the *Generalized Stam inequality* from Section 2. We start with the defining relation (13), i.e.,

$$N_q(\mathcal{Y}) = \frac{1}{2\pi}q^{1/(1-q)}\|\mathcal{G}\|_q^{2q/[(1-q)D]}, \quad (A20)$$

and consider $q \in [1/2, 1]$ so that $q/(1-q) > 0$. For the q-norm, we can write

$$\|\mathcal{G}\|_q = \left(\int_{\mathbb{R}^D} d\mathbf{y}\,|\psi_{\mathcal{G}}(\mathbf{y})|^{2q}\right)^{1/q} = \|\psi_{\mathcal{G}}\|_{2q}^2 \geq \|\hat{\psi}_{\mathcal{G}}\|_{2r}^2 = \|\psi_{\mathcal{F}}\|_{2r}^2 = \|\mathcal{F}\|_r. \quad (A21)$$

Here, $2r$ and $2q$ are Hölder conjugates so that $r \in [1, \infty]$. The inequality employed is due to the Hausdorff–Young inequality (which in turn is a simple consequence of the Hölder inequality [64]). We further have

$$
\begin{aligned}
\|\mathcal{F}\|_r &= \left(\int_{\mathbb{R}^D} d\mathbf{x} |\psi_\mathcal{F}(\mathbf{x})|^{2r}\right)^{1/r} \geq \left|\int_{\mathbb{R}^D} d\mathbf{x} |\psi_\mathcal{F}(\mathbf{x})|^{2r} \frac{\nabla_i \nabla_i e^{i\mathbf{a}\cdot\mathbf{x}}}{a_i^2}\right|^{1/r} \\
&= \left|r \int_{\mathbb{R}^D} d\mathbf{x} \left[(r-1)\rho_r(\mathbf{x}) V_i(\mathbf{x}) V_i(\mathbf{x}) + \rho_r(\mathbf{x})\frac{\nabla_i \nabla_i \mathcal{F}(\mathbf{x})}{\mathcal{F}(\mathbf{x})}\right] e^{i\mathbf{a}\cdot\mathbf{x}}\right|^{1/r} \\
&\quad \times \frac{\left(\int_{\mathbb{R}^D} d\mathbf{x} |\psi_\mathcal{F}(\mathbf{x})|^{2r}\right)^{1/r}}{a_i^{2/r}} \\
&\geq \left|r \int_{\mathbb{R}^D} d\mathbf{x} \left[(r-1)\rho_r(\mathbf{x}) V_i(\mathbf{x}) V_i(\mathbf{x}) + \rho_r(\mathbf{x})\frac{\nabla_i \nabla_i \mathcal{F}(\mathbf{x})}{\mathcal{F}(\mathbf{x})}\right] \cos(\mathbf{a}\cdot\mathbf{x})\right|^{1/r} \\
&\quad \times \frac{\left(\int_{\mathbb{R}^D} d\mathbf{x} |\psi_\mathcal{F}(\mathbf{x})|^{2r}\right)^{1/r}}{a_i^{2/r}} \\
&\geq \left|r \int_{V_D} d\mathbf{x} \rho_r(\mathbf{x}) \frac{\nabla_i \nabla_i \mathcal{F}(\mathbf{x})}{\mathcal{F}(\mathbf{x})} \cos(\mathbf{a}\cdot\mathbf{x})\right|^{1/r} \frac{\left(\int_{\mathbb{R}^D} d\mathbf{x} |\psi_\mathcal{F}(\mathbf{x})|^{2r}\right)^{1/r}}{a_i^{2/r}}, \quad (A22)
\end{aligned}
$$

where $\mathbf{a} \in \mathbb{R}^D$ is an arbitrary x-independent vector, $\nabla_i \equiv \partial/\partial x_i$ and V_D denotes a regularized volume of $\mathbb{R}^D$—D-dimensional ball of a very large (but finite) radius R. In the first line of (A22), we have employed the triangle inequality $|\mathbb{E}_r(e^{i\mathbf{a}\cdot\mathbf{x}})| \leq 1$ (with equality if and only if $\mathbf{a} = \mathbf{0}$), namely

$$
\left|\int_{\mathbb{R}^D} d\mathbf{x} |\psi_\mathcal{F}(\mathbf{x})|^{2r} e^{i\mathbf{a}\cdot\mathbf{x}}\right| = \left|\int_{\mathbb{R}^D} d\mathbf{x} \rho_r(\mathbf{x}) e^{i\mathbf{a}\cdot\mathbf{x}} \int_{\mathbb{R}^D} d\mathbf{x} |\psi_\mathcal{F}(\mathbf{x})|^{2r}\right| \leq \int_{\mathbb{R}^D} d\mathbf{x} |\psi_\mathcal{F}(\mathbf{x})|^{2r}. \quad (A23)
$$

The inequality in the last line holds for $a_i = \pi/(2R)$ (for all i), since, in this case, $\cos(\mathbf{a}\cdot\mathbf{x}) \geq 0$ for all $\mathbf{x}$ from the D-dimensional ball. In this case, one may further estimate the integral from below by neglecting the positive integrand $(r-1)\rho_r(\mathbf{x})[V_i(\mathbf{x})]^2$.

Note that (A22) implies

$$
\frac{r|\mathbb{E}_r[\mathcal{F}^{-1}\nabla_i\nabla_i\mathcal{F}\cos(\mathbf{a}\cdot\mathbf{x})]|}{a_i^2} \leq 1, \quad (A24)
$$

with equality if and only if $\mathbf{a} \to \mathbf{0}$ (to see this, one should apply L'Hospital's rule). Equation (A24) allows for writing

$$
\begin{aligned}
\|\mathcal{F}\|_r &\geq \frac{r^\gamma |\mathbb{E}_r[\mathcal{F}^{-1}\nabla_i\nabla_i\mathcal{F}\cos(\mathbf{a}\cdot\mathbf{x})]|^\gamma}{a_i^{2\gamma}} \left(\int_{\mathbb{R}^D} d\mathbf{x} |\psi_\mathcal{F}(\mathbf{x})|^{2r}\right)^{1/r} \\
&\geq \frac{r^\gamma |\mathbb{E}_r[\mathcal{F}^{-1}\nabla_i\nabla_i\mathcal{F}\cos(\mathbf{a}\cdot\mathbf{x})]|^\gamma}{a_i^{2\gamma}} \frac{1}{V_D^{1-1/r}} \\
&= \frac{r^\gamma |\mathbb{E}_r[\mathcal{F}^{-1}\nabla_i\nabla_i\mathcal{F}\cos(\mathbf{a}\cdot\mathbf{x})]|^\gamma}{a_i^{2\gamma}} \frac{1}{C_D^{1-1/r} R^{D-D/r}}, \quad (A25)
\end{aligned}
$$

where $\gamma > 0$ is some as yet unspecified constant and $C_D = \pi^{D/2}/\Gamma(D/2+1)$. In deriving (A25), we have used the Hölder inequality

$$\begin{aligned} 1 &= \left(\int_{\mathbb{R}^D} d\boldsymbol{x}\, 1 \cdot |\psi_{\mathcal{F}}(\boldsymbol{x})|^2\right) \leq \left(\int_{\mathbb{R}^D} d\boldsymbol{x}\, 1^{r'}\right)^{1/r'} \left(\int_{\mathbb{R}^D} d\boldsymbol{x}\, |\psi_{\mathcal{F}}(\boldsymbol{x})|^{2r}\right)^{1/r} \\ &= V_D^{1-1/r} \left(\int_{\mathbb{R}^D} d\boldsymbol{x}\, |\psi_{\mathcal{F}}(\boldsymbol{x})|^{2r}\right)^{1/r}. \end{aligned} \quad (A26)$$

Here, and also in (A22) and (A25), $V_D = C_D R^D$ denotes the regularized volume of $\mathbb{R}^D$.

As already mentioned, the best estimate of the inequality (A25) is obtained for $a \to 0$. As we have seen, a_i goes to zero as $\pi/(2R)$ which allows for choosing the constant γ so that the denominator in (A25) stays finite in the limit $R \to \infty$. This implies that $\gamma = D/2 - D/(2r)$. Consequently, (A25) acquires in the large R limit the form

$$\|\mathcal{F}\|_r \geq \frac{[4(r-1)/r]^{D/2-D/2r}[\Gamma(D/2+1)]^{1-1/r}}{\pi^{3D/2-3D/2r}} [(\mathbb{J}_r)_{ii}(\mathcal{X})]^{D/2-D/2r}, \quad (A27)$$

With this, we can write [see Equations (A20)–(A21)]

$$N_q(\mathcal{Y}) \geq \frac{1}{(2\pi)^2} q^{1/(1-q)} [(\mathbb{J}_r)_{ii}(\mathcal{X})] \geq \frac{1}{16\pi^2} [(\mathbb{J}_r)_{ii}(\mathcal{X})], \quad (A28)$$

where, in the last inequality, we have used the fact that $q^{1/(1-q)} \geq 1/4$ for $q \in [1/2, 1]$ and that $[\Gamma(D/2+1)]^{2/D} \geq \pi/4$. As a final step, we employ Equations (A18) and (A28) to write

$$N_q(\mathcal{Y}) \geq \frac{1}{16\pi^2 D} \operatorname{Tr}(\mathbb{J}_r(\mathcal{X})) \geq \frac{1}{16\pi^2} [\det(\mathbb{J}_r(\mathcal{X}))]^{1/D}, \quad (A29)$$

which completes the proof of the generalized Stam's inequality.

References

1. Bennaim, A. *Information, Entropy, Life in addition, the Universe: What We Know Amnd What We Do Not Know*; World Scientific: Singapore, 2015.
2. Jaynes, E.T. *Papers on Probability and Statistics and Statistical Physics*; D. Reidel Publishing Company: Boston, MA, USA, 1983.
3. Millar, R.B. *Maximum Likelihood Estimation and Infrence*; John Wiley and Soms Ltd.: Chichester, UK, 2011.
4. Leff, H.S.; Rex, A.F. (Eds.) *Maxwell's Demon 2: Entropy, Classical and Quantum Information, Computing*; Institute of Physics: London, UK, 2002.
5. Shannon, C.E. A mathematical theory of communication. *Bell Syst. Tech. J.* **1948**, *27*, 379–423; 623–656. [CrossRef]
6. Shannon, C.E.; Weaver, W. *The Mathematical Theory of Communication*; University of Illinois Press: New York, NY, USA, 1949.
7. Feinstein, A. *Foundations of Information Theory*; McGraw Hill: New York, NY, USA, 1958.
8. Campbell, L.L. A Coding Theorem and Rényi's Entropy. *Inf. Control* **1965**, *8*, 423–429. [CrossRef]
9. Bercher, J.-F. Source Coding Escort Distributions Rényi Entropy Bounds. *Phys. Lett. A* **2009**, *373*, 3235–3238. [CrossRef]
10. Thurner, S.; Hanel, R.; Klimek, P. *Introduction to the Theory of Complex Systems*; Oxford University Press: Oxford, UK, 2018.
11. Tsallis, C. *Introduction to Nonextensive Statistical Mechanics; Approaching a Complex World*; Springer: New York, NY, USA, 2009.
12. Bialynicki-Birula, I. Rényi Entropy and the Uncertainty Relations. *AIP Conf. Proc.* **2007**, *889*, 52–61.
13. Jizba, P.; Ma, Y.; Hayes, A.; Dunningham J.A. One-parameter class of uncertainty relations based on entropy power. *Phys. Rev. E* **2016**, *93*, 060104-1(R)–060104-5(R). [CrossRef] [PubMed]
14. Maassen, H.; Uffink, J.B.M. Generalized entropic uncertainty relations. *Phys. Rev. Lett.* **1988**, *60*, 1103–1106. [CrossRef] [PubMed]
15. Bialynicki-Birula, I.; Mycielski, J. Uncertainty relations for information entropy in wave mechanics. *Commun. Math. Phys.* **1975**, *44*, 129–132. [CrossRef]
16. Dang, P.; Deng, G.-T.; Qian, T. A sharper uncertainty principle. *J. Funct. Anal.* **2013**, *265*, 2239–2266. [CrossRef]
17. Ozawa, T.; Yuasa, K. Uncertainty relations in the framework of equalities. *J. Math. Anal. Appl.* **2017**, *445*, 998–1012. [CrossRef]
18. Zeng, B.; Chen, X.; Zhou, D.-L.; Wen, X.-G. *Quantum Information MeetsQuantum Matter: From Quantum Entanglement to Topological Phase in Many-Body Systems*; Springer: New York, NY, USA, 2018.
19. Melcher, B.; Gulyak, B.; Wiersig, J. Information-theoretical approach to the many-particle hierarchy problem. *Phys. Rev. A* **2019**, *100*, 013854-1–013854-5. [CrossRef]

20. Ryu, S.; Takayanagi, T. Holographic derivation of entanglement entropy from AdS/CFT. *Phys. Rev. Lett.* **2006**, *96*, 181602-1–181602-4. [CrossRef]
21. Eisert, J.; Cramer, M.; Plenio, M.B. Area laws for the entanglement entropy—A review. *Rev. Mod. Phys.* **2010**, *82*, 277–306. [CrossRef]
22. Pikovski, I.; Vanner, M.R.; Aspelmeyer, M.; Kim, M.S.; Brukner, Č. Probing Planck-Scale physics Quantum Optics. *Nat. Phys.* **2012**, *8*, 393–397. [CrossRef]
23. Marin, F.; Marino, F.; Bonaldi, M.; Cerdonio, M.; Conti, L.; Falferi, P.; Mezzena, R.; Ortolan, A.; Prodi, G.A.; Taffarello, L.; et al. Gravitational bar detectors set limits to Planck-scale physics on macroscopic variables *Nat. Phys.* **2013**, *9*, 71–73.
24. An, S.; Zhang, J.-N.; Um, M.; Lv, D.; Lu, Y.; Zhang, J.; Yin, Z.-Q.; Quan, H.T.; Kim, K. Experimental test of the quantum Jarzynski equality with a trapped-ion system. *Nat. Phys.* **2014**, *11*, 193–199. [CrossRef]
25. Campisi, M.; Hänggi, P.; Talkner, P. Quantum fluctuation relations: Foundations and applications. *Rev. Mod. Phys.* **2011**, *83*, 771–791. [CrossRef]
26. Erhart, J.; Sponar, S.; Sulyok, G.; Badurek, G.; Ozawa, M.; Hasegawa, Y. Experimental demonstration of a universally valid error—Disturbance uncertainty relation in spin measurements. *Nat. Phys.* **2012**, *8*, 185–189. [CrossRef]
27. Sulyok, G.; Sponar, S.; Erhart, J.; Badurek, G.; Ozawa, M.; Hasegawa, Y. Violation of Heisenberg's error-disturbance uncertainty relation in neutron-spin measurements *Phys. Rev. A* **2013**, *88*, 022110-1–022110-15. [CrossRef]
28. Baek, S.Y.; Kaneda, F.; Ozawa, M.; Edamatsu, K. Experimental violation and reformulation of the Heisenberg's error-disturbance uncertainty relation . *Sci. Rep.* **2013**, *3*, 2221-1–2221-5. [CrossRef] [PubMed]
29. Dressel, J.; Nori, F. Certainty in Heisenberg's uncertainty principle: Revisiting definitions for estimation errors and disturbance. *Phys. Rev. A* **2014**, *89*, 022106-1–022106-14. [CrossRef]
30. Busch, P.; Lahti, P.; Werner, R.F. Proof of Heisenberg's Error-Disturbance Relation. *Phys. Rev. Lett.* **2013**, *111*, 160405-1–160405-5. [CrossRef]
31. Jizba, P.; Arimitsu, T. The world according to Rényi: Thermodynamics of multifractal systems. *Ann. Phys.* **2004**, *312*, 17–59. [CrossRef]
32. Liu, R.; Liu, T.; Poor, H.V.; Shamai, S. A Vector Generalization of Costa's Entropy-Power Inequality with Applications. *IEEE Trans. Inf. Theory* **2010**, *56*, 1865–1879.
33. Costa, M.H.M. On the Gaussian interference channel. *IEEE Trans. Inf. Theory* **1985**, *31*, 607–615. [CrossRef]
34. Polyanskiy, Y.; Wu, Y. Wasserstein continuity of entropy and outer bounds for interference channels. *arXiv* **2015**, arXiv:1504.04419.
35. Bagherikaram, G.; Motahari, A.S.; Khandani, A.K. The Secrecy Capacity Region of the Gaussian MIMO Broadcast Channel. *IEEE Trans. Inf. Theory* **2013**, *59*, 2673–2682. [CrossRef]
36. De Palma, G.; Mari, A.; Lloyd, S.; Giovannetti, V. Multimode quantum entropy power inequality. *Phys. Rev. A* **2015**, *91*, 032320-1–032320-6. [CrossRef]
37. Costa, M.H. A new entropy power inequality. *IEEE Trans. Inf. Theory* **1985**, *31*, 751–760. [CrossRef]
38. Frieden, B.R. *Science from Fisher Information: A Unification*; Cambridge University Press: Cambridge, UK, 2004.
39. Courtade, T.A. Strengthening the Entropy Power Inequality. *arXiv* **2016**, arXiv:1602.03033.
40. Barron, A.R. Entropy and the Central Limit Theorem. *Ann. Probab.* **1986**, *14*, 336–342. [CrossRef]
41. Pardo, L. New Developments in Statistical Information Theory Based on Entropy and Divergence Measures. *Entropy* **2019**, *21*, 391. [CrossRef]
42. Biró, T.; Barnaföldi, G.; Ván, P. New entropy formula with fluctuating reservoir. *Physics A* **2015**, *417*, 215–220. [CrossRef]
43. Bíró, G.; Barnaföldi, G.G.; Biró, T.S.; Ürmössy, K.; Takács, Á. Systematic Analysis of the Non-Extensive Statistical Approach in High Energy Particle Collisions—Experiment vs. Theory. *Entropy* **2017**, *19*, 88. [CrossRef]
44. Hanel, R.; Thurner, S. When do generalized entropies apply? How phase space volume determines entropy. *Europhys. Lett.* **2011**, *96*, 50003-1–50003-6. [CrossRef]
45. Hanel, R.; Thurner, S.; Gell-Mann, M. How multiplicity determines entropy and the derivation of the maximum entropy principle for complex systems. *Proc. Natl. Acad. Sci. USA* **2014**, *111*, 6905–6910. [CrossRef] [PubMed]
46. Burg, J.P. The Relationship Between Maximum Entropy Spectra In addition, Maximum Likelihood Spectra. *Geophysics* **1972**, *37*, 375–376. [CrossRef]
47. Tsallis, C. Possible generalization of Boltzmann-Gibbs statistics. *J. Stat. Phys.* **1988**, *52*, 479–487. [CrossRef]
48. Havrda, J.; Charvát, F. Quantification Method of Classification Processes: Concept of Structural α-Entropy. *Kybernetika* **1967**, *3* 30–35.
49. Frank, T.; Daffertshofer, A. Exact time-dependent solutions of the Renyi Fokker–Planck equation and the Fokker–Planck equations related to the entropies proposed by Sharma and Mittal. *Physics A* **2000**, *285*, 351–366. [CrossRef]
50. Sharma, B.D.; Mitter, J.; Mohan, M. On measures of "useful" information. *Inf. Control* **1978**, *39*, 323–336. [CrossRef]
51. Jizba, P.; Korbel, J. On q-non-extensive statistics with non-Tsallisian entropy. *Physics A* **2016**, *444*, 808–827. [CrossRef]
52. Jizba, P.; Arimitsu, T. Generalized statistics: Yet another generalization. *Physics A* **2004**, *340*, 110–116. [CrossRef]
53. Vos, G. Generalized additivity in unitary conformal field theories. *Nucl. Phys. B* **2015**, *899*, 91–111. [CrossRef]
54. Uffink, J. Can the maximum entropy principle be explained as a consistency requirement? *Stud. Hist. Phil. Mod. Phys.* **1995**, *26*, 223–261. [CrossRef]

55. Jizba, P.; Korbel, J. Maximum Entropy Principle in Statistical Inference: Case for Non-Shannonian Entropies. *Phys. Rev. Lett.* **2019**, *122*, 120601-1–120601-6. [CrossRef] [PubMed]
56. Jizba, P.; Arimitsu, T. Observability of Rényi's entropy. *Phys. Rev. E* **2004**, *69*, 026128-1–026128-12. [CrossRef]
57. Elben, A.; Vermersch, B.; Dalmonte, M.; Cirac, J.I.; Zoller, P. Rényi Entropies from Random Quenches in Atomic Hubbard and Spin Models. *Phys. Rev. Lett.* **2018**, *120*, 050406-1–050406-6. [CrossRef]
58. Bacco, D.; Canale, M.; Laurenti, N.; Vallone, G.; Villoresi, P. Experimental quantum key distribution with finite-key security analysis for noisy channels. *Nat. Commun.* **2013**, *4*, 2363-1–2363-8. [CrossRef] [PubMed]
59. Müller-Lennert, M.; Dupuis, F.; Szehr, O.; Fehr, S.; Tomamichel, M. On quantum Renyi entropies: A new generalization and some properties. *J. Math. Phys.* **2013**, *54*, 122203-1–122203-20. [CrossRef]
60. Coles, P.J.; Colbeck, R.; Yu, L.; Zwolak, M. Uncertainty Relations from Simple Entropic Properties. *Phys. Rev. Lett.* **2012**, *108*, 210405-1–210405-5. [CrossRef]
61. Minter, F.; Kuś, M.; Buchleitner, A. Concurrence of Mixed Bipartite Quantum States in Arbitrary Dimensions. *Phys. Rev. Lett.* **2004**, *92*, 167902-1–167902-4.
62. Vidal, G.; Tarrach, R. Robustness of entanglement. *Phys. Rev. A* **1999**, *59*, 141–155. [CrossRef]
63. Bengtsson, I.; Życzkowski, K. *Geometry of Quantum States. An Introduction to Quantum Entanglement*; Cambridge University Press: Cambridge, UK, 2006.
64. Jizba, P.; Dunningham, J.A.; Joo, J. Role of information theoretic uncertainty relations in quantum theory. *Ann. Phys.* **2015**, *355*, 87–114. [CrossRef]
65. Toranzo, I.V.; Zozor, S.; Brossier, J.-M. Generalization of the de Bruijn Identity to General ϕ-Entropies and ϕ-Fisher Informations. *IEEE Trans. Inf. Theory* **2018**, *64*, 6743–6758. [CrossRef]
66. Rioul, O. Information Theoretic Proofs of Entropy Power Inequalities. *IEEE Trans. Inf. Theory* **2011**, *57*, 33–55. [CrossRef]
67. Dembo, A.; Cover, T.M. Information Theoretic Inequalitis. *IEEE Trans. Inf. Theory* **1991**, *37*, 1501–1517. [CrossRef]
68. Lutwak, E.; Lv, S.; Yang, D.; Zhang, G. Extensions of Fisher Information and Stam's Inequality. *IEEE Trans. Inf. Theory* **2012**, *58*, 1319–1327. [CrossRef]
69. Widder, D.V. *The Laplace Transform*; Princeton University Press: Princeton, NJ, USA, 1946.
70. Knott, P.A.; Proctor, T.J.; Hayes, A.J.; Ralph, J.F.; Kok, P.; Dunningham, J.A. Local versus Global Strategies in Multi-parameter Estimation. *Phys. Rev. A* **2016**, *94*, 062312-1– 062312-7. [CrossRef]
71. Beck, C.; Schlögl, F. *Thermodynamics of Chaotic Systems*; Cambridge University Press: Cambridge, UK, 1993.
72. Gardner, R.J. The Brunn-Minkowski inequality. *Bull. Am. Math. Soc.* **2002**, *39*, 355–405. [CrossRef]
73. Cover, T.M.; Thomas, J.A. *Elements of Information Theory*; Wiley-Interscience: Hoboken, NJ, USA, 2006.
74. Einstein, A. Theorie der Opaleszenz von homogenen Flüssigkeiten und Flüssigkeitsgemischen in der Nähe des kritischen Zustandes. *Ann. Phys.* **1910**, *33*, 1275–1298. [CrossRef]
75. De Palma, G. The entropy power inequality with quantum conditioning. *J. Phys. A Math. Theor.* **2019**, *52*, 08LT03-1–08LT03-12. [CrossRef]
76. Ram, E.; Sason, I. On Rényi Entropy Power Inequalities. *IEEE Trans. Inf. Theory* **2016**, *62*, 6800–6815. [CrossRef]
77. Stam, A. Some inequalities satisfied by the quantities of information of Fisher and Shannon. *Inform. Control* **1959**, *2*, 101–112. [CrossRef]
78. Rényi, A. *Probability Theory*; Selected Papers of Alfred Rényi; Akadémia Kiado: Budapest, Hungary, 1976; Volume 2.
79. Cramér, H. *Mathematical Methods of Statistics*; Princeton University Press: Princeton, NJ, USA, 1946.
80. Wilk, G.; Włodarczyk, Z. Uncertainty relations in terms of the Tsallis entropy. *Phys. Rev. A* **2009**, *79*, 062108-1–062108-6. [CrossRef]
81. Schrödinger, E. About Heisenberg Uncertainty Relation. *Sitzungsber. Preuss. Akad. Wiss.* **1930**, *24*, 296–303.
82. Robertson, H.P. The Uncertainty Principle. *Phys. Rev.* **1929**, *34*, 163–164. [CrossRef]
83. Hirschman, I.I., Jr. A Note on Entropy . *Am. J. Math.* **1957**, *79*, 152–156. [CrossRef]
84. D'Ariano, M.G.; De Laurentis, M.; Paris, M.G.A.; Porzio, A.; Solimeno, S. Quantum tomography as a tool for the characterization of optical devices. *J. Opt. B* **2002**, *4*, 127–132. [CrossRef]
85. Lvovsky, A.I.; Raymer, M.G. Continuous-variable optical quantum-state tomography. *Rev. Mod. Phys.* **2009**, *81*, 299–332. [CrossRef]
86. Gross, D.; Liu, Y.-K.; Flammia, S.T.; Becker, S.; Eisert, J. Quantum State Tomography via Compressed Sensing. *Phys. Rev. Lett.* **2010**, *105*, 150401-1–150401-4. [CrossRef] [PubMed]
87. Beckner, W. Inequalities in Fourier Analysis. *Ann. Math.* **1975**, *102*, 159–182. [CrossRef]
88. Babenko, K.I. An inequality in the theory of Fourier integrals. *Am. Math. Soc. Transl.* **1962**, *44*, 115–128.
89. Samko, S.G.; Kilbas, A.A.; Marichev, O.I. *Fractional Integrals and Derivatives: Theory and Applications*; Gordon and Breach: New York, NY, USA, 1993.
90. Reed, M.; Simon, B. *Methods of Modern Mathematical Physics*; Academic Press: New York, NY, USA, 1975; Volume XI.
91. Wallace, D.L. Asymptotic Approximations to Distributions. *Ann. Math. Stat.* **1958**, *29*, 635–654. [CrossRef]
92. Zolotarev, V.M. Mellin—Stieltjes Transforms in Probability Theory. *Theory Probab. Appl.* **1957**, *2*, 444–469. [CrossRef]
93. Tagliani, A. Inverse two-sided Laplace transform for probability density functions. *J. Comp. Appl. Math.* **1998**, *90*, 157–170. [CrossRef]
94. Lukacs, E. *Characteristic Functions*; Charles Griffin: London, UK, 1970.

95. Pal, N.; Jin, C.; Lim, W.K. *Handbook of Exponential and Related Distributions for Engineers and Scientists*; Taylor & Francis Group: New York, NY, USA, 2005.
96. Kira, M.; Koch, S.W.; Smith, R.P.; Hunter, A.E.; Cundiff, S.T. Quantum spectroscopy with Schrödinger-cat states *Nat. Phys.* **2011**, *7*, 799–804.
97. Knott, P.A.; Cooling, J.P.; Hayes, A.; Proctor, T.J.; Dunningham, J.A. Practical quantum metrology with large precision gains in the low-photon-number regime. *Phys. Rev. A* **2016**, *93*, 033859-1–033859-7. [CrossRef]
98. Wei, L. On the Exact Variance of Tsallis Entanglement Entropy in a Random Pure State. *Entropy* **2019**, *21*, 539. [CrossRef] [PubMed]
99. Marcinkiewicz, J. On a Property of the Gauss law. *Math. Z.* **1939**, *44*, 612–618. [CrossRef]

Article

Estimation for Entropy and Parameters of Generalized Bilal Distribution under Adaptive Type II Progressive Hybrid Censoring Scheme

Xiaolin Shi [1], Yimin Shi [2,*] and Kuang Zhou [2]

[1] School of Electronics Engineering, Xi'an University of Posts and Telecommunications, Xi'an 710121, China; linda20016@163.com
[2] School of Mathematics and Statistics, Northwestern Polytechnical University, Xi'an 710072, China; kzhoumath@nwpu.edu.cn
* Correspondence: ymshi@nwpu.edu.cn

Abstract: Entropy measures the uncertainty associated with a random variable. It has important applications in cybernetics, probability theory, astrophysics, life sciences and other fields. Recently, many authors focused on the estimation of entropy with different life distributions. However, the estimation of entropy for the generalized Bilal (GB) distribution has not yet been involved. In this paper, we consider the estimation of the entropy and the parameters with GB distribution based on adaptive Type-II progressive hybrid censored data. Maximum likelihood estimation of the entropy and the parameters are obtained using the Newton–Raphson iteration method. Bayesian estimations under different loss functions are provided with the help of Lindley's approximation. The approximate confidence interval and the Bayesian credible interval of the parameters and entropy are obtained by using the delta and Markov chain Monte Carlo (MCMC) methods, respectively. Monte Carlo simulation studies are carried out to observe the performances of the different point and interval estimations. Finally, a real data set has been analyzed for illustrative purposes.

Keywords: entropy; generalized Bilal distribution; adaptive Type-II progressive hybrid censoring scheme; maximum likelihood estimation; Bayesian estimation; Lindley's approximation; confidence interval; Markov chain Monte Carlo method

1. Introduction

To analyze and evaluate the reliability of products, life tests are often carried out. For products with long lives and high reliability, a censoring scheme is often adopted during the test to save on time and costs. Two commonly used censoring schemes are Type-I and Type-II censoring, but these two censoring schemes do not have the flexibility of allowing the removal of units at points other than the terminal point of the experiment. To allow for more flexibility in removing surviving units from the test, more general censoring approaches are required. The progressive Type-II censoring scheme is appealing and has attracted much attention in the literature. This topic can be found in [1]. One may also refer to [2] for a comprehensive review on progressive censoring. One drawback of the Type-II progressive censoring scheme is that the length of the experiment may be quite long for long-life products. Therefore, Kundu and Joarder [3] proposed a Type-II progressive hybrid censoring scheme where the experiment terminates at a pre-specified time. However, for the Type-II progressive hybrid censoring scheme, the drawback is that the effective sample size is a random variable, which may be very small or even zero. To strike a balance between the total testing time and the efficiency in statistical inference, Ng et al. [4] introduced an adaptive Type-II progressive hybrid censoring scheme (ATII-PHCS). This censoring scheme is described as follows. Suppose that n units are placed on test and $X_1, X_2, \ldots, X_n$ denote the corresponding lifetimes from a distribution with the cumulative distribution function

(CDF) $F(x)$ and the probability density function (PDF) $f(x)$. The number of observed failures m and time T are specified in advance and $m < n$. At the first failure time $X_{1:m:n}$, R_1 units are randomly removed from the remaining $n-1$ units. Similarly, at the second failure time $X_{2:m:n}$, R_2 units from the remaining $n-2-R_1$ units are randomly removed, and so on. If the mth failure occurs before time T (i.e., $X_{m:m;n} < T$), the test terminates at time $X_{m:m:n}$ and all remaining R_m units are removed, where $R_m = n - m - \sum_{i=1}^{m-1} R_i$ and R_i is specified in advance ($i = 1, 2, \ldots, m$). If the Jth failure occurs before time T (i.e., $X_{J:m:n} < T < X_{J+1:m:n}$ where $J + 1 < m$), then we will not withdraw any units from the test by setting $R_{J+1} = R_{J+2} = \ldots = R_{m-1} = 0$, and the test will continue until the failure unit number reaches the prefixed number m. At the time of the mth failure, all remaining R_m units are removed and the test terminates, where $R_m = n - m - \sum_{i=1}^{J} R_i$.

The main advantage of ATII-PHCS is that it speeds up the test when the test duration exceeds the predetermined time T and ensures we get the effective number of failures m. It also illustrates how an experimenter can control the experiment. If one is interested in getting observations early, one will remove fewer units (or even none). For convenience, we let $X_i = X_{i:m:n}, i = 1, 2, \ldots, m$. After the above test, we get one of the following observation data cases:

Case I: $(X_1, R_1), (X_2, R_2), \ldots, (X_m, R_m)$ if $X_m < T$, where $R_m = n - \sum_{i=1}^{m-1} R_i - m$.
Case II: $(X_1, R_1), (X_2, R_2), \ldots, (X_J, R_J), (X_{J+1}, 0), \ldots, (X_{m-1}, 0), (X_m, R_m)$ if $X_J < T < X_{J+1}$ and $J < m$, where $R_m = n - m - \sum_{i=1}^{J} R_i$.

The ATII-PHCS has been studied in recent years. Mazen et al. [5] discussed the statistical analysis of the Weibull distribution under an adaptive Type-II progressive hybrid censoring scheme. Zhang et al. [6] investigated the maximum likelihood estimations (MLEs) of the unknown parameters and acceleration factors in the step-stress accelerated life test, based on the tampered failure rate model with ATII-PHC samples. Cui et al. [7] studied the point and interval estimates of the parameters from the Weibull distribution, based on adaptive Type-II progressive hybrid censored data in a constant-stress accelerated life test. Ismail [8] proposed that the MLE of the Weibull distribution parameters and the acceleration factor were derived based on ATII-PHC schemes under a step-stress partially accelerated life test model. The statistical inference of the dependent competitive failure system under the constant-stress accelerated life test with ATII-PHC data was studied by Zhang et al. [9]. Under an adaptive Type-II progressive censoring scheme, Ye et al. [10] investigated the general statistical properties and then used the maximum likelihood technique to estimate the parameters of the extreme value distribution. Some other studies on the statistical inference of life models using ATII-PHCS were presented by Sobhi and Soliman [11] and Nassar et al. [12]. Xu and Gui [13] studied entropy estimation for the two-parameter inverse Weibull distribution under adaptive type-II progressive hybrid censoring schemes.

Entropy measures the uncertainty associated with a random variable. Let X be a random variable having a continuous CDF $F(x)$ and PDF $f(x)$. Then, the Shannon entropy is defined as

$$H(f) = -\int_{-\infty}^{+\infty} f(x) \ln f(x) dx. \qquad (1)$$

In recent years, several scholars have studied the entropy estimation of different life distributions. Kang et al. [14] investigated the entropy estimators of a double exponential distribution based on multiply Type-II censored samples. Cho et al. [15] derived an estimation for the entropy function of a Rayleigh distribution based on doubly generalized Type-II hybrid censored samples. Baratpour et al. [16] developed the entropy of the upper record values and provided several upper and lower bounds for this entropy by using the hazard rate function. Cramer and Bagh [17] discussed the entropy of the Weibull distribution under progressive censoring. Cho et al. [18] obtained estimators for the entropy function of the Weibull distribution based on a generalized Type-II hybrid censored sample. Yu et al. [19] studied statistical inference in the Shannon entropy of the inverse Weibull distribution under progressive first-failure censoring.

In addition to the above-mentioned life distributions, the generalized Bilal (GB) distribution is also an important life distribution for analyzing lifetime data. The PDF and the CDF of the GB distribution, respectively, are given as

$$f(x;\beta,\lambda) = 6\beta\lambda x^{\lambda-1}\exp(-2\beta x^{\lambda})[1-\exp(-\beta x^{\lambda})], x>0, \beta>0, \lambda>0, \qquad (2)$$

$$F(x;\beta,\lambda) = 1 - \exp(-2\beta x^{\lambda})[3 - 2\exp(-\beta x^{\lambda})], x>0, \beta>0, \lambda>0, \qquad (3)$$

The Shannon entropy of the GB distribution is given by

$$H(f) = H(\beta,\lambda) = 2.5 + \gamma - \ln(27/4) - \ln(\lambda\beta^{\frac{1}{\lambda}}) + \frac{1}{\lambda}(\ln(9/8) - \gamma), \beta>0, \lambda>0,$$

where γ denotes the Euler–Mascheroni constant and $\gamma = 0.5772$.

The GB distribution was first introduced by Abd-Elrahman [20]. He investigated the properties of the probability density and failure rate function of this distribution. A comprehensive mathematical treatment of the GB distribution was provided, and the maximum likelihood estimations of unknown parameters were derived under the complete sample. Abd-Elrahman [21] provided the MLEs and Bayesian estimations of the unknown parameters and the reliability function based on a Type-II censored sample. Since the failure rate function of GB distribution has an upside-down bathtub shape, and it can also be monotonically decreasing or monotonically increasing at some selected values of the shape parameters λ, the GB model is very useful in survival analysis and reliability studies.

To the best of our knowledge, there has been no published work on the estimation of the entropy and parameters of GB distribution under an ATII-PHCS. As such, these issues are considered in this paper. The main objective of this paper is to provide the estimation of the entropy and unknown parameters of GB distribution under an ATII-PHCS by using the frequency and Bayesian methods.

The rest of this paper is organized as follows. In Section 2, the MLEs of the parameters and entropy of GB distribution are obtained, and approximate confidence intervals are constructed using the ATII-PHC data. In Section 3, the Bayesian estimation of the parameters and entropy under three different loss functions are provided using Lindley's approximation method. In addition, the Bayesian credible intervals of the parameters and entropy are also obtained by using the Markov chain Monte Carlo (MCMC) method. In Section 4, Monte Carlo simulations are carried out to investigate the performance of different point estimates and interval estimates. In Section 5, a real data set is analyzed for illustrative purposes. Some conclusions are presented in Section 6.

2. Maximum Likelihood Estimation

In this section, the MLE and approximate confidence intervals of the parameters and entropy of GB distribution will be discussed under the ATII-PHCS. Based on the data in Case I and Case II, the likelihood functions can be respectively written as

$$\text{Case I}: L_I(\beta,\lambda|\vec{x}) \propto \prod_{i=1}^{m} f(x_i;\beta,\lambda)[1 - F(x_i;\beta,\lambda))]^{R_i}, \qquad (4)$$

$$\text{Case II}: L_{II}(\beta,\lambda|\vec{x}) \propto \prod_{i=1}^{m} f(x_i;\beta,\lambda))\prod_{i=1}^{J}[1 - F(x_i;\beta,\lambda)]^{R_i}[1 - F(x_m;\beta,\lambda)]^{n-m-\sum_{i=1}^{J}R_i}, \qquad (5)$$

where $\vec{x} = (x_1, x_2, \ldots, x_m)$.

By combining $L_I(\beta,\lambda|\vec{x})$ and $L_{II}(\beta,\lambda|\vec{x})$, the likelihood functions can be written uniformly as

$$L(\beta,\lambda|\vec{x}) \propto \prod_{i=1}^{m} f(x_i;\beta,\lambda)) \prod_{i=1}^{D} [1-F(x_i;\beta,\lambda)]^{R_i} [1-F(x_m;\beta,\lambda)]^{R^*} =$$
$$= \prod_{i=1}^{m} 6\beta\lambda x_i^{\lambda-1} \exp(-2\beta x_i^\lambda)[1-\exp(-\beta x_i^\lambda)] \prod_{i=1}^{D} [\exp(-2\beta x_i^\lambda)(3-2\exp(-\beta x_i^\lambda))]^{R_i} \times [\exp(-2\beta x_m^\lambda)(3-2\exp(-\beta x_m^\lambda))]^{R^*}, \quad (6)$$

where $R^* = n - m - \sum_{i=1}^{D} R_i$ and, for Case I, $D = m$, $R^* = 0$, and for Case II, $D = J$, $R^* = n - m - \sum_{i=1}^{J} R_i$.

The log-likelihood function is given by

$$l = \ln L(\beta,\lambda|\vec{x}) \propto m\ln(6\beta\lambda) + \sum_{i=1}^{m}[(\lambda-1)\ln x_i - 2\beta x_i^\lambda + \ln(1-\exp(-\beta x_i^\lambda))] + \\ + \sum_{i=1}^{D}[-2R_i\beta x_i^\lambda + R_i\ln(3-2\exp(-\beta x_i^\lambda))] - 2R^*\beta x_m^\lambda + R^*\ln(3-2\exp(-\beta x_m^\lambda)). \quad (7)$$

By taking the first partial derivative of the log-likelihood function with regard to β and λ and equating them to zero, the following results can be obtained:

$$\frac{\partial l}{\partial \beta} = \frac{m}{\beta} + \sum_{i=1}^{m}[-3x_i^\lambda + x_i^\lambda[y_1(\theta)]^{-1}] + \sum_{i=1}^{D}[-3R_ix_i^\lambda + 3R_ix_i^\lambda[y_2(\theta)]^{-1}] - 3R^*x_m^\lambda + 3R^*x_m^\lambda[y_3(\theta)]^{-1} = 0, \quad (8)$$

$$\frac{\partial l}{\partial \lambda} = \frac{m}{\lambda} + \sum_{i=1}^{m}[\ln x_i - 3\beta x_i^\lambda \ln x_i + \beta x_i^\lambda \ln x_i[y_1(\theta)]^{-1}] + \sum_{i=1}^{D}[-3R_i\beta x_i^\lambda \ln x_i + 3R_i\beta x_i^\lambda \ln x_i[y_2(\theta)]^{-1}] - \\ - 3R^*\beta x_m^\lambda \ln x_m + 3R^*\beta x_m^\lambda \ln x_m[y_3(\theta)]^{-1} = 0, \quad (9)$$

where $\theta = (\beta,\lambda)$, $y_1(\theta) = 1 - \exp(-\beta x_i^\lambda)$, $y_2(\theta) = 3 - 2\exp(-\beta x_i^\lambda)$, $y_3(\theta) = 3 - 2\exp(-\beta x_m^\lambda)$

The MLEs of β and λ can be obtained by solving Equations (7) and (8), but the above two equations do not yield an analytical solution. Thus, we use the Newton–Raphson iteration method to obtain the MLEs of the parameters. For this purpose, we firstly calculate the second partial derivatives of the log-likelihood function with regard to β and λ:

$$\frac{\partial^2 l}{\partial \beta^2} = -\frac{m}{\beta^2} - \sum_{i=1}^{m}[x_i^{2\lambda}\exp(-\beta x_i^\lambda)][y_1(\theta)]^{-2} - \sum_{i=1}^{D} 6R_i x_i^{2\lambda}\exp(-\beta x_i^\lambda)[y_2(\theta)]^{-2} - 6R^*x_m^{2\lambda}\exp(-\beta x_m^\lambda)[y_3(\theta)]^{-2}, \quad (10)$$

$$\frac{\partial^2 l}{\partial \beta \partial \lambda} = \sum_{i=1}^{m}[-3x_i^\lambda \ln x_i + x_i^\lambda \ln x_i(y_1(\theta))^{-1}[1-\beta x_i^\lambda \exp(-\beta x_i^\lambda)(y_1(\theta))^{-1}] + \\ + \sum_{i=1}^{D}[-3R_i x_i^\lambda \ln x_i + 3R_i x_i^\lambda \ln x_i(y_2(\theta))^{-1}(1-2\beta x_i^\lambda \exp(-\beta x_i^\lambda)(y_2(\theta))^{-1})] - \\ -3R^*x_m^\lambda + 3R^*x_m^\lambda \ln x_m[y_3(\theta)]^{-1}[1-2\beta x_m^\lambda \exp(-\beta x_m^\lambda)(y_3(\theta))^{-1}], \quad (11)$$

$$\frac{\partial^2 l}{\partial \lambda^2} = -\frac{m}{\lambda^2} + \sum_{i=1}^{m}\left[\beta x_i^\lambda (\ln x_i)^2[-3+(y_1(\theta))^{-1}] - \beta^2 x_i^{2\lambda}(\ln x_i)^2 \exp(-\beta x_i^\lambda)(y_1(\theta))^{-2}\right] \\ + \sum_{i=1}^{D}\left[-3R_i\beta x_i^\lambda(\ln x_i)^2(1-(y_2(\theta))^{-1}) - 6R_i\beta^2 x_i^{2\lambda}(\ln x_i)^2 \exp(-\beta x_i^\lambda)(y_2(\theta))^{-2}\right] \\ -3R^*\beta x_m^\lambda(\ln x_m)^2(1-(y_3(\theta))^{-1}) - 6R^*\beta^2 x_m^{2\lambda}(\ln x_m)^2 \exp(-\beta x_m^\lambda)(y_3(\theta))^{-2}. \quad (12)$$

Let $I(\beta,\lambda) = \begin{bmatrix} I_{11} & I_{12} \\ I_{21} & I_{22} \end{bmatrix}$, where

$$I_{11} = -\frac{\partial^2 l}{\partial \beta^2}, \quad I_{22} = -\frac{\partial^2 l}{\partial \lambda^2}, \quad I_{12} = I_{21} = -\frac{\partial^2 l}{\partial \beta \partial \lambda}. \quad (13)$$

On the basis of the above calculation results, we can implement the Newton–Raphson iteration method to obtain the MLEs of unknown parameters. The specific steps of this iteration method can be seen in Appendix B. After obtaining the MLE $\hat{\beta}$ and $\hat{\lambda}$ of the parameters β and λ, using the invariant property of MLEs, the MLE of the entropy H (f) for the generalized Bilal distribution is given by

$$\hat{H}(f) = 2.5 + \gamma - \ln(27/4) - \frac{1}{\hat{\lambda}}\ln\hat{\beta} - \ln\hat{\lambda} + \frac{1}{\hat{\lambda}}(\ln(9/8) - \gamma). \quad (14)$$

Approximate Confidence Interval

In this subsection, the approximate confidence intervals of the parameters β, λ and the Shannon entropy $H(f)$ are derived. Based on regularity conditions, the MLEs $(\hat{\beta}, \hat{\lambda})$ are an approximately bivariate normal distribution $N((\beta, \lambda), I^{-1}(\hat{\beta}, \hat{\lambda}))$, where the covariance matrix $I^{-1}(\beta, \lambda)$ is an estimation of $I^{-1}(\beta, \lambda)$ and $I^{-1}(\hat{\beta}, \hat{\lambda}) = \begin{bmatrix} I_{11} & I_{12} \\ I_{21} & I_{22} \end{bmatrix}^{-1}_{(\beta,\lambda)=(\hat{\beta},\hat{\lambda})}$,

I_{11}, I_{22}, I_{12} and I_{21} are given by Equations (10)–(13), respectively.

Thus, the approximate $100(1 - \alpha)\%$ two-sided confidence intervals (CIs) for parameters β, λ are given by

$$\left(\hat{\beta} \pm z_{\alpha/2}\sqrt{Var(\hat{\beta})}\right), \left(\hat{\lambda} \pm z_{\alpha/2}\sqrt{Var(\hat{\lambda})}\right), \tag{15}$$

where $z_{\alpha/2}$ is the upper $\alpha/2$ percentile of the standard normal distribution and $Var(\hat{\beta})$, $Var(\hat{\lambda})$ are the main diagonal elements of the matrix $I^{-1}(\hat{\beta}, \hat{\lambda})$.

Next, we use the delta method to obtain the asymptotic confidence interval of the entropy $H(f)$. The delta method is a general approach to compute CIs for functions of MLEs. Under a progressive Type-II censored sample, the authors of [22] used the delta method to study the estimation of a new Weibull–Pareto distribution. The authors of [23] also used this method to investigate the estimation of the two-parameter bathtub lifetime model.

Let $M^T = (\frac{\partial H(f)}{\partial \beta}, \frac{\partial H(f)}{\partial \lambda})$, where $\frac{\partial H(f)}{\partial \beta} = -\frac{1}{\beta\lambda}$, $\frac{\partial H(f)}{\partial \lambda} = \frac{1}{\lambda^2}\ln\beta - \frac{1}{\lambda} - \frac{1}{\lambda^2}(\ln\frac{9}{8} - \gamma)$.

Then, the approximate estimates of $var(\hat{H}(f))$ is given by

$$v\hat{a}r(\hat{H}(f)) = [M^T I^{-1}(\beta, \lambda) M]|_{(\beta,\lambda)=(\hat{\beta},\hat{\lambda})},$$

where $\hat{\beta}$ and $\hat{\lambda}$ are the MLEs of β and λ, respectively, and $I^{-1}(\beta, \lambda)$ denotes the inverse of the matrix $I(\beta, \lambda) = \begin{bmatrix} I_{11} & I_{12} \\ I_{21} & I_{22} \end{bmatrix}$. The elements of the matrix $I(\beta, \lambda)$ are given by Equations (10)–(13), respectively. Thus, $\frac{\hat{H}(f) - H(f)}{\sqrt{v\hat{a}r(\hat{H}(f))}}$ is asymptotically distributed as $N(0, 1)$. The asymptotic $100(1 - \alpha)\%$ CI for the entropy $H(f)$ is given by

$$\left(\hat{H}(f) \pm Z_{\alpha/2}\sqrt{v\hat{a}r(\hat{H}(f))}\right)$$

where $z_{\alpha/2}$ is the upper $\alpha/2$ percentile of the standard normal distribution.

3. Bayesian Estimation

In this section, we discuss the Bayesian point estimation of the parameters and entropy $H(f)$ for generalized Bilal distribution using Lindley's approximation method under symmetric as well as asymmetric loss functions. Furthermore, the Bayesian CI of the parameters and entropy are also derived by using the Markov chain Monte Carlo method.

3.1. Loss Functions and Posterior Distribution

Choosing the loss function is an important part in the Bayesian inference. The commonly used symmetric loss function is the squared error loss (SEL) function, which is defined as

$$L_1(U, \hat{U}) = (\hat{U} - U)^2. \tag{16}$$

Two popular asymmetric loss functions are the Linex loss (LL) and general entropy loss (EL) functions, which are respectively given by

$$L_2(U, \hat{U}) = \exp(h(\hat{U} - U)) - h(\hat{U} - U) - 1, \ h \neq 0, \tag{17}$$

$$L_3(U,\hat{U}) \propto \left(\frac{\hat{U}}{U}\right)^q - q\ln\left(\frac{\hat{U}}{U}\right) - 1, \; q \neq 0. \tag{18}$$

Here, $U = U(\beta, \lambda)$ is any function of β and λ, and $\hat{U}$ is an estimate of U. The constant h and q represent the weight of errors on different decisions. Under the above loss functions, the Bayesian estimate of function U can be calculated by

$$\hat{U}_S = E(U|\vec{x}). \tag{19}$$

$$\hat{U}_L = -\frac{1}{h}\ln[E(\exp(-hU)|\vec{x})], \quad h \neq 0. \tag{20}$$

$$\hat{U}_E = [E(U^{-q}|\vec{x})]^{-1/q}, \quad q \neq 0. \tag{21}$$

To derive the Bayesian estimates of the function $U(\beta, \lambda)$, we consider prior distributions of the unknown parameters β and λ as independent Gamma distributions $Ga(a,b)$ and $Ga(c,d)$, respectively. Therefore, the joint prior distribution of β and λ becomes

$$\pi(\beta,\lambda) = \frac{b^a \beta^{a-1}}{\Gamma(a)}\exp(-b\beta)\frac{d^c \lambda^{c-1}}{\Gamma(c)}\exp(-d\lambda), \; (\beta,\lambda,a,b,c,d > 0).$$

Based on the likelihood function $L(\beta,\lambda|\vec{x})$ and the joint prior distribution of β and λ, the joint posterior density of parameters β and λ can be written as

$$\pi(\beta,\lambda|\vec{x}) = \frac{\pi(\beta,\lambda)L(\beta,\lambda|\vec{x})}{\int_0^\infty \int_0^\infty \pi(\beta,\lambda)L(\beta,\lambda|\vec{x})d\beta d\lambda}$$

$$\propto \pi(\beta,\lambda)L(\beta,\lambda|\vec{x})$$
$$= \beta^{a-1}\exp(-b\beta)\lambda^{c-1}\exp(-d\lambda)A_1(\beta,\lambda)A_2(\beta,\lambda)A_3(\beta,\lambda), \tag{22}$$

where

$$A_1(\beta,\lambda) = \prod_{i=1}^{m} 6\beta\lambda x_i^{\lambda-1}\exp(-2\beta x_i^\lambda)[1-\exp(-\beta x_i^\lambda)],$$

$$A_2(\beta,\lambda) = \prod_{i=1}^{D}[\exp(-2\beta x_i^\lambda)(3-2\exp(-\beta x_i^\lambda))]^{R_i},$$

$$A_3(\beta,\lambda) = [\exp(-2\beta x_m^\lambda)(3-2\exp(-\beta x_m^\lambda))]^{R^*}.$$

Therefore, the Bayesian estimate of $U(\beta,\lambda)$ under the SEL, LL and GEL functions are respectively given by

$$\hat{U}_S(\beta,\lambda) = \frac{\int_0^\infty \int_0^\infty U(\beta,\lambda)\pi(\beta,\lambda)L(\beta,\lambda|\vec{x})d\beta d\lambda}{\int_0^\infty \int_0^\infty \pi(\beta,\lambda)L(\beta,\lambda|\vec{x})d\beta d\lambda}, \tag{23}$$

$$\hat{U}_L(\beta,\lambda) = -\frac{1}{h}\ln\left[\frac{\int_0^\infty \int_0^\infty \exp(-hU(\beta,\lambda))\pi(\beta,\lambda)L(\beta,\lambda|\vec{x})d\beta d\lambda}{\int_0^\infty \int_0^\infty \pi(\beta,\lambda)L(\beta,\lambda|\vec{x})d\beta d\lambda}\right], \tag{24}$$

$$\hat{U}_E(\beta,\lambda) = \left[\frac{\int_0^\infty \int_0^\infty (U(\beta,\lambda))^{-q}\pi(\beta,\lambda)L(\beta,\lambda|\vec{x})d\beta d\lambda}{\int_0^\infty \int_0^\infty \pi(\beta,\lambda)L(\beta,\lambda|\vec{x})d\beta d\lambda}\right]^{-\frac{1}{q}}. \tag{25}$$

3.2. Lindley's Approximation

From Equations (23)–(25), it is observed that all of these estimates of the $U(\beta,\lambda)$ are in the form of the ratio of two integrals which cannot be reduced to a closed form. Therefore, we use Lindley's approximation method to obtain the Bayesian estimates. If we let $\theta = (\theta_1, \theta_2)$, then the posterior expectation of a function U (θ_1, θ_2) can be approximated as in [18]:

$$\hat{U} = U(\hat{\theta}_1, \hat{\theta}_2) + 0.5(A + z_{30}B_{12} + z_{03}B_{21} + z_{21}C_{12} + z_{12}C_{21}) + p_1 A_{12} + p_2 A_{21}, \quad (26)$$

where $U(\hat{\theta}_1, \hat{\theta}_2)$ is the MLE of $U(\theta_1, \theta_2)$ and

$$A = \sum_{i=1}^{2} \sum_{j=1}^{2} u_{ij} \tau_{ij}, B_{ij} = (u_i \tau_{ii} + u_j \tau_{ij}) \tau_{ii}, C_{ij} = 3 u_i \tau_{ii} \tau_{ij} + u_j (\tau_{ii} \tau_{jj} + 2 \tau_{ij}^2),$$

$$p_i = \frac{\partial p}{\partial \theta_i}, u_i = \frac{\partial U}{\partial \theta_i}, u_{ij} = \frac{\partial^2 U}{\partial \theta_i \partial \theta_j}, p = \ln \pi(\theta_1, \theta_2), A_{ij} = u_i \tau_{ii} + u_j \tau_{ji},$$

$$z_{ij} = \frac{\partial^{i+j} l(\theta_1, \theta_2)}{\partial \theta_1^i \partial \theta_2^j}, i, j = 0, 1, 2, 3, i + j = 3,$$

where l denotes the log-likelihood function and $\tau_{ij}(i,j)$ denotes the (i,j)–th element of the matrix $[-\partial^2 l / \partial \theta_1^i \partial \theta_2^j]^{-1}$. All terms are estimated by MLEs of the parameters θ_1 and θ_2.

Based on the above equations, we have

$$z_{30} = \frac{\partial^3 l}{\partial \beta^3} = \frac{2m}{\beta^3} + \sum_{i=1}^{m} \{x_i^{3\lambda} \exp(-\beta x_i^\lambda)(y_1(\theta))^{-2}[1 + 2(y_1(\theta))^{-1} \exp(-\beta x_i^\lambda)]\}$$
$$+ \sum_{i=1}^{D} \{6 R_i x_i^{3\lambda} \exp(-\beta x_i^\lambda)(y_2(\theta))^{-2}[1 + 4 \exp(-\beta x_i^\lambda)(y_2(\theta))^{-1}]\} \quad (27)$$
$$+ 6 R^* x_m^{3\lambda} \exp(-\beta x_m^\lambda)(y_3(\theta))^{-2}[1 + 4 \exp(-\beta x_m^\lambda)(y_3(\theta))^{-1}].$$

$$z_{03} = \frac{\partial^3 l}{\partial \lambda^3} = \frac{2m}{\lambda^3} + \sum_{i=1}^{m} \{\beta x_i^\lambda (\ln x_i)^3 (-3 + (y_1(\theta))^{-1}) - \beta^2 x_i^{2\lambda} (\ln x_i)^3 \exp(-\beta x_i^\lambda)(y_1(\theta))^{-2}$$
$$\times [3 - \beta x_i^\lambda - 2\beta x_i^\lambda \exp(-\beta x_i^\lambda)(y_1(\theta))^{-1}]\} + \sum_{i=1}^{D} \{[-3 R_i \beta x_i^\lambda (\ln x_i)^3 [1 - (y_2(\theta))^{-1}] +$$
$$+ 6 R_i \beta^2 x_i^{2\lambda} (\ln x_i)^3 \exp(-\beta x_i^\lambda)(y_2(\theta))^{-2}(-3 + \beta x_i^\lambda + 4\beta x_i^\lambda (y_2(\theta))^{-1} \exp(-\beta x_i^\lambda))\} + \quad (28)$$
$$+ 3 R^* \beta x_m^\lambda (\ln x_m)^3 [-1 + (y_3(\theta))^{-1}] + 6 R^* \beta^2 x_m^{2\lambda} (\ln x_m)^3 \exp(-\beta x_m^\lambda)(y_3(\theta))^{-2}[-3 + \beta x_m^\lambda + 4\beta x_m^\lambda (y_3(\theta))^{-1} \exp(-\beta x_m^\lambda)].$$

$$z_{21} = \frac{\partial^3 l}{\partial \beta^2 \partial \lambda} = \sum_{i=1}^{m} \left[-x_i^{2\lambda} \ln x_i \exp(-\beta x_i^\lambda)(y_1(\theta))^{-2}[2 - \beta x_i^\lambda - 2\beta x_i^\lambda \exp(-\beta x_i^\lambda)(y_1(\theta))^{-1}] \right]$$
$$- \sum_{i=1}^{D} [6 R_i x_i^{2\lambda} \ln x_i \exp(-\beta x_i^\lambda)(y_2(\theta))^{-2}[2 - \beta x_i^\lambda - 4\beta x_i^\lambda \exp(-\beta x_i^\lambda)(y_2(\theta))^{-1}] \quad (29)$$
$$- 6 R^* x_m^{2\lambda} \ln x_m \exp(-\beta x_m^\lambda)(y_3(\theta))^{-2}[2 - \beta x_m^\lambda - 4\beta x_m^\lambda (y_3(\theta))^{-1} \exp(-\beta x_m^\lambda)].$$

$$z_{12} = \frac{\partial^3 l}{\partial \beta \partial \lambda^2} = \sum_{i=1}^{m} [-3 x_i^\lambda (\ln x_i)^2 + x_i^\lambda (\ln x_i)^2 (y_1(\theta))^{-1}$$
$$+ \beta x_i^{2\lambda} (\ln x_i)^2 \exp(-\beta x_i^\lambda)(y_1(\theta))^{-2}[-3 + \beta x_i^\lambda + y_1(\theta))^{-1} \beta x_i^\lambda \exp(-\beta x_i^\lambda)]] + \sum_{i=1}^{D} \{-3 R_i x_i^\lambda (\ln x_i)^2 + 3 R_i x_i^\lambda (\ln x_i)^2 (y_2(\theta))^{-1}$$
$$+ 6 \beta R_i x_i^{2\lambda} (\ln x_i)^2 \exp(-\beta x_i^\lambda)(y_2(\theta))^{-2}[-3 + \beta x_i^\lambda \exp(-\beta x_i^\lambda) + 4(y_2(\theta))^{-1} \beta x_i^\lambda \exp(-\beta x_i^\lambda)] \quad (30)$$
$$- 3 R^* x_m^\lambda (\ln x_m)^2 + 3 R^* x_m^\lambda (\ln x_m)^2 (y_3(\theta))^{-1}\}$$
$$+ 6 \beta R^* x_m^{2\lambda} (\ln x_m)^2 \exp(-\beta x_m^\lambda)(y_3(\theta))^{-2}[-3 + \beta x_m^\lambda \exp(-\beta x_m^\lambda) + 4(y_3(\theta))^{-1} \beta x_m^\lambda \exp(-\beta x_m^\lambda)].$$

$$p_1 = \frac{a-1}{\beta} - b, p_2 = \frac{c-1}{\lambda} - d,$$

$$\tau_{11} = -\frac{z_{02}}{z_{20} z_{02} - z_{11}^2}, \tau_{22} = -\frac{z_{20}}{z_{20} z_{02} - z_{11}^2}, \tau_{12} = \tau_{21} = \frac{z_{11}}{z_{20} z_{02} - z_{11}^2},$$

$$z_{20} = \frac{\partial^2 l}{\partial \beta^2}, z_{11} = \frac{\partial^2 l}{\partial \beta \partial \lambda}, z_{02} = \frac{\partial^2 l}{\partial \lambda^2},$$

where z_{20}, z_{11}, z_{02} are given by Equations (10)–(12), respectively.

Based on Lindley's approximation, we can derive the Bayesian estimation of the two parameters, β and λ, and the entropy under different loss functions.

3.2.1. Squared Error Loss Function

When U $(\beta, \lambda) = \beta$ or λ, the Bayesian estimations of the parameters β and λ under the SEL function are given by, respectively,

$$\hat{\beta}_S = \hat{\beta} + 0.5[\tau_{11}^2 z_{30} + \tau_{21}\tau_{22}z_{03} + 3\tau_{11}\tau_{12}z_{21} + (\tau_{11}\tau_{22} + 2\tau_{21}^2)z_{12}] + \tau_{11}p_1 + \tau_{12}p_2,$$

$$\hat{\lambda}_S = \hat{\lambda} + 0.5[\tau_{11}\tau_{12}z_{30} + \tau_{22}^2 z_{03} + 3\tau_{22}\tau_{21}z_{12} + (\tau_{11}\tau_{22} + 2\tau_{21}^2)z_{21}] + \tau_{21}p_1 + \tau_{22}p_2,$$

where $\hat{\beta}$ and $\hat{\lambda}$ are the MLEs of the parameters β and λ, respectively.

Similarly, the Bayesian estimation of the entropy can be derived. We notice that

$$U(\beta, \lambda) = H(\beta, \lambda) = 2.5 + \gamma - \ln(27/4) - \ln \lambda - \frac{1}{\lambda}\ln \beta + \frac{1}{\lambda}(\ln(9/8) - \gamma),$$
$$u_1 = -\frac{1}{\beta\lambda}, u_2 = -\frac{1}{\lambda} + \frac{1}{\lambda^2}(\ln \beta - \ln(9/8) + \gamma),$$
$$u_{11} = \frac{1}{\beta^2\lambda}, u_{22} = \frac{1}{\lambda^2} - \frac{2}{\lambda^3}(\ln \beta - \ln(9/8) + \gamma), u_{12} = u_{21} = \frac{1}{\beta\lambda^2}.$$

Thus, the Bayesian estimation of the entropy H (f) under the SEL function is given by

$$\hat{H}_S(f) = \hat{H}(f) + 0.5[u_{11}\tau_{11} + 2u_{12}\tau_{12} + u_{22}\tau_{22} + z_{30}(u_1\tau_{11} + u_2\tau_{12})\tau_{11} + z_{03}(u_2\tau_{22} + u_1\tau_{12})\tau_{22}$$
$$+ z_{21}(3u_1\tau_{11}\tau_{12} + u_2(\tau_{11}\tau_{22} + 2\tau_{12}^2)) + z_{12}(3u_2\tau_{22}\tau_{21} + u_1(\tau_{11}\tau_{22} + 2\tau_{21}^2))] \quad (31)$$
$$+ p_1(u_1\tau_{11} + u_2\tau_{21}) + p_2(u_2\tau_{22} + u_1\tau_{12}),$$

where $\hat{H}(f)$ represents the maximum likelihood estimate of H (f).

3.2.2. Linex Loss Function

Based on Lindley's approximation, the Bayesian estimations of two parameters, β and λ, and the entropy under the LL function can, respectively, be given by

$$\hat{\beta}_L = -\frac{1}{h}\ln\{\exp(-h\hat{\beta}) + 0.5[u_{11}\tau_{11} + u_1\tau_{11}^2 z_{30} + u_1\tau_{21}\tau_{22}z_{03} + 3u_1\tau_{11}\tau_{12}z_{21}$$
$$+ (\tau_{11}\tau_{22} + 2u_1\tau_{21}^2)u_1 z_{12}] + u_1\tau_{11}p_1 + u_1\tau_{12}p_2\}$$

$$\hat{\lambda}_L = -\frac{1}{h}\ln\{\exp(-h\hat{\lambda}) + 0.5[u_{22}\tau_{22} + u_2\tau_{11}\tau_{12}z_{30} + u_2\tau_{22}^2 z_{03} + (\tau_{11}\tau_{22} + 2\tau_{12}^2)u_2 z_{21}$$
$$+ 3u_2\tau_{22}\tau_{21}z_{21}] + u_2\tau_{12}p_1 + u_2\tau_{22}p_2\}$$

$$\hat{H}_L(f) = -\frac{1}{h}\ln\{\exp[-h\hat{H}(f)] + 0.5[u_{11}\tau_{11} + 2u_{12}\tau_{12} + u_{22}\tau_{22} + z_{30}(u_1\tau_{11} + u_2\tau_{12})\tau_{11} + z_{03}(u_2\tau_{22} + u_1\tau_{21})\tau_{22}$$
$$+ z_{21}(3u_1\tau_{11}\tau_{12} + u_2(\tau_{11}\tau_{22} + 2\tau_{12}^2)) + z_{12}(3u_2\tau_{22}\tau_{21} + u_1(\tau_{11}\tau_{22} + 2\tau_{21}^2))] \quad (32)$$
$$+ p_1(u_1\tau_{11} + u_2\tau_{21}) + p_2(u_2\tau_{22} + u_1\tau_{12})\}.$$

Here, $\hat{\beta}$ and $\hat{\lambda}$ are the MLEs of the parameters β and λ, and $\hat{H}(f)$ represents the MLE of H (f). The detailed derivation of these Bayesian estimates is shown in Appendix C.

3.2.3. General Entropy Loss Function

Using Lindley's approximation method, the Bayesian estimations of two parameters, β and λ, and the entropy under the GEL function can, respectively, be given by

$$\hat{\beta}_E = \{\hat{\beta}^{-q} + 0.5[u_{11}\tau_{11} + u_1\tau_{11}^2 z_{30} + u_1\tau_{21}\tau_{22}z_{03} + 3u_1\tau_{11}\tau_{12}z_{21} + (\tau_{11}\tau_{22} + 2u_1\tau_{21}^2)u_1 z_{12}] + u_1\tau_{11}p_1 + u_1\tau_{12}p_2\}^{-1/q}$$

$$\hat{\lambda}_L = \{\hat{\lambda}^{-q} + 0.5[u_{22}\tau_{22} + u_2\tau_{11}\tau_{12}z_{30} + u_2\tau_{22}^2 z_{03} + (\tau_{11}\tau_{22} + 2\tau_{12}^2)u_2 z_{21} + 3u_2\tau_{22}\tau_{21}z_{21}] + u_2\tau_{21}p_1 + u_2\tau_{22}p_2\}^{-1/q}$$

$$\hat{H}_E(f) = \{[\hat{H}(f)]^{-q} + 0.5[(u_{11}\tau_{11} + 2u_{12}\tau_{12} + u_{22}\tau_{22}) + z_{30}(u_1\tau_{11} + u_2\tau_{12})\tau_{11} + z_{03}(u_2\tau_{22} + u_1\tau_{12})\tau_{22}$$
$$+ z_{21}(3u_1\tau_{11}\tau_{12} + u_2(\tau_{11}\tau_{22} + 2\tau_{12}^2)) + z_{12}(3u_2\tau_{22}\tau_{21} + u_1(\tau_{11}\tau_{22} + 2\tau_{21}^2))] \quad (33)$$
$$+ p_1(u_1\tau_{11} + u_2\tau_{21}) + p_2(u_2\tau_{22} + u_1\tau_{12})\}^{-1/q}.$$

Here, $\hat{\beta}$ and $\hat{\lambda}$ are the MLEs of the parameters β and λ, and $\hat{H}(f)$ represents the MLE of H(f). The detailed derivation of these Bayesian estimates is shown in Appendix D.

3.3. Bayesian Credible Interval

In the previous subsection, we used the Lindley's approximation method to obtain the Bayesian point estimation of the parameters and entropy. However, this approximation method cannot determine the Bayesian CIs. Thus, the MCMC method is applied to obtain the Bayesian CI for the parameters and entropy. The MCMC method is a useful technique for estimating complex Bayesian models. The Gibbs sampling and Metropolis–Hastings algorithm are the two most frequently applied MCMC methods which are used in reliability analysis, statistical physics and machine learning, among other applications. Due to their practicality, they have gained some attention among researchers, and interesting results have been obtained. For example, Gilks and Wild [24] proposed adaptive rejection sampling to handle non-conjugacy in applications of Gibbs sampling. Koch [25] studied the Gibbs sampler by means of the sampling–importance resampling algorithm. Martino et al. [26] established a new approach, namely by recycling the Gibbs sampler to improve the efficiency without adding any extra computational cost. Panahi and Moradi [27] developed a hybrid strategy, combining the Metropolis–Hastings [28,29] algorithm with the Gibbs sampler to generate samples from the respective posterior, arising from the inverted, exponentiated Rayleigh distribution. In this paper, we adopt the method proposed in [27] to generate samples from the respective posterior arising from the GB distribution. From Equations (6) and (22), the joint posterior of the parameters β, λ can be written as

$$\pi(\beta,\lambda|\vec{x}) \propto \pi(\beta,\lambda)L(\beta,\lambda|\vec{x}) \propto [V(\lambda)]^{m+a}\beta^{m+a-1}\exp[-\beta V(\lambda)]\prod_{i=1}^{m}[1-\exp(-\beta x_i^\lambda)] \quad (34)$$

$$\times \frac{1}{[V(\lambda)]^{m+a}}\prod_{i=1}^{D}(3-2\exp(-\beta x_i^\lambda))^{R_i}(3-2\exp(-\beta x_m^\lambda))^{R^*}\lambda^{m+c-1}\exp(-d\lambda)\prod_{i=1}^{m}x_i^{\lambda-1}$$

Here, $V(\lambda) = (b + 2\sum_{i=1}^{m}x_i^\lambda + 2\sum_{i=1}^{D}R_i x_i^\lambda + 2R^* x_m^\lambda)$. Therefore, we have

$$\pi(\beta,\lambda|\vec{x}) \propto \pi_1(\beta|\lambda,\vec{x})\pi_2(\lambda|\beta,\vec{x}), \quad (35)$$

where

$$\pi_1(\beta|\lambda,\vec{x}) \propto [V(\lambda)]^{m+a}\beta^{m+a-1}\exp[-\beta V(\lambda)] \quad (36)$$

$$\pi_2(\lambda|\beta,\vec{x}) \propto \frac{\lambda^{m+c-1}}{[V(\lambda)]^{m+a}}\exp(-d\lambda)\exp[-\beta(2\sum_{i=1}^{m}x_i^\lambda + 2\sum_{i=1}^{D}R_i x_i^\lambda + 2R^* x_m^\lambda)] \quad (37)$$
$$\times \prod_{i=1}^{m}[1-\exp(-\beta x_i^\lambda)]\prod_{i=1}^{D}(3-2\exp(-\beta x_i^\lambda))^{R_i}(3-2\exp(-\beta x_m^\lambda))^{R^*}\prod_{i=1}^{m}x_i^{\lambda-1}.$$

It is observed that the posterior density $\pi_1(\beta|\lambda,\vec{x})$ of β, given λ, is the PDF of the Gamma distribution $Gamma(m+a, b+2\sum_{i=1}^{m}x_i^\lambda+2\sum_{i=1}^{D}R_i x_i^\lambda+2R^* x_m^\lambda)$. However, the posterior density $\pi_2(\lambda|\beta,\vec{x})$ of λ, given β, cannot be reduced analytically to a known distribution. Therefore, we use the Metropolis–Hastings method with normal proposal distribution to generate random numbers from Equation (37). We use the next algorithm (Algorithm 1), proposed in [27], to generate random numbers from Equation (34) and construct the Bayesian credible interval of λ, β and the entropy H(f).

Algorithm 1 The MCMC method

Step 1: Choose the initial value $(\beta^{(0)}, \lambda^{(0)})$.
Step 2: At stage i and for the given m, n and ATII-PH censored data, generate $\beta^{(i)}$ from the following:
$$Gamma(m+a,\ b+2\sum_{i=1}^{m} x_i^{\lambda} + 2\sum_{i=1}^{D} R_i x_i^{\lambda} + 2R^* x_m^{\lambda}).$$
Step 3: Generate $\lambda^{(i)}$ from $\pi_2(\lambda^{(i-1)} | \beta^{(i)}, \vec{x})$ using the following steps.
Step 3-1: Generate λ' from $N(\lambda^{(i-1)}, var(\lambda))$.
Step 3-2: Generate the ω from the uniform distribution U(0, 1).
Step 3-3: Set $\lambda^{(i)} = \begin{cases} \lambda', & if\,\omega \leq r^* \\ \lambda^{(i-1)}, & if\,\omega > r^* \end{cases}$, where $r^* = \min\{1, \frac{\pi_2(\lambda' | \beta^{(i)}, \vec{x})}{\pi_2(\lambda^{(i-1)} | \beta^{(i)}, \vec{x})}\}$.
Step 4: Set $i = i + 1$.
Step 5: By repeating Steps 2–4 N times, we get $(\beta_1, \lambda_1), (\beta_2, \lambda_2), \ldots, (\beta_N, \lambda_N)$. Furthermore, we compute $H_1, H_2, \ldots, H_N$, where $H_i = H(\beta_i, \lambda_i)$, $i = 1, 2, \ldots, N$ and $H(\beta, \lambda)$ is the Shannon entropy of the GB distribution.

Rearrange $(\beta_1, \beta_2, \ldots, \beta_N)$, and $(H_1, H_2, \ldots, H_N)$ into $(\beta_{(1)}, \beta_{(2)}, \ldots, \beta_{(N)})$, $(\lambda_{(1)}, \lambda_{(2)}, \ldots, \lambda_{(N)})$ and $(H_{(1)} H_{(2)}, \ldots, H_{(N)})$, where $(\beta_{(1)} < \beta_{(2)} < \ldots < \beta_{(N)})$, $(\lambda_{(1)} < \lambda_{(2)} < \ldots < \lambda_{(N)})$ and $(H_{(1)} < H_{(2)} < \ldots < H_{(N)})$.

Then, the $100(1-\alpha)\%$ Bayesian credible interval of the two parameters β, λ and the entropy are given by $(\beta_{(N\alpha/2)}, \beta_{(N(1-\alpha/2))})$, $(\lambda_{(N\alpha/2)}, \lambda_{(N(1-\alpha/2))})$ and $(H_{(N\alpha/2)}, H_{(N(1-\alpha/2))})$.

4. Simulation Study

In this section, a Monte Carlo simulation study is carried out to observe the performance of different estimators of the entropy, in terms of the MSEs for different values of n, m, T and censoring schemes. In addition, the average 95% asymptotic confidence intervals (ACIs), Bayesian credible intervals (BCIs) of β, λ and the entropy, as well as the average interval length (IL), are computed, and the performances are also compared. We consider the following three different progressive censoring schemes (CSs):

- **CS I:** $R_m = n - m$, $R_i = 0, i \neq m$;
- **CS II:** $R_1 = n - m$, $R_i = 0, i \neq 1$;
- **CS III:** $R_{m/2} = n - m$, $R_i = 0$, for $i \neq \frac{m}{2}$, if m is even or $R_{(m+1)/2} = n - m$, $R_i = 0$, for $i \neq \frac{m+1}{2}$, if m is odd.

Based on the following algorithm proposed by Balakrishnan and Sandhu [30] (Algorithm 2), we can generate an adaptive Type-II progressive hybrid censored sample from the GB distribution.

Algorithm 2. Generating a adaptive Type-II progressive hybrid censored sample from the GB distribution.

Step1: Generate m independent observations $Z_1, Z_2, \ldots, Z_m$, where Z_i follows the uniform distribution $U(0, 1)$, $i = 1, 2, \ldots, m$.
Step 2: For the known censoring scheme $(R_1, R_2, \ldots, R_m)$, let
$\xi_i = Z_i^{1/(i+R_m+R_{m-1}+\ldots+R_{m-i+1})}, i = 1, 2, \ldots, m$.
Step 3: By setting $U_i = 1 - \xi_m \xi_{m-1} \cdots \xi_{m-i+1}$, then $U_1, U_2, \ldots, U_m$ is a Type-II progressive censored sample from the uniform distribution $U(0, 1)$.
Step 4: Using the inverse transformation $X_{i:m:n} = F^{-1}(U_i)$, $i = 1, 2, \ldots, m$, we obtain a Type-II progressive censored sample from the GB distribution; that is, $X_{1:m:n}, X_{2:m:n}, \ldots, X_{m:m:n}$, where $F^{-1}(\cdot)$ denotes the GB distribution's inverse cumulative functional expression with the parameter (β, λ). The following theorem1 gives the uniqueness of the solution for the equation $X_{i:m:n} = F^{-1}(U_i)$, $i = 1, 2, \ldots, m$.
Step 5: If there exists a real number J satisfying $X_{J:m:n} < T \leq X_{J+1:m:n}$, then we set index J and record $X_{1:m:n}, X_{2:m:n}, \ldots, X_{J+1:m:n}$.
Step 6: Generate the first $m - J - 1$ order statistics $X_{J+2:m:n}, X_{J+3:m:n}, \ldots, X_{m:m:n}$ from the truncated distribution $f(x; \beta, \lambda) / [1 - F(x_{J+1}; \beta, \lambda)]$ with a sample size $n - J - 1 - \sum_{i=1}^{J} R_i$.

Theorem 1. *The equation $X_{i:m:n} = F^{-1}(U_i)$ has a unique solution, $i = 1, 2, \ldots, m$.*

Proof. See Appendix A. □

In the simulation study, we took the values of the parameters of the GB distribution as $\beta = 1, \lambda = 2$. In this case, H(f) = 0.2448. The hyperparameter values of the prior distribution were taken as $a = 1, b = 3, c = 2, d = 3$. For the Linex loss function and general entropy loss function, we set $h = -1.0$, 1.0 and $q = -1.0$, 1.0, respectively. In the Newton iterative algorithm and MCMC sampling algorithm, we chose the initial values of β and λ as $\beta^{(0)} = 0.9, \lambda^{(0)} = 1.9$; the value of ε was taken as 10^{-6}. For different sample sizes n and different effective samples m and time T, we used 3000 simulated samples in each case. The average values and mean square errors (MSEs) of the MLEs and Bayesian estimations (BEs) for β, λ and the entropy were calculated. These results are reported in Tables 1–6.

Table 1. The average maximum likelihood estimations (MLEs) and mean square errors (MSEs) of β, λ and the entropy ($\beta = 1, \lambda = 2$, H(f) = 0.2448).

(n, m)	SC	T = 0.6			T = 1.5		
		$\hat{\beta}$ MSE	$\hat{\lambda}$ MSE	$\hat{H}$ MSE	$\hat{\beta}$ MSE	$\hat{\lambda}$ MSE	$\hat{H}$ MSE
(40, 15)	I	1.1850 0.1224	2.2096 0.1428	0.1903 0.0979	1.1875 0.1213	2.2848 0.1521	0.1950 0.0963
	II	1.0727 0.0709	2.1448 0.1258	0.2015 0.0376	1.0619 0.0609	2.1541 0.1336	0.2017 0.0279
	III	1.1819 0.1217	2.2354 0.1413	0.1947 0.0910	1.1864 0.1208	2.2362 0.1514	0.1968 0.0902
(50, 15)	I	1.1326 0.1053	2.1803 0.1398	0.2086 0.0797	1.0905 0.0741	2.1931 0.1483	0.1997 0.0750
	II	1.0498 0.0390	2.1017 0.1243	0.2281 0.0280	1.0390 0.0374	2.1076 0.1263	0.2169 0.0197
	III	1.1184 0.1013	2.1817 0.1345	0.2035 0.0742	1.0740 0.0602	2.1284 0.1448	0.2013 0.0598
(60, 30)	I	1.1006 0.0889	2.1758 0.1374	0.2029 0.0625	1.0689 0.0683	2.1795 0.1368	0.2033 0.0547
	II	1.0451 0.0363	2.0847 0.1066	0.2260 0.0231	1.0476 0.0383	2.0877 0.1048	0.2170 0.0158
	III	1.0860 0.0653	2.1528 0.1368	0.2086 0.0601	1.0583 0.0592	2.1571 0.1335	0.2090 0.0418
(70, 30)	I	1.0641 0.0704	2.1296 0.1202	0.2163 0.0516	1.0581 0.0597	2.1197 0.1278	0.2134 0.0417
	II	1.0246 0.0265	2.0785 0.0849	0.2294 0.0198	1.0231 0.0317	2.0715 0.0946	0.2245 0.0148
	III	1.0517 0.0580	2.1483 0.1203	0.2199 0.0591	1.0468 0.0485	2.1132 0.1203	0.2195 0.0361

From Tables 1–6, the following observations can be made:

1. For the fixed m and T values, the MSEs of the MLEs and Bayesian estimations of the two parameters and the entropy decreased when n increased. As such, we tended to get better estimation results with an increase in the test sample size;
2. For the fixed n and m values, when T increased, the MSEs of the MLEs and Bayesian estimations of the two parameters and the entropy did not show any specific trend. This could be due to the fact that the number of observed failures was preplanned, and no additional failures were observed when T increased;
3. In most cases, the MSEs of the Bayesian estimations under a squared error loss function were smaller than those of the MLEs. There was no significant difference in the MSEs between the Linex loss and general entropy loss functions;
4. For fixed values of n, m and T, Scheme II was smaller than Scheme I and Scheme III in terms of the MSE.

Table 2. The average Bayesian estimations and MSEs of β, λ and the entropy under the squared error loss functon ($\beta = 1$, $\lambda = 2$; $\beta = 1$, $\lambda = 2$, H(f) = 0.2448).

(n, m)	SC	T = 0.6 $\hat{\beta}$ MSE	$\hat{\lambda}$ MSE	$\hat{H}$ MSE	T = 1.5 $\hat{\beta}$ MSE	$\hat{\lambda}$ MSE	$\hat{H}$ MSE
(40, 15)	I	0.8625 0.0353	1.8735 0.1325	0.3357 0.0930	0.8687 0.0337	1.8761 0.1317	0.3301 0.0920
	II	0.9480 0.0235	1.9583 0.0954	0.2630 0.0342	0.9546 0.0255	1.9531 0.0938	0.2616 0.0217
	III	0.8795 0.0340	1.8041 0.1314	0.3264 0.0948	0.8837 0.0310	1.8996 0.1299	0.3034 0.0902
(50, 15)	I	0.9325 0.0297	1.8917 0.1185	0.3189 0.0796	0.8973 0.0289	1.8345 0.0975	0.2732 0.0741
	II	0.9645 0.0218	1.9907 0.0827	0.2580 0.0260	0.9694 0.0223	1.9763 0.0812	0.2303 0.0198
	III	0.9475 0.0253	1.9013 0.1072	0.3016 0.0546	0.9824 0.0234	1.9314 0.0972	0.2661 0.0486
(60, 30)	I	0.9274 0.0224	1.8445 0.1151	0.2357 0.0575	0.9457 0.0263	1.8781 0.0919	0.2674 0.0508
	II	0.9671 0.0202	1.9932 0.0728	0.2398 0.0235	0.9688 0.0207	2.0176 0.0741	0.2235 0.0179
	III	0.9185 0.0211	1.8525 0.1072	0.2301 0.0534	0.9316 0.0227	1.9427 0.0954	0.2652 0.0504
(70, 30)	I	0.9742 0.0198	1.9360 0.0775	0.2538 0.0404	0.9515 0.0213	1.9504 0.0892	0.2553 0.0401
	II	0.9895 0.0174	2.0413 0.0613	0.2506 0.0195	0.9804 0.0186	2.0378 0.0537	0.2260 0.0105
	III	0.9787 0.0182	1.9746 0.0761	0.2512 0.0397	0.9713 0.0194	1.9714 0.0683	0.2537 0.0346

Table 3. The average Bayesian estimations and MSEs of β, λ and the entropy under the Linex loss function ($\beta = 1$, $\lambda = 2$, T = 0.6, H(f) = 0.2448).

(n, m)	SC	h=−1 $\hat{\beta}$ MSE	$\hat{\lambda}$ MSE	$\hat{H}$ MSE	h=1 $\hat{\beta}$ MSE	$\hat{\lambda}$ MSE	$\hat{H}$ MSE
(40, 15)	I	0.8835 0.0355	1.8558 0.1261	0.3583 0.0964	0.8531 0.0366	1.8248 0.1343	0.2802 0.0904
	II	0.9740 0.0255	1.9161 0.0885	0.2587 0.0721	0.9308 0.0246	1.9092 0.1008	0.2469 0.0304
	III	0.9047 0.0308	1.8768 0.1249	0.3343 0.0929	0.8670 0.0335	1.8405 0.1889	0.2638 0.0884
(50, 15)	I	0.9047 0.0301	1.9415 0.1238	0.3158 0.0939	0.8704 0.0337	1.9175 0.1329	0.2736 0.0764
	II	0.9852 0.0218	2.0538 0.0789	0.2502 0.0623	0.9674 0.0213	1.9201 0.0912	0.2358 0.0265
	III	0.9105 0.0284	1.9771 0.0986	0.3046 0.0904	0.8924 0.0293	1.9203 0.1257	0.2604 0.0654
(60, 30)	I	0.9341 0.0223	1.9788 0.1127	0.2792 0.0836	0.9035 0.0238	1.9221 0.1308	0.2520 0.0543
	II	0.9834 0.0198	2.0465 0.0664	0.3743 0.0365	0.9609 0.0211	1.9447 0.0791	0.2118 0.0220
	III	0.9498 0.0204	1.9837 0.0973	0.3424 0.0829	0.9258 0.0207	1.9253 0.1227	0.2319 0.0425
(70, 30)	I	0.9561 0.0197	1.9889 0.0768	0.2546 0.0579	0.9378 0.0184	1.9543 0.0975	0.2407 0.0403
	II	0.9957 0.0174	2.0312 0.0572	0.2371 0.0281	0.9798 0.0159	2.0164 0.0614	0.2410 0.0187
	III	0.9687 0.0185	2.0024 0.0746	0.2265 0.0536	0.9451 0.0120	1.9623 0.0784	0.2409 0.0354

Table 4. The average Bayesian estimations and MSEs of β, λ and the entropy under the Linex loss function ($\beta = 1$, $\lambda = 2$, T = 1.5, H(f) = 0.2448).

(n, m)	SC	h = −1			h = 1		
		$\hat{\beta}$ MSE	$\hat{\lambda}$ MSE	$\hat{H}$ MSE	$\hat{\beta}$ MSE	$\hat{\lambda}$ MSE	$\hat{H}$ MSE
(40, 15)	I	0.8896 0.0330	1.8328 0.1359	0.3492 0.1025	0.8510 0.0375	1.8127 0.1396	0.3381 0.0947
	II	0.9638 0.0248	1.9177 0.0863	0.2743 0.0365	0.9272 0.0265	1.9167 0.0982	0.2657 0.0301
	III	0.8922 0.0321	1.8691 0.1306	0.3424 0.0948	0.8631 0.0334	1.8430 0.1328	0.3343 0.0803
(50, 15)	I	0.9024 0.0234	1.8678 0.1094	0.3217 0.0921	0.8823 0.0315	1.8874 0.1173	0.3216 0.0810
	II	0.9713 0.0221	1.9401 0.0731	0.2601 0.0262	0.9418 0.0217	1.9824 0.0884	0.2632 0.0223
	III	0.9135 0.0231	1.8792 0.090	0.3383 0.0921	0.8975 0.0314	1.8845 0.1121	0.3210 0.0693
(60, 30)	I	0.9470 0.0219	1.8946 0.0951	0.3222 0.0727	0.9080 0.0234	1.9012 0.1075	0.3251 0.0536
	II	0.9795 0.0209	1.9452 0.0719	0.2518 0.0246	0.9548 0.0199	1.9616 0.0776	0.2513 0.0219
	III	0.9425 0.0213	1.8978 0.0906	0.3197 0.0648	0.9253 0.0213	1.9041 0.1069	0.3218 0.0412
(70, 30)	I	0.9583 0.0184	1.9562 0.0748	0.3165 0.0473	0.9491 0.0179	1.9493 0.0861	0.3314 0.0392
	II	0.9901 0.0163	2.0576 0.0652	0.2318 0.0168	0.9814 0.0153	2.0997 0.0608	0.2459 0.0161
	III	0.9711 0.0175	1.9230 0.0697	0.3027 0.0389	0.9502 0.0162	1.9894 0.0841	0.3267 0.0304

Table 5. The average Bayesian estimations and MSEs of β, λ and the entropy under the general entropy loss function ($\beta = 1$, $\lambda = 2$, T = 0.6, H(f) = 0.2448).

(n, m)	SC	q = −1			q = 1		
		$\hat{\beta}$ MSE	$\hat{\lambda}$ MSE	$\hat{H}$ MSE	$\hat{\beta}$ MSE	$\hat{\lambda}$ MSE	$\hat{H}$ MSE
(40, 15)	I	0.8739 0.0341	1.8380 0.1348	0.3181 0.0891	0.8288 0.0437	1.8173 0.1381	0.3558 0.1091
	II	0.9546 0.0239	1.9184 0.0966	0.2832 0.0234	0.9169 0.0265	1.9081 0.1084	0.2628 0.0315
	III	0.8828 0.0324	1.8422 0.1306	0.3097 0.0863	0.8494 0.0389	1.8266 0.1361	0.3207 0.1063
(50, 15)	I	0.9013 0.0305	1.8948 0.1191	0.3017 0.0463	0.8972 0.0380	1.8728 0.1231	0.3423 0.0598
	II	0.9701 0.0214	1.9386 0.0803	0.2695 0.0186	0.9430 0.0236	1.9471 0.0962	0.2268 0.0271
	III	0.9251 0.0263	1.8984 0.1093	0.3023 0.0486	0.8613 0.0308	1.8498 0.1176	0.3287 0.0525
(60, 30)	I	0.9270 0.0232	1.9089 0.0824	0.2776 0.0390	0.8975 0.0276	1.8785 0.1127	0.3270 0.0477
	II	0.9610 0.0190	2.0351 0.0686	0.2318 0.0197	0.9481 0.0210	2.0453 0.0791	0.2391 0.0245
	III	0.9406 0.0210	1.9105 0.0874	0.2698 0.0375	0.9116 0.0231	1.8938 0.1109	0.3168 0.0418
(70, 30)	I	0.9501 0.0171	1.9492 0.0778	0.2536 0.0265	0.9213 0.0202	1.9308 0.0840	0.2924 0.0392
	II	0.9817 0.0158	2.0147 0.0436	0.2325 0.0148	0.9681 0.0151	2, 1489 0.0526	0.2410 0.0272
	III	0.9546 0.0174	1.9602 0.0738	0.2513 0.0168	0.9467 0.0173	1.9436 0.0724	0.2902 0.0312

Table 6. The average Bayesian estimations and MSEs of β, λ and the entropy under the general entropy loss function ($\beta = 1$, $\lambda = 2$, $T = 1.5$, $H(f) = 0.2448$).

(n, m)	SC	\multicolumn{3}{c}{$q=-1$}			\multicolumn{3}{c}{$q=1$}		
		$\hat{\beta}$ MSE	$\hat{\lambda}$ MSE	$\hat{H}$ MSE	$\hat{\beta}$ MSE	$\hat{\lambda}$ MSE	$\hat{H}$ MSE
(40, 15)	I	0.8770 0.0335	1.8569 0.1332	0.3564 0.0903	0.8224 0.0455	1.7924 0.1331	0.3598 0.1075
	II	0.9560 0.0218	1.9221 0.0914	0.2729 0.0198	0.9112 0.0257	1.9038 0.0913	0.2786 0.0294
	III	0.8836 0.0315	1.8297 0.1217	0.3519 0.0841	0.8453 0.0348	1.8374 0.1224	0.3547 0.1024
(50, 15)	I	0.8947 0.0298	1.8979 0.0981	0.3028 0.0372	0.8631 0.0362	1.8308 0.1134	0.3143 0.0483
	II	0.9685 0.0206	1.9793 0.0801	0.2610 0.0164	0.9377 0.0216	1.9467 0.0910	0.2656 0.0283
	III	0.8984 0.0278	1.9078 0.0931	0.3012 0.0416	0.8702 0.0302	1.8547 0.1086	0.3125 0.0502
(60, 30)	I	0.9244 0.0221	1.8446 0.0772	0.2731 0.0283	0.8930 0.0267	1.9208 0.1041	0.2812 0.0421
	II	0.9767 0.0188	2.0526 0.0614	0.2554 0.0164	0.9440 0.0202	2.0658 0.0718	0.2627 0.0238
	III	0.9387 0.0198	1.9541 0.0824	0.2709 0.0346	0.9125 0.0210	1.9435 0.0983	0.2801 0.0431
(70, 30)	I	0.9531 0.0167	1.9578 0.0738	0.2501 0.0247	0.9230 0.0188	1.9447 0.0814	0.2523 0.0370
	II	0.9814 0.0140	2.2263 0.0394	0.2309 0.0135	0.9675 0.0140	2.2680 0.0338	0.2352 0.0247
	III	0.9624 0.0163	1.9795 0.0745	0.2486 0.0216	0.9457 0.0164	1.9539 0.0718	0.2501 0.0306

To further demonstrate the conclusions, the MSEs are plotted when the sample size increases under different censoring schemes. The trends are shown in Figure 1 (values come from Tables 1–6).

Furthermore, the average 95% ACIs and BCIs of β, λ and the entropy, as well as the average lengths (ALs) and coverage probabilities of the confidence intervals, were computed. These results are displayed in Tables A1–A4 (See Appendix E).

From Tables A1–A4, the following can be observed:

1. The coverage probability of the approximate confidence intervals and Bayes credible intervals became bigger when n increased while m and T remain fixed;
2. For fixed values of n and m, when T increased, we did not observe any specific trend in the coverage probability of the approximate confidence intervals and Bayesian credible intervals;
3. For fixed values of n and T, the average length of the approximate confidence intervals and Bayesian credible intervals were narrowed down when n increased;
4. The average length of the Bayesian credible intervals was smaller than that of the asymptotic confidence intervals in most cases;
5. For fixed values of n and m, when T increased, we did not observe any specific trend in the average length of the confidence intervals;
6. For fixed values of n, m and T, Scheme II was smaller than Scheme I and Scheme III in terms of the average length of the credible interval;
7. For fixed values of n, m and T, the coverage probability of the approximate confidence intervals and Bayesian credible intervals were bigger than Scheme I and Scheme III.

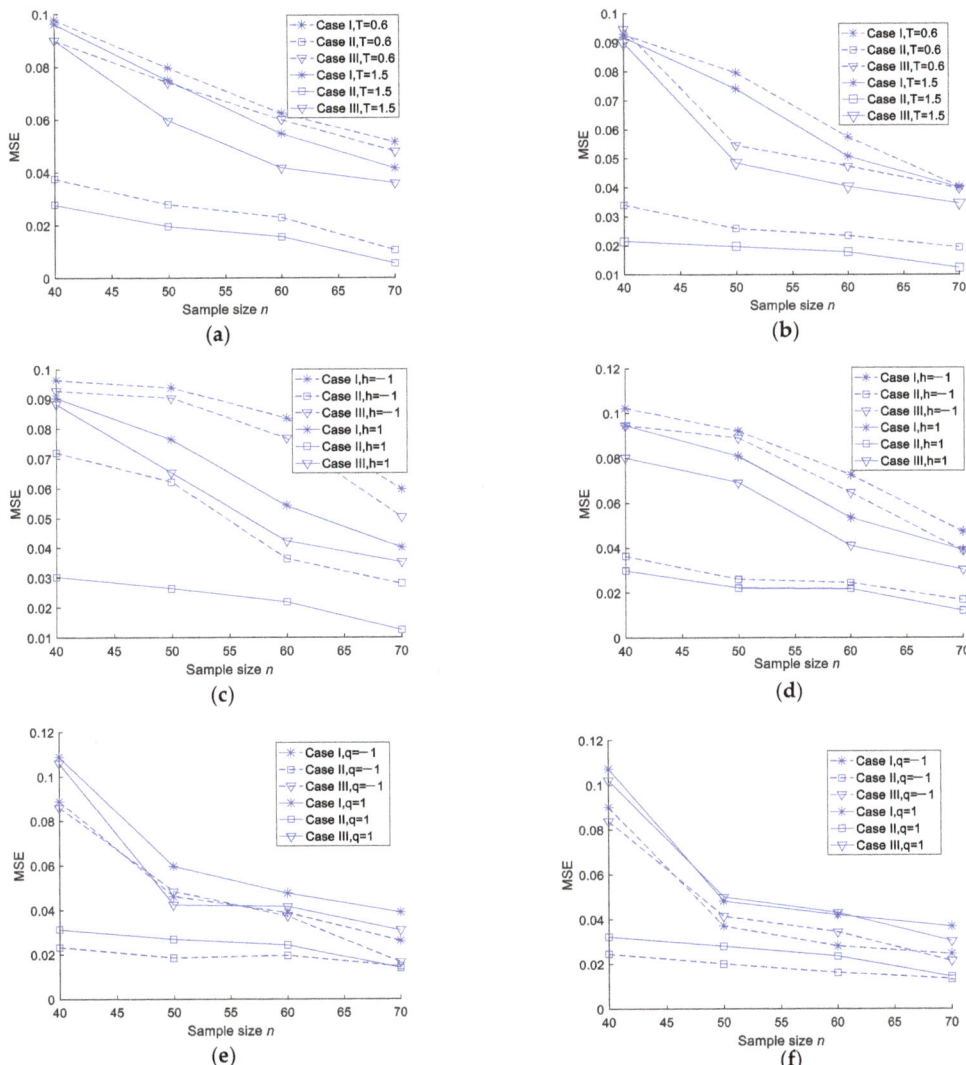

Figure 1. MSEs of different entropy estimations. (**a**) MSEs of MLEs of entropy in the case of T = 0.6 and T = 1.5. (**b**) MSEs of Bayesian estimations of entropy under a squared error loss function in the case of T = 0.6 and T = 1.5. (**c**) MSEs of Bayesian estimations of entropy under a Linex loss function in the case of T = 0.6. (**d**) MSEs of Bayesian estimations of entropy under a Linex loss function in the case of T = 1.5. (**e**) MSEs of Bayesian estimations of entropy under a general entropy loss function in the case of T = 0.6. (**f**) MSEs of Bayesian estimations of entropy under a general entropy loss function in the case of T = 1.5.

5. Real Data Analysis

In this subsection, a real data set is considered to illustrate the use of the inference procedures discussed in this paper. This data set consisted of 30 successive values of March precipitation in Minneapolis–Saint Paul, which were reported by Hinkley [31]. The data set points are expressed in inches as follows: 0.32, 0.47, 0.52, 0.59, 0.77, 0.81, 0.81, 0.9, 0.96, 1.18, 1.20, 1.20, 1.31, 1.35, 1.43, 1.51, 1.62, 1.74, 1.87, 1.89, 1.95, 2.05, 2.10, 2.20, 2.48, 2.81, 3.0, 3.09, 3.37 and 4.75 in.

This data was used by Barreto-Souza and Cribari-Neto [32] for fitting the generalized exponential-Poisson (GEP) distribution and by Abd-Elrahman [20] for fitting the Bilal and GB distributions. In the complete sample case, the MLEs of β and λ were 0.4168 and 1.2486, respectively. In this case, we calculated the maximum likelihood estimate of the entropy as H(f) = 1.2786. For the above data set, Abd-Elrahman [20] pointed out that the negative of the log likelihood, Kolmogorov–Smirnov (K–S) test statistics and its corresponding p value related to these MLEs were 38.1763, 0.0532 and 1.0, respectively. Based on the value of p, it is clear that the GB distribution was found to fit the data very well. Using the above data set, we generated an adaptive Type-II progressive hybrid censoring scheme with an effective failure number m (m = 20).

When we took T = 4.0 and $R_1 = R_2 = \ldots = R_5 = 1$, $R_6 = R_7 = \ldots = R_{15} = 0$, $R_{16} = R_{17} = \ldots = R_{20} = 1$, the obtained data in Case I were as follows:

Case I: 0.32, 0.52, 0.77, 0.81, 0.96, 1.18, 1.20, 1.31, 1.35, 1.43, 1.51, 1.62, 1.74, 1.87, 1.89, 1.95, 2.10, 2.48, 2.81 and 3.37.

When we took T = 2.0, $R_1 = 1$, $R_2 = R_3 = \ldots = R_8 = 0$, $R_9 = R_{10} = \ldots R_{15} = 1$, $R_{16} = R_{17} = \ldots = R_{19} = 0$ and $R_{20} = 2$, the obtained data in Case II were as follows:

Case II: 0.32, 0.47, 0.52, 0.59, 0.77, 0.81, 0.9, 0.96, 1.18, 1.20, 1.35, 1.43, 1.74, 1.87, 1.95, 2.10, 2.20, 2.48, 2.81 and 3.09.

Based on the above data, the maximum likelihood estimation and Bayesian estimation of the entropy and the two parameters could be calculated. For the Bayesian estimation, since we had no prior information about the unknown parameters, we considered the noninformative gamma priors of the unknown parameters as a = b = c = d = 0. For the Linex loss and general entropy functions, we set $h = -1.0$, 1.0 and $q = -1.0$, 1.0, respectively. The MLEs and Bayesian estimations of the entropy and the two parameters were calculated by using the Newton–Raphson iteration and Lindley's approximation method. These results are tabulated in Tables 7 and 8. In addition, the 95% asymptotic confidence intervals (ACIs) and Bayesian credible intervals (BCs) of the two parameters and the entropy were calculated using the Newton–Raphson iteration, delta method and MCMC method. These results are displayed in Table 9.

Table 7. MLEs and Bayesian estimations of the parameters and the entropy.

MLEs	Case I	Case II	BEs (Squared Loss)	Case I	Case II
$\hat{\beta}_M$	0.3289	0.3948	$\hat{\beta}_S$	0.3428	0.4044
$\hat{\lambda}_M$	1.0408	1.3373	$\hat{\lambda}_S$	0.9974	1.2410
$\hat{H}_M$	1.5890	1.3881	$\hat{H}_S$	1.6230	1.4701

Table 8. Bayesian estimations of the parameters and the entropy under two loss functions.

BEs Linex Loss	$h = -1$		$h = 1$		BEs Entropy Loss	$q = -1$		$q = 1$	
	Case I	Case II	Case I	Case II		Case I	Case II	Case I	Case II
$\hat{\beta}_L$	0.3406	0.4031	0.3330	0.3958	$\hat{\beta}_E$	0.3369	0.4025	0.3273	0.3852
$\hat{\lambda}_L$	1.2893	1.0217	1.2442	0.9898	$\hat{\lambda}_E$	1.2618	1.0060	1.2173	0.9765
$\hat{H}_L$	1.4714	1.6681	1.4385	1.6276	$\hat{H}_E$	1.4608	1.6340	1.4370	1.6249

From Tables 7–9, we can observe that the MLEs and Bayesian estimations of the parameters and the entropy were close to the estimations in the complete sample case. In most cases, the length of the Bayesian credible intervals was smaller than that of the asymptotic confidence intervals.

Table 9. The 95% asymptotic confidence intervals (ACIs) and Bayesian credible intervals (BCIs) with the corresponding interval lengths (ILs) of the two parameters and the entropy.

Parameter	ACIs IL		Parameter	BCIs IL	
	Case I	Case II		Case I	Case II
β	(0.2406, 0.5409) 0.3003	(0.1812, 0.4564) 0.2752	β	(0.2760, 0.5625) 0.2865	(0.2210, 0.4923) 0.2713
λ	(0.6899, 1.3918) 0.7019	(0.9884, 1.7863) 0.7979	λ	(0.7021, 1.3566) 0.6545	(0.8776, 1.6743) 0.7967
H	(1.2012, 1.9314) 0.7302	(1.0299, 1.7863) 0.7164	H	(1.2487, 1.9707) 0.7220	(1.1266, 1.8671) 0.7405

6. Conclusions

In this paper, we considered the estimation of parameters and entropy for generalized Bilal distribution using adaptive Type-II progressive hybrid censored data. Using an iterative procedure and asymptotic normality theory, we developed the MLEs and approximate confidence intervals of the unknown parameters and the entropy. The Bayesian estimates were derived by Lindley's approximation under the square, Linex and general entropy loss functions. Since Lindley's method failed to construct the intervals, we utilized Gibbs sampling together with the Metropolis–Hastings sampling procedure to construct the Bayesian credence intervals of the unknown parameters and the entropy. A Monte Carlo simulation was provided to show all the estimation results. The results illustrate that the proposed methods performed well. The applicability of the considered model in a real situation was illustrated, based on the data of March precipitation in Minneapolis–Saint Paul. It was observed that the considered model could be utilized to analyze this real data appropriately.

Author Contributions: Methodology and writing, X.S.; supervision, Y.S.; simulation study, K.Z. All authors have read and agreed to the published version of the manuscript.

Funding: This work is supported by the National Natural Science Foundation of China (71571144, 71401134, 71171164, 11701406) and the Program of International Cooperation and Exchanges in Science and Technology funded by Shaanxi Province (2016KW-033).

Institutional Review Board Statement: Not applicable.

Informed Consent Statement: Not applicable.

Data Availability Statement: Not available.

Acknowledgments: The authors would like to thank the editors and the anonymous reviewers.

Conflicts of Interest: The authors declare no conflict of interest.

Appendix A. Proof of Theorem 1

We set $y = \exp(-\beta x^\lambda)$, then $0 < y < 1$. The cumulative distribution function of GB distribution can be written as

$$F(x; \beta, \lambda) = 1 - 3y^2 + 2y^3, 0 < y < 1$$

By setting $u = 1 - 3y^2 + 2y^3, 0 < u < 1$, then we get $3y^2 - 2y^3 + u - 1 = 0, 0 < y < 1$.

Set $\rho(y) = 3y^2 - 2y^3 + u - 1$, take the first derivative of $\rho(y)$ with respect to y, and we have $\frac{d\rho(y)}{dy} = 6y - 6y^2 > 0$, as $0 < y < 1$.

Notice that $\rho(y)$ is a monotonically increasing function when $0 < y < 1$. Thus, there is a unique solution to the equation $3y^2 - 2y^3 + u - 1 = 0$ when $0 < y < 1$. As such, we have proven that the equation $X_{i:m:n} = F^{-1}(U_i)$ has a unique solution ($i = 1, 2, \ldots, m$).

Appendix B. The Specific Steps of the Newton–Raphson Iteration Method

Step 1: Give the initial values of $\theta = (\beta, \lambda)$; that is, $\theta^{(0)} = (\beta^{(0)}, \lambda^{(0)})$.

Step 2: In the kth iteration, calculate $\left(\frac{\partial l}{\partial \beta}, \frac{\partial l}{\partial \lambda}\right)\Big|_{\substack{\beta = \beta^{(k)} \\ \lambda = \lambda^{(k)}}}$ and $I(\beta^{(k)}, \lambda^{(k)})$, where

$$I(\beta^{(k)}, \lambda^{(k)}) = \begin{bmatrix} I_{11} & I_{12} \\ I_{21} & I_{22} \end{bmatrix}\Big|_{\substack{\beta = \beta^{(k)} \\ \lambda = \lambda^{(k)}}}$$ is the observed information matrix of the parameters β and λ, and $I_{ij}, i = 1, 2, 3$ are given by Equations (10)–(13).

Step 3: Update $(\beta, \lambda)^T$ with

$$(\beta^{(k+1)}, \lambda^{(k+1)})^T = (\beta^{(k)}, \lambda^{(k)})^T + I^{-1}(\beta^{(k)}, \lambda^{(k)}) \times \left(\frac{\partial l}{\partial \beta}, \frac{\partial l}{\partial \lambda}\right)^T\Big|_{\substack{\beta = \beta^{(k)} \\ \lambda = \lambda^{(k)}}}.$$

Here, $(\beta, \lambda)^T$ is the transpose of vector (β, λ), and $I^{-1}(\beta^{(k)}, \lambda^{(k)})$ represents the inverse of the matrix $I(\beta^{(k)}, \lambda^{(k)})$.

Step 4: Setting $k = k+1$, the MLEs of the parameters (denoted by $\hat{\beta}$ and $\hat{\lambda}$) can be obtained by repeating Steps 2 and 3 until $|(\beta^{(k+1)}, \lambda^{(k+1)})^T - (\beta^{(k)}, \lambda^{(k)})^T| < \varepsilon$, where ε is a threshold value that is fixed in advance.

Appendix C. The Detailed Derivation of Bayesian Estimates of two Parameters (β, λ) and the Entropy under the LL Function

In this case, we take $U(\beta, \lambda) = \exp(-h\beta)$, and then

$$u_1 = -h\exp(-h\beta), u_{11} = h^2 \exp(-h\beta), u_{12} = u_{21} = u_{22} = u_2 = 0.$$

Using Equation (26), the Bayesian estimation of parameter β is given by

$$\hat{\beta}_L = -\frac{1}{h}\ln\{\exp(-h\hat{\beta}) + 0.5[u_{11}\tau_{11} + u_1\tau_{11}^2 z_{30} + u_1\tau_{21}\tau_{22}z_{03} + 3u_1\tau_{11}\tau_{12}z_{21} + (\tau_{11}\tau_{22} + 2u_1\tau_{21}^2)u_1 z_{12}] + u_1\tau_{11}p_1 + u_1\tau_{12}p_2\}$$

Similarly, the Bayesian estimation of parameter λ is obtained by

$$\hat{\lambda}_L = -\frac{1}{h}\ln\{\exp(-h\hat{\lambda}) + 0.5[u_{22}\tau_{22} + u_2\tau_{11}\tau_{12}z_{30} + u_2\tau_{22}^2 z_{03} + (\tau_{11}\tau_{22} + 2\tau_{12}^2)u_2 z_{21} + 3u_2\tau_{22}\tau_{21}z_{21}] + u_2\tau_{12}p_1 + u_2\tau_{22}p_2\}$$

For the Bayesian estimation of the entropy, we have

$$U(\beta, \lambda) = \exp[-hH(f)], u_1 = \frac{h}{\beta\lambda}\exp[-hH(f)],$$
$$u_2 = -h[-\frac{1}{\lambda} + \frac{1}{\lambda^2}(\ln\beta - \ln(9/8) + \gamma)]\exp[-hH(f)],$$
$$u_{11} = h[\frac{-1}{\beta^2\lambda} + \frac{h}{\beta^2\lambda^2}]\exp[-hH(f)],$$

$$u_{22} = \{-h[\frac{1}{\lambda^2} - \frac{2}{\lambda^3}(\ln\beta - \ln(9/8) + \gamma)] + h^2[-\frac{1}{\lambda} + \frac{1}{\lambda^2}(\ln\beta - \ln(9/8) + \gamma)]^2\}\exp[-hH(f)],$$

$$u_{12} = u_{21} == h[\frac{h-1}{\beta\lambda^2} - h\frac{1}{\beta\lambda^3}(\ln\beta - \ln(9/8) + \gamma)]\exp[-hH(f)].$$

The Bayesian estimation of the entropy under the LL function is given by

$$\hat{H}_L(f) = -\frac{1}{h}\ln\{\exp[-h\hat{H}(f)] + 0.5[u_{11}\tau_{11} + 2u_{12}\tau_{12} + u_{22}\tau_{22} + z_{30}(u_1\tau_{11} + u_2\tau_{12})\tau_{11} + z_{03}(u_2\tau_{22} + u_1\tau_{21})\tau_{22}$$
$$+ z_{21}(3u_1\tau_{11}\tau_{12} + u_2(\tau_{11}\tau_{22} + 2\tau_{12}^2)) + z_{12}(3u_2\tau_{22}\tau_{21} + u_1(\tau_{11}\tau_{22} + 2\tau_{21}^2))]$$
$$+ p_1(u_1\tau_{11} + u_2\tau_{21}) + p_2(u_2\tau_{22} + u_1\tau_{12})\}$$

Appendix D. The Derivation of Bayesian Estimates of two Parameters (β, λ) and the Entropy under the GEL Function

In this case, we take $U(\beta, \lambda) = \beta^{-q}$ and then $u_1 = -q\beta^{-q-1}$, $u_{11} = q(q+1)\beta^{-q-2}$, and $u_{12} = u_{21} = u_{22} = u_2 = 0$.

Using Equation (26), the Bayesian estimation of parameter β is given by

$$\hat{\beta}_E = \{\hat{\beta}^{-q} + 0.5[u_{11}\tau_{11} + u_1\tau_{11}^2 z_{30} + u_1\tau_{21}\tau_{22}z_{03} + 3u_1\tau_{11}\tau_{12}z_{21} + (\tau_{11}\tau_{22} + 2u_1\tau_{21}^2)u_1 z_{12}] + u_1\tau_{11}p_1 + u_1\tau_{12}p_2\}^{-1/q}$$

Similarly, the Bayesian estimation of parameter λ is obtained by

$$\hat{\lambda}_L = \{\hat{\lambda}^{-q} + 0.5[u_{22}\tau_{22} + u_2\tau_{11}\tau_{12}z_{30} + u_2\tau_{22}^2 z_{03} + (\tau_{11}\tau_{22} + 2\tau_{12}^2)u_2 z_{21} + 3u_2\tau_{22}\tau_{21}z_{21}] + u_2\tau_{21}p_1 + u_2\tau_{22}p_2\}^{-1/q}$$

For the Bayesian estimation of the entropy under the general EL function, we take $U(\beta, \lambda) = [H(f)]^{-q}$, and then

$$u_1 = \tfrac{q}{\beta\lambda}[H(f)]^{-q-1},\ u_2 = [\tfrac{q}{\lambda} - \tfrac{q}{\lambda^2}(\ln\beta - \ln(9/8) + \gamma)][H(f)]^{-q-1},$$

$$u_2 = [\tfrac{q}{\lambda} - \tfrac{q}{\lambda^2}(\ln\beta - \ln(9/8) + \gamma)][H(f)]^{-q-1},\ u_{11} = \tfrac{q(q+1)}{\beta^2\lambda^2}[H(f)]^{-q-2} - \tfrac{q}{\beta^2\lambda}[H(f)]^{-q-1},$$

$$u_{22} = [\tfrac{-q}{\lambda^2} + \tfrac{2q}{\lambda^3}(\ln\beta - \ln(9/8) + \gamma)][H(f)]^{-q-1} + q(q+1)[\tfrac{1}{\lambda} - \tfrac{1}{\lambda^2}(\ln\beta - \ln(9/8) + \gamma)]^2[H(f)]^{-q-2},$$

$$u_{12} = u_{21} = q(q+1)[\tfrac{1}{\beta\lambda^2} - \tfrac{1}{\beta\lambda^3}(\ln\beta - \ln(9/8) + \gamma)][H(f)]^{-q-2} - \tfrac{q}{\beta\lambda^2}[H(f)]^{-q-1}.$$

Using Equation (26), the approximate Bayesian estimation of the entropy is given by

$$\hat{H}_E(f) = \{[\hat{H}(f)]^{-q} + 0.5[(u_{11}\tau_{11} + 2u_{12}\tau_{12} + u_{22}\tau_{22}) + z_{30}(u_1\tau_{11} + u_2\tau_{12})\tau_{11} + z_{03}(u_2\tau_{22} + u_1\tau_{12})\tau_{22}$$
$$+ z_{21}(3u_1\tau_{11}\tau_{12} + u_2(\tau_{11}\tau_{22} + 2\tau_{12}^2)) + z_{12}(3u_2\tau_{22}\tau_{21} + u_1(\tau_{11}\tau_{22} + 2\tau_{21}^2))]$$
$$+ p_1(u_1\tau_{11} + u_2\tau_{21}) + p_2(u_2\tau_{22} + u_1\tau_{12})\}^{-1/q}.$$

Appendix E.

Table A1. The average 95% approximate confidence intervals and average lengths and coverage probabilities of β, λ and the entropy ($\beta = 1$, $\lambda = 2$, $H(f) = 0.2448$, $T = 0.6$).

(n, m)	SC	β AL	CP	λ AL	CP	H AL	CP
(40, 15)	I	(0.6598, 1.5736) 0.9138	0.9042	(1.2220, 3.1773) 1.9573	0.9216	(0.0293, 1.1866) 1.1573	0.9184
	II	(0.6711, 1.4742) 0.8031	0.9253	(1.4238, 2.8658) 1.4420	0.9361	(0.0393, 0.7733) 0.7340	0.929
	III	(0.6343, 1.5347) 0.9004	0.9130	(1.2645, 3.1064) 1.9319	0.9281	(0.0254, 1.1244) 1.0990	0.9174
(50, 15)	I	(0.6421, 1.5458) 0.9037	0.9162	(1.2837, 3.0913) 1.8076	0.9314	(0.0203, 1.0469) 1.0266	0.9216
	II	(0.7102, 1.3884) 0.6782	0.9394	(1.4416, 2.7246) 1.2830	0.9406	(0.0438, 0.6924) 0.6486	0.9392
	III	(0.6914, 1.5147) 0.8233	0.9253	(1.3021, 2.9705) 1.6684	0.9370	(0.0264, 1.0759) 1.0495	0.9261
(60, 30)	I	(0.6377, 1.5335) 0.8958	0.9374	(1.3388, 3.0191) 1.6803	0.9487	(0.0151, 0.9112) 0.8959	0.9393
	II	(0.7093, 1.3769) 0.6676	0.9516	(1.4807, 2.6886) 1.2069	0.9542	(0.0536, 0.6667) 0.6131	0.9461
	III	(0.6934, 1.4786) 0.7852	0.9405	(1.3955, 2.9630) 1.5675	0.9506	(0.0325, 0.8630) 0.8305	0.9428
(70, 30)	I	(0.7329, 1.4293) 0.6964	0.9472	(1.4068, 2.8432) 1.34364	0.9534	(0.0298, 0.7943) 0.7645	0.9446
	II	(0.7247, 1.2859) 0.5602	0.9651	(1.5369, 2.5891) 1.0522	0.9680	(0.0614, 0.5498) 0.4884	0.9632
	III	(0.7392, 1.3486) 0.6154	0.9514	(1.4476, 2.7845) 1.3361	0.9573	(0.0498, 0.7185) 0.6687	0.9521

Table A2. The average 95% approximate confidence intervals and average lengths and coverage probabilities of β, λ and the entropy ($\beta = 1$, $\lambda = 2$, $H(f) = 0.2448$, $T = 1.5$).

(n, m)	SC	β AL	CP	λ AL	CP	H AL	CP
(40, 15)	I	(0.5234, 1.8717) 1.3483	0.9231	(1.2469, 3.2287) 1.9818	0.9274	(0.0284, 1.1887) 1.1603	0.9267
	II	(0.6662, 1.4576) 0.7914	0.9372	(1.4322, 2.8760) 1.4438	0.9405	(0.0436, 0.7887) 0.7451	0.9393
	III	(0.5619, 1.8110) 1.2491	0.9252	(1.2679, 3.2045) 1.9364	0.9364	(0.0212, 1.1173) 1.0961	0.9340
(50, 15)	I	(0.5601, 1.6810) 1.1209	0.9230	(1.3076, 3.0214) 1.7136	0.9363	(0.0245, 0.9304) 0.9059	0.9347
	II	(0.7124, 1.3705) 0.6581	0.9418	(1.4548, 2.7213) 1.2665	0.9462	(0.0458, 0.6740) 0.6282	0.9515
	III	(0.6103, 1.5868) 0.9765	0.9336	(1.3320, 2.9769) 1.6449	0.9372	(0.0259, 0.8461) 0.8202	0.9347
(60, 30)	I	(0.6659, 1.5135) 0.8476	0.9418	(1.3454, 3.0335) 1.6881	0.9521	(0.0206, 1.0400) 1.0194	0.9464
	II	(0.7051, 1.3680) 0.6619	0.9592	(1.4812, 2.6942) 1.2130	0.9574	(0.0456, 0.6604) 0.6148	0.9531
	III	(0.6913, 1.4513) 0.7600	0.9431	(1.3775, 2.8768) 1.4983	0.9520	(0.0237, 0.9934) 0.9697	0.9506
(70, 30)	I	(0.7381, 1.3951) 0.6570	0.9492	(1.4501, 2.7820) 1.3319	0.9582	(0.0321, 0.7553) 0.7232	0.9523
	II	(0.7573, 1.2850) 0.5277	0.9704	(1.5514, 2.5845) 1.0331	0.9726	(0.0647, 0.5680) 0.5033	0.9741
	III	(0.7554, 1.3492) 0.5938	0.9546	(1.4967, 2.7071) 1.2104	0.9615	(0.0410, 0.7147) 0.6737	0.9591

Table A3. The average 95% Bayesian credible intervals and average lengths and coverage probabilities of β, λ and the entropy ($\beta = 1$, $\lambda = 2$ $H(f) = 0.2448$, $T = 0.6$).

(n, m)	SC	β AL	CP	λ AL	CP	H AL	CP
(40, 15)	I	(0.5521, 1.2841) 0.7320	0.9194	(1.0215, 2.4593) 1.4378	0.9241	(0.0213, 1.1750) 1.1537	0.9263
	II	(0.6378, 1.3228) 0.6850	0.9433	(1.2854, 2.5238) 1.2384	0.9472	(0.0395, 0.7752) 0.7357	0.9380
	III	(0.5670, 1.2953) 0.7283	0.9253	(1.0579, 2.4762) 1.4183	0.9294	(0.0224, 1.1192) 1.0968	0.9308
(50, 15)	I	(0.5924, 1.2871) 0.6947	0.9312	(1.1731, 2.5054) 1.3323	0.9397	(0.0298, 0.9231) 0.8933	0.9386
	II	(0.6897, 1.2921) 0.6024	0.9491	(1.3580, 2.4935) 1.1355	0.9465	(0.0548, 0.6751) 0.6203	0.9507
	III	(0.6067, 1.2854) 0.6787	0.9342	(1.2051, 2.4718) 1.2667	0.9354	(0.0278, 0.8553) 0.8275	0.9326
(60, 30)	I	(0.6450, 1.2925) 0.6475	0.9481	(1.1389, 2.4565) 1.3176	0.9536	(0.0397, 1.0509) 1.0112	0.9394
	II	(0.6870, 1.2905) 0.6035	0.9614	(1.3883, 2.4740) 1.0857	0.9656	(0.0578, 0.6717) 0.6139	0.9562
	III	(0.6565, 1.2812) 0.6247	0.9532	(1.1919, 2.4423) 1.2504	0.9561	(0.0319, 0.8408) 0.8029	0.9528
(70, 30)	I	(0.7062, 1.2494) 0.5432	0.9512	(1.3068, 2.4374) 1.1306	0.9563	(0.0324, 0.7516) 0.7192	0.9536
	II	(0.7451, 1.2449) 0.4998	0.9711	(1.4821, 2.4494) 0.9673	0.9744	(0.0701, 0.5672) 0.4971	0.9783
	III	(0.7162, 1.2359) 0.5197	0.9583	(1.3597, 2.4443) 1.0846	0.9604	(0.0440, 0.7067) 0.6627	0.9578

Table A4. The average 95% Bayesian credible intervals and average lengths and coverage probabilities of β, λ and the entropy ($\beta = 1$, $\lambda = 2$ H(f) = = 0.2448, T = 1.5).

(n, m)	SC	β AL	CP	λ AL	CP	H AL	CP
(40, 15)	I	(0.5554, 1.2954) 0.7400	0.9218	(1.0243, 2.4612) 1.4369	0.9354	(0.0251, 1.1801) 1.1550	0.9258
	II	(0.6417, 1.3339) 0.6922	0.9439	(1.2824, 2.5169) 1.2345	0.9485	(0.0372, 0.7728) 0.7356	0.9394
	III	(0.5696, 1.3033) 0.7337	0.9275	(1.0556, 2.4672) 1.4116	0.9318	(0.0241, 1.1200) 1.0959	0.9337
(50, 15)	I	(0.5954, 1.2947) 0.6993	0.9417	(1.1722, 2.4804) 1.3002	0.9420	(0.0224, 1.0231) 1.0007	0.9418
	II	(0.68902, 1.2954) 0.6062	0.9506	(1.3599, 2.5034) 1.1435	0.9525	(0.0479, 0.6710) 0.6239	0.9526
	III	(0.6045, 1.2801) 0.6756	0.9359	(1.2337, 2.5094) 1.2757	0.9364	(0.0324, 1.0047) 0.9723	0.9371
(60, 30)	I	(0.6418, 1.2835) 0.6417	0.9494	(1.1349, 2.4455) 1.3106	0.9548	(0.0250, 0.9212) 0.8960	0.9417
	II	(0.6896, 1.2970) 0.6074	0.9628	(1.3987, 2.4911) 1.0924	0.9662	(0.0479, 0.6608) 0.6129	0.9573
	III	(0.6600, 1.2856) 0.6256	0.9556	(1.1549, 2.4283) 1.2734	0.9571	(0.0217, 0.8359) 0.8142	0.9538
(70, 30)	I	(0.7061, 1.2472) 0.5411	0.9526	(1.3179, 2.4521) 1.1342	0.9571	(0.0363, 0.7509) 0.7146	0.9548
	II	(0.7451, 1.2413) 0.4962	0.9725	(1.4663, 2.4268) 0.9605	0.9757	(0.0778, 0.5701) 0.4923	0.9793
	III	(0.7154, 1.2267) 0.5113	0.9594	(1.3542, 2.4118) 1.0576	0.9624	(0.0604, 0.7108) 0.6504	0.9585

References

1. Balakrishnan, N.; Aggarwala, R. *Progressive Censoring: Theory, Methods, and Applications*; Birkhauser: Boston, MA, USA, 2000.
2. Balakrishnan, N. Progressive censoring methodology: An appraisal. *Test* **2007**, *16*, 211–259. [CrossRef]
3. Kundu, D.; Joarder, A. Analysis of type-II progressively hybrid censored data. *Comput. Stat. Data Anal.* **2006**, *50*, 2509–2528. [CrossRef]
4. Ng, H.K.T.; Kundu, D.; Chan, P.S. Statistical analysis of exponential lifetimes under an adaptive Type II progressive censoring scheme. *Naval Res. Logist.* **2010**, *56*, 687–698. [CrossRef]
5. Nassar, M.; Abo-Kasem, O.; Zhang, C.; Dey, S. Analysis of weibull distribution under adaptive Type-II progressive hybrid censoring scheme. *J. Indian Soc. Probab. Stat.* **2018**, *19*, 25–65. [CrossRef]
6. Zhang, C.; Shi, Y. Estimation of the extended Weibull parameters and acceleration factors in the step-stress accelerated life tests under an adaptive progressively hybrid censoring data. *J. Stat Comput. Simulat.* **2016**, *86*, 3303–3314. [CrossRef]
7. Cui, W.; Yan, Z.; Peng, X. Statistical analysis for constant-stress accelerated life test with Weibull distribution under adaptive Type-II hybrid censored data. *IEEE Access* **2019**. [CrossRef]
8. Ismail, A.A. Inference for a step-stress partially accelerated life test model with an adaptive Type-II progressively hybrid censored data from Weibull distribution. *J. Comput. Appl. Math.* **2014**, *260*, 533–542. [CrossRef]
9. Zhang, C.; Shi, Y. Inference for constant-stress accelerated life tests with dependent competing risks from bivariate Birnbaum-Saunders distribution based on adaptive progressively hybrid censoring. *IEEE Trans. Reliab.* **2017**, *66*, 111–122. [CrossRef]
10. Ye, Z.S.; Chan, P.S.; Xie, M. Statistical inference for the extreme value distribution under adaptive Type-II progressive censoring schemes. *J. Stat. Comput. Simulat.* **2014**, *84*, 1099–1114. [CrossRef]
11. Sobhi, M.M.; Soliman, A.A. Estimation for the exponentiated Weibull model with adaptive Type-II progressive censored schemes. *Appl. Math. Model.* **2016**, *40*, 1180–1192. [CrossRef]
12. Nassar, M.; Abo-Kasem, O.E. Estimation of the inverse Weibull parameters under adaptive type-II progressive hybrid censoring scheme. *J. Comput. Appl. Math.* **2017**, *315*, 228–239. [CrossRef]
13. Xu, R.; Gui, W.H. Entropy estimation of inverse Weibull Distribution under adaptive Type-II progressive hybrid censoring schemes. *Symmetry* **2019**, *11*, 1463. [CrossRef]
14. Kang, S.B.; Cho, Y.S.; Han, J.T.; Kim, J. An estimation of the entropy for a double exponential distribution based on multiply Type-II censored samples. *Entropy* **2012**, *14*, 161–173. [CrossRef]
15. Cho, Y.; Sun, H.; Lee, K. An estimation of the entropy for a Rayleigh distribution based on doubly-generalized Type-II hybrid censored samples. *Entropy* **2014**, *16*, 3655–3669. [CrossRef]
16. Baratpour, S.; Ahmadi, J.; Arghami, N.R. Entropy properties of record statistics. *Stat. Pap.* **2017**, *48*, 197–213. [CrossRef]

17. Cramer, E.; Bagh, C. Minimum and maximum information censoring plans in progressive censoring. *Commun. Stat. Theory Methods* **2011**, *40*, 2511–2527. [CrossRef]
18. Cho, Y.; Sun, H.; Lee, K. Estimating the entropy of a weibull distribution under generalized progressive hybrid censoring. *Entropy* **2015**, *17*, 102–122. [CrossRef]
19. Yu, J.; Gui, W.H.; Shan, Y.Q. Statistical inference on the Shannon entropy of inverse Weibull distribution under the progressive first-failure censoring. *Entropy* **2019**, *21*, 1209. [CrossRef]
20. Abd-Elrahman, A.M. A new two-parameter lifetime distribution with decreasing, increasing or upside-down bathtub-shaped failure rate. *Commun. Stat. Theory Methods* **2017**, *46*, 8865–8880. [CrossRef]
21. Abd-Elrahman, A.M. Reliability estimation under type-II censored data from the generalized Bilal distribution. *J. Egypt. Math. Soc.* **2019**, *27*, 1–15. [CrossRef]
22. Mahmoud, M.; EL-Sagheer, R.M.; Abdallah, S. Inferences for new Weibull–Pareto distribution based on progressively Type-II censored data. *J. Stat. Appl. Probab.* **2016**, *5*, 501–514. [CrossRef]
23. Ahmed, E.A. Bayesian estimation based on progressive Type-II censoring from two-parameter bathtub-shaped lifetime model: An Markov chain Monte Carlo approach. *J. Appl. Stat.* **2014**, *41*, 752–768. [CrossRef]
24. Gilks, W.R.; Wild, P. Adaptive rejection sampling for Gibbs sampling. *J. R. Stat. Soc.* **1992**, *C41*, 337–348. [CrossRef]
25. Koch, K.R. Gibbs sampler by sampling-importance-resampling. *J. Geod.* **2007**, *81*, 581–591. [CrossRef]
26. Martino, L.; Elvira, V.; Camps-Valls, G. The recycling gibbs sampler for efficient learning. *Digit. Signal Process.* **2018**, *74*, 1–13. [CrossRef]
27. Panahi, H.; Moradi, N. Estimation of the inverted exponentiated Rayleigh distribution based on adaptive Type II progressive hybrid censored sample. *J. Comput. Appl. Math.* **2020**, *364*, 112345. [CrossRef]
28. Metropolis, N.; Rosenbluth, A.W.; Rosenbluth, M.N.; Teller, A.H.; Teller, E. Equations of state calculations by fast computing machines. *J. Chem. Phys.* **1953**, *21*, 1087–1092. [CrossRef]
29. Hastings, W.K. Monte Carlo sampling methods using Markov chains and their applications. *Biometrika* **1970**, *57*, 97–109. [CrossRef]
30. Balakrishnan, N.; Sandhu, R.A. A simple simulational algorithm for generating progressive Type-II censored samples. *Am. Stat.* **1995**, *49*, 229–230.
31. Hinkley, D. On quick choice of power transformations. *Appl. Stat.* **1977**, *26*, 67–96. [CrossRef]
32. Barreto-Souza, W.; Cribari-Neto, F. A generalization of the exponential-Poisson distribution. *Stat. Probab. Lett.* **2009**, *79*, 2493–2500. [CrossRef]

Article

Calibration Invariance of the MaxEnt Distribution in the Maximum Entropy Principle

Jan Korbel [1,2,3]

[1] Section for the Science of Complex Systems, Center for Medical Statistics, Informatics, and Intelligent Systems (CeMSIIS), Medical University of Vienna, Spitalgasse 23, 1090 Vienna, Austria; jan.korbel@meduniwien.ac.at
[2] Complexity Science Hub Vienna, Josefstädterstrasse 39, 1080 Vienna, Austria
[3] Faculty of Nuclear Sciences and Physical Engineering, Czech Technical University, 11519 Prague, Czech Republic

Abstract: The maximum entropy principle consists of two steps: The first step is to find the distribution which maximizes entropy under given constraints. The second step is to calculate the corresponding thermodynamic quantities. The second part is determined by Lagrange multipliers' relation to the measurable physical quantities as temperature or Helmholtz free energy/free entropy. We show that for a given MaxEnt distribution, the whole class of entropies and constraints leads to the same distribution but generally different thermodynamics. Two simple classes of transformations that preserve the MaxEnt distributions are studied: The first case is a transform of the entropy to an arbitrary increasing function of that entropy. The second case is the transform of the energetic constraint to a combination of the normalization and energetic constraints. We derive group transformations of the Lagrange multipliers corresponding to these transformations and determine their connections to thermodynamic quantities. For each case, we provide a simple example of this transformation.

Keywords: maximum entropy principle; MaxEnt distribution; calibration invariance; Lagrange multipliers

Citation: Korbel, J. Calibration Invariance of the MaxEnt Distribution in the Maximum Entropy Principle. *Entropy* **2021**, *23*, 96. https://doi.org/10.3390/e23010096

Received: 11 December 2020
Accepted: 9 January 2021
Published: 11 January 2021

Publisher's Note: MDPI stays neutral with regard to jurisdictional clai-ms in published maps and institutio-nal affiliations.

Copyright: © 2021 by the author. Licensee MDPI, Basel, Switzerland. This article is an open access article distributed under the terms and conditions of the Creative Commons Attribution (CC BY) license (https://creativecommons.org/licenses/by/4.0/).

1. Introduction

The maximum entropy principle (MEP) is one of the most fundamental concepts in equilibrium statistical mechanics. It was originally proposed by Jaynes [1,2] in order to connect information entropy introduced by Shannon and thermodynamic entropy introduced by Clausius, Boltzmann, and Gibbs. Although the MEP was originally introduced for the case of Shannon entropy, with the advent of generalized entropies [3–17] the natural effort was to apply the maximum entropy principle beyond the case of Shannon entropy. Another question that arose naturally is whether the MEP can be applied to other than ordinary linear constraints. Examples of the constraints that might be considered in connection with the MEP are *escort constraints* [18–20], *Kolmogorov–Nagumo means* [21,22], or more exotic types of constraints [23]. It brought some discussion about the applicability of the principle for the case of generalized entropies [24,25] and nonlinear constraints and its thermodynamic interpretation [26–30]. Indeed, MEP is not the only one extremal principle in statistical physics, let us mention, e.g., the *principle of maximum caliber* [31] which is useful in non-equilibrium physics. In this paper, we stick, however, to MEP, as it is the most widespread principle and the theory of generalized thermostatistics has been mainly focused on MEP. For a recent review of other principles, see also in [32]. For the discussion between entropy arising from information theory and thermodynamics, see in [33]. For the sake of simplicity, let us consider canonical ensemble, i.e., fluctuations in internal energy. For the case of the grand-canonical ensemble, one can obtain similar results to the ones presented in this paper for the case of a chemical potential μ.

In order to grasp the debate about the applicability of the MEP, let us emphasize that the MEP consists of two main parts:

(I) Finding a distribution (*MaxEnt* distribution) that maximizes entropy under given constraints.
(II) Plugging the distribution into the entropic functional and calculating physical quantities as thermodynamic potentials, temperature, or response coefficients (specific heat, compressibility, etc.).

The first part is rather a mathematical procedure of finding a maximum subject to constraints. This is done by the *method of Lagrange multipliers*, by defining a Lagrange function in the form

$$Lagrange\ function = entropy - (Lagrange\ multiplier) \cdot (constraint)$$

The Lagrange multipliers' role at this stage is to ensure fulfillment of constraints as they are determined from the set of equations obtained from the maximization of the Lagrange function. This procedure is known in statistics as *Softmax*, a method used to infer distribution from given data. Shore and Johnson [34,35] therefore studied MEP as a statistical inference procedure and established a set of consistency axioms. Shore and Johnson's work heated a debate about whether MEP for generalized entropies can be also understood as a statistical inference method satisfying the consistency requirements [24,36–41]. In [42], it was shown that the class of entropies satisfying the original Shore–Johnson axioms is wider than previously thought. Moreover, in [43], the connection between Shore–Johnson axioms and Shannon–Khinchin axioms was investigated and the equivalence of information theory and statistical inference axiomatics was established.

In the second part, the physical interpretation of entropy starts to arise. Similar to the case of Lagrangian mechanics, where the Lagrangian is the difference between kinetic and potential energy and the Lagrange multipliers play the role of the normal force to the constraints, here the entropy becomes a thermodynamic state variable. For Shannon entropy and linear constraints, the Lagrange multipliers become inverse temperature and free entropy, respectively.

The main aim of this paper is to discuss the relation between points (I) and (II). In the first part, it is possible to find a class of entropic functionals and constraints leading to the same MaxEnt distribution. However, in the second part, different entropy and/or constraints lead to different thermodynamics and different relations between physical quantities and Lagrange multipliers. The two main messages of this paper are listed below.

(i) For each MaxEnt distribution, there exists the whole class of entropies and constraints leading to generally different thermodynamics.
(ii) It is possible to establish transformation relations of Lagrange parameters (and subsequently the thermodynamic quantities) for classes of entropies and constraints giving the same MaxEnt distribution.

We call the latter transformation relation *calibration invariance* of the MaxEnt distribution. A straightforward consequence is that in order to fully determine the statistical properties of a thermal system in equilibrium, it is not enough to measure the statistical distribution of energies.

The rest of the paper is organized as follows. In the next section, we briefly discuss the main aspects of MEP for the case of general entropic functional and general constraints. In the following two sections, we introduce two simple transformations of entropic functional (Section 3) and constraints (Section 4) that lead to the same MaxEnt distribution and derive transformations between the Lagrange multipliers. These transformations form a group. After the general derivation, we provide a few simple examples for each case. The last section is devoted to conclusions.

2. Maximum Entropy Principle in Statistical Physics

Maximum entropy principle is the way of obtaining the representing probability distribution from the limited amount of information. Our aim is to find the probability distribution of the system $P = \{p_i\}_{i=1}^n$ under the set of given constraints. In the simplest case, the principle can be formulated as follows.

Maximum entropy principle: *Maximize entropy $S(P)$ under the normalization constraint $f_0(P) = 0$, and energy constraint $f_E(P) = 0$.*

The normalization condition is considered in the regular form, i.e., $f_0(P) = \sum_i p_i - 1 = \langle 1 \rangle - 1$. Moreover, we have a class of constraints, which originally described the average energy of the system. Therefore, we call them *energy constraints*. We consider only one energy constraint, for simplicity, although there can be more constraints, and they do not have to consider only internal energy but also other thermodynamic quantities. In the original formulation, the energy constraint is linear in probabilities, i.e.,

$$f_E(P) = \sum_i p_i E_i - E = \langle E \rangle - E, \qquad (1)$$

but it can be generally any nonlinear function of probabilities—escort means provide an example. A large class of energy constraints can be written in a *separable form*, which means that $f_E(P) = \mathcal{E}(P) - E$, i.e., in the form expressing the "expected" internal energy (macroscopic variable) as a function of probability distribution (microscopic variable). This class of constraints plays a dominant role in the thermodynamic systems.

In order to find a solution of the Maximum entropy principle, we use a common method of Lagrange multipliers, which can be done through maximization of Lagrange function:

$$\mathcal{L}(P; \alpha, \beta) = S(P) - \alpha f_0(P) - \beta f_E(P) \qquad (2)$$

The maximization procedure leads to the set of equations

$$\begin{aligned}
\frac{\partial \mathcal{L}(P; \alpha, \beta)}{\partial p_i} &= 0 \quad \forall i \in \{1, \ldots, n\} \\
\frac{\partial \mathcal{L}(P; \alpha, \beta)}{\partial \alpha} &= f_0(P) = 0 \\
\frac{\partial \mathcal{L}(P; \alpha, \beta)}{\partial \beta} &= f_E(P) = 0
\end{aligned} \qquad (3)$$

from which we determine the resulting MaxEnt distribution. In order to obtain a unique solution, we require that the entropic functional should be a Schur-concave symmetric function [42].

As a consequence, we obtain the values of Lagrange multipliers α and β. From the strictly mathematical point of view, Lagrange multipliers are just auxiliary parameters to be solved from the set of Equation (3). However, in physics, Lagrange parameters also have a physical interpretation. In Lagrangian mechanics, Lagrange parameters play the role of normal force to the constraints. Similarly, in ordinary statistical mechanics based on Shannon entropy $H(P) = -\sum_i p_i \log p_i$ and linear constraints (1), the Lagrange multipliers have the particular physical interpretation:

$$\beta = \frac{1}{T} \quad \text{(inverse temperature)}, \qquad (4)$$

$$\alpha = S - \frac{1}{T} E \quad \text{(free entropy)}. \qquad (5)$$

Note that the free entropy is, similarly to Helmholtz free energy, a Legendre transform of entropy w.r.t. internal energy. For the case of ordinary thermodynamics (Shannon entropy and linear constraints), it is equal to the logarithm of the partition function.

This interpretation is valid only in this case. In the case, when we use different entropy functional or different constraints, these relation between Lagrange multipliers and thermodynamic quantities are no longer valid. This is even the case, when the resulting MaxEnt distribution is the same.

The main aim of this paper is to show how the invariance of MaxEnt distribution affects the Lagrange multipliers and their relations to thermodynamic quantities. Let us now solve Equation (3). The first set of equations leads to

$$\frac{\partial S(P)}{\partial p_i} - \alpha \frac{\partial f_0(P)}{\partial p_i} - \beta \frac{\partial f_E(P)}{\partial p_i} = 0. \tag{6}$$

Let us assume the normalization in the usual way which leads to $\frac{\partial f_0(P)}{\partial p_i} = 1$. Moreover, let us consider separable energy constraint, so $\frac{\partial f_E(P)}{\partial p_i} = \frac{\partial \mathcal{E}(P)}{\partial p_i}$. The resulting probability distribution can be expressed as

$$p_i^\star = \frac{\partial S}{\partial p_i}^{(-1)} \left[\alpha + \beta \frac{\partial \mathcal{E}(P)}{\partial p_i} \right]. \tag{7}$$

where (-1) denotes inverse function of $\partial S / \partial p_i$ (provided it exists and is unique). We can express α by multiplying the equation by p_i and summing over i, which leads to

$$\alpha = \langle \nabla_P S(P) \rangle - \beta \langle \nabla_P \mathcal{E}(P) \rangle \tag{8}$$

where $\langle X \rangle = \sum_i x_i p_i$ and $\nabla_P = (\frac{\partial}{\partial p_1}, \ldots, \frac{\partial}{\partial p_n})$. By plugging back to the previous equation, we can get β as

$$\beta = \frac{\Delta_i(\nabla S(P))}{\Delta_i(\nabla \mathcal{E}(P))} \tag{9}$$

where $\Delta_i(X) = x_i - \langle X \rangle$ is the difference from the average.

The solution of Equation (3) depends on the internal energy E. However, in thermodynamics it is natural to invert the relation $\beta = \beta(E)$ and express the relevant quantities in terms of β, so $E = E(\beta)$. With that, we can calculate dependence of entropy on β:

$$\frac{\partial S}{\partial \beta} = \sum_i \frac{\partial S}{\partial p_i} \frac{\partial p_i}{\partial \beta} = \sum_i \left(\alpha + \beta \frac{\partial \mathcal{E}(P)}{\partial p_i} \right) \frac{\partial p_i}{\partial \beta} = \beta \sum_i \frac{\partial f_E}{\partial p_i} \frac{\partial p_i}{\partial \beta} = \beta \left(-\frac{\partial f_E}{\partial E} \frac{\partial E}{\partial \beta} \right) \tag{10}$$

For separable energy constraints, $\frac{\partial f_E}{\partial E} = -1$, so we obtain the well-known relation

$$\frac{\partial S}{\partial \beta} = \beta \frac{\partial E}{\partial \beta} \Rightarrow \beta = \frac{\partial S}{\partial E}. \tag{11}$$

Let us now define the Legendre conjugate of entropy called *free entropy* (also called Jaynes parameter [44] or Massieu function [45]):

$$\psi = S - \frac{\partial S}{\partial E} E = S - \beta E \tag{12}$$

Free entropy is connected to Helmholtz free energy as $\psi = -\beta F$. The difference between α and ψ can be expressed as

$$\psi - \alpha = (S - \langle \nabla_P S \rangle) - \beta(E - \langle \nabla_P \mathcal{E} \rangle) \tag{13}$$

Therefore, we can understand the difference $\psi - \alpha$ as the Legendre transform of ψ with respect to P. From this, we see that the difference between ψ and α is a constant (not depending on thermodynamic quantities), if two independent conditions are fulfilled, i.e., $E = \langle \nabla_P \mathcal{E}(P) \rangle$ and $S = \langle \nabla_P S \rangle + a$. The former constraint leads to linear energy constraints, while the latter one leads to the the conclusion that the entropy must be in trace form $S(P) = \sum_i g(p_i)$. Moreover, the function g has to fulfill the following equation,

$$g(x) - ax = xg'(x) \tag{14}$$

leading to $g(x) = -ax \log(x) + bx$ which is equivalent to Shannon entropy.

In the next sections, we will explore how the transformation of the entropy and the energy constraint that leaves the MaxEnt distribution invariant affects the Lagrange multipliers and their relation to thermodynamic quantities.

3. Calibration Invariance of MaxEnt Distribution with Entropy Transformation

The simplest transformation of Lagrange functional that leaves the MaxEnt distribution invariant is to consider an arbitrary increasing function of entropy, i.e., we replace $S(P)$ by $c(S(P))$, where $c'(x) > 0$. Let us note that this transform preserves the uniqueness of the MEP because it is easy to show that if $S(P)$ is Schur-concave, $c(S(P))$ is also Schur-concave [42] which is a sufficient condition for uniqueness of the MaxEnt distribution.

In this case, the Lagrange equations are adjusted as follows,

$$c'(S(P))\frac{\partial S(P)}{\partial p_i} - \alpha_c \frac{\partial f_0(P)}{\partial p_i} - \beta_c \frac{\partial \mathcal{E}(P)}{\partial p_i} = 0 \tag{15}$$

leading to

$$\alpha_c = c'(S(P))\langle \nabla_P S(P) \rangle - \beta_c \langle \nabla_P \mathcal{E}(P) \rangle \tag{16}$$

and

$$\beta_c = c'(S(P))\frac{\Delta_i(\nabla_P S(P))}{\Delta_i(\nabla_P \mathcal{E}(P))} \tag{17}$$

so we get that the function c causes *rescaling* of α and β, so

$$\alpha_c = c'(S(P))\,\alpha \tag{18}$$
$$\beta_c = c'(S(P))\,\beta \tag{19}$$

while its ratio remains unchanged, i.e., $\alpha_c/\beta_c = \alpha/\beta$. Actually, the set of increasing functions conform a group of Lagrange multipliers, because it is easy to show that the Lagrange parameters related to the entropy $c_1(c_2(S(P))$

$$\beta_{c_1 \circ c_2} = c_1'(c_2(S(P)) \cdot c_2'(S(P))\,\beta = c_1'(c_2(S(P))\beta_{c_2} \tag{20}$$

which can be described as the group operation $(c_1 \circ c_2) \mapsto c_1'(c_2) \cdot c_2'$.

An important property of this transformation is that it changes the extensive–intensive duality of the conjugated pair of thermodynamic variables and the respective forces while it maintains the distribution. Notably, by changing the entropic functional from extensive (i.e., $S(n) \sim U(n)$) to non-extensive, it changes β from intensive (i.e., size-independent, at least in the thermodynamic limit) to non-intensive, i.e., explicitly size-dependent. This point has been discussed in connection with q-non-extensive statistical physics of [29,30] and the relation to the zeroth law of thermodynamics was shown in [46]. As one can see from the example below, although Rényi entropy and Tsallis entropy have the same maximizer, the corresponding thermodynamics is different. While Rényi entropy is additive (and therefore extensive for systems where $U(n) \sim n$) and the temperature is intensive, Tsallis entropy is non-extensive, and the corresponding temperature explicitly depends on the size of the system.

Let us finally mention that the difference between free entropy and Lagrange parameter α transforms as

$$\psi_c - \alpha_c = (c(S) - c'(S)\langle \nabla_P S(P) \rangle - c'(S)\beta(E - \langle \nabla_P \mathcal{E}(P) \rangle) = c'(S)(\psi - \alpha) + (c(S) - c'(S) \cdot S). \tag{21}$$

While free entropy and other thermodynamic potentials are transformed, the heat change remains invariant under this transformation:

$$đQ_c = T_c d\, c(S) = \frac{T}{c'(S)} c'(S) dS = T dS = đQ. \qquad (22)$$

Example 1. *We exemplify the calibration invariance on two popular examples of closely related entropies.*

- **Rényi entropy and Tsallis entropy**: *Two most famous examples of generalized entropies are Rényi entropy $R_q(P) = \frac{1}{1-q} \ln\left(\sum_i p_i^q\right)$ and Tsallis entropy $S_q(P) = \frac{1}{1-q}\left(\sum_i p_i^q - 1\right)$. Their relation can be expressed as*

$$R_q(P) = c_q(S_q(P)) = \frac{1}{1-q} \ln\bigl[(1-q)S_q(P) + 1\bigr] \qquad (23)$$

and therefore we obtain that

$$c'_q(S_q(P)) = \frac{1}{1+(1-q)S_q} = \frac{1}{\sum_i p_i^q}. \qquad (24)$$

The difference between free entropy and α can be obtained as

$$\psi_R - \alpha_R = \frac{1}{\sum_i p_i^q}(\psi_S - \alpha_S) + \left(R_q(P) - \frac{S_q(P)}{\sum_i p_i^q}\right). \qquad (25)$$

One can therefore see that even though Rényi and Tsallis entropy lead to the same MaxEnt distribution, their thermodynamic quantities, such as temperature or free entropy, are different. Whether the system follows Rényi or Tsallis entropy depends on additional facts, as e.g., (non)-extensitivity and (non)-intensivity of thermodynamic quantities.

- **Shannon entropy and Entropy power**: *A similar example is provided with Shannon entropy $H(P) = \sum_i p_i \ln 1/p_i$ and entropy power $\mathcal{P}(P) = \prod_i (1/p_i)^{p_i}$. The relation between them is simply*

$$H(P) = c(\mathcal{P}(P)) = \log(\mathcal{P}(P)), \qquad (26)$$

so we obtain that

$$c'(\mathcal{P}(P)) = 1/(\mathcal{P}(P)) = \exp(-H(P)). \qquad (27)$$

For the difference between free entropy and α, we obtain that

$$0 = \psi_H - \alpha_H = \frac{1}{\mathcal{P}(P)}(\psi_\mathcal{P} - \alpha_\mathcal{P}) + (H(P) - 1) \qquad (28)$$

from which we get that

$$\psi_\mathcal{P} - \alpha_\mathcal{P} = \mathcal{P}(P)(1 - \log \mathcal{P}(P)). \qquad (29)$$

Therefore, we see that even that the MaxEnt distribution remains unchanged, the relation between α and free energy is different.

4. Calibration Invariance of MaxEnt Distribution with Constraints Transformation

Similarly, one can uncover the invariance of the MaxEnt distribution when the constraints are transformed in a certain way. Generally, if two sets of constraints define the same domain, the resulting Maximum entropy principle should lead to equivalent results. We will not be so general, but we focus on a specific situation, which might be quite interesting for thermodynamic applications. Let us remind two conditions, which we assume: normalization $f_0(P) = 0$ and energy constraint $f_E(P) = 0$. Let us investigate the latter. Similarly to the previous case, it is possible to take any function g of $f_E(P)$, for which $g(y) = 0$

if $y = 0$. More generally, we can also take into account the normalization constraint and replace the original energy condition by

$$g(f_0(P), f_E(P)) = 0 \qquad (30)$$

for any $g(x,y)$, for which $g(x,y) = 0 \Rightarrow y = 0$. Let us investigate the Maximum entropy principle for this case. We can express the Lagrange function as

$$\mathcal{L}(P) = S(P) - \alpha_g f_0(P) - \beta_g g(f_0(P), f_E(P)) \qquad (31)$$

which leads to a set of equations

$$\frac{\partial S(P)}{\partial p_i} - \alpha_g \frac{\partial f_0(P)}{\partial p_i} - \beta_g \left[G^{(1,0)} \frac{\partial f_0(P)}{\partial p_i} + G^{(0,1)} \frac{\partial \mathcal{E}(P)}{\partial p_i} \right] = 0 \qquad (32)$$

where $G^{(1,0)} = \frac{\partial g(x,y)}{\partial x}|_{(0,0)}$ and $G^{(0,1)} = \frac{\partial g(y,x)}{\partial x}|_{(0,0)}$. We take again into account that $\frac{\partial f_0(P)}{\partial p_i} = 1$, multiply the equations by p_i and some over i. This gives us

$$\alpha_g = \langle \nabla_P S(P) \rangle - \beta_g \left[G^{(1,0)} + G^{(0,1)} \langle \nabla_P \mathcal{E}(P) \rangle \right]. \qquad (33)$$

By plugging α_g back, we end with relation for β_g:

$$\beta_g = \frac{1}{G^{(0,1)}} \frac{\Delta_i(\nabla_P S(P))}{\Delta_i(\nabla_P \mathcal{E}(P))}. \qquad (34)$$

For α_g we end with

$$\alpha_g = \langle \nabla_P S(P) \rangle - \frac{\Delta_i(\nabla_P S(P))}{\Delta_i(\nabla_P \mathcal{E}(P))} \langle \nabla f_E(P) \rangle \left[1 + \frac{G^{(1,0)}}{G^{(0,1)}} \frac{1}{\langle \nabla_P \mathcal{E}(P) \rangle} \right]. \qquad (35)$$

Thus, we end again with rescaling of α_g and β_g, which reads

$$\alpha_g(\alpha, \beta) = \alpha - \frac{G^{(1,0)}}{G^{(0,1)}} \beta, \qquad (36)$$

$$\beta_g(\beta) = \frac{\beta}{G^{(0,1)}}. \qquad (37)$$

The ratio of Lagrange multipliers is also transformed, so we get

$$\frac{\alpha_g}{\beta_g} = G^{(0,1)} \frac{\alpha}{\beta} - G^{(1,0)}. \qquad (38)$$

Again, the set of all functions fulfilling the aforementioned condition conform a group. The group operation can be described by the relation between coefficients $G^{(1,0)}$ and $G^{(0,1)}$ for the composite function $g(x,y) = g_1(x, g_2(x,y))$. We obtain that

$$G^{(1,0)} = G_1^{(1,0)} + G_1^{(0,1)} G_2^{(1,0)} \qquad (39)$$

$$G^{(0,1)} = G_1^{(0,1)} G_2^{(0,1)} \qquad (40)$$

which leads to group relations

$$\alpha_g(\alpha, \beta) = \alpha_{g_1}(\alpha_{g_2}(\alpha, \beta), \beta_{g_2}(\beta)) - \frac{G_1^{(1,0)}}{G_1^{(0,1)}} \beta_{g_2}(\beta) \qquad (41)$$

$$\beta_g(\beta) = \frac{\beta_{g_2}(\beta)}{G_1^{(0,1)}}. \qquad (42)$$

Example 2. *Here we mention two simple examples of the aforementioned transformation.*

- **Energy shift:** *Under this scheme, we can assume the constant shift in the energy spectrum. Let us rewrite the constraint $f(P)$ in the following form,*

$$f_E(P) = \sum p_i E_i - E = \sum p_i (E_i - E') - (E - E') \tag{43}$$

which allows us to identify the function $g(x,y)$ as

$$g(x,y) = y - E'x + E' \tag{44}$$

We obtain $G^{(1,0)} = -E'$ and $G^{(0,1)} = 1$, which means that $\alpha' = \alpha - \beta E'$.

- **Latent escort means:** *Apart from linear means, it is possible to use some generalized approaches. One of these examples is provided by so-called escort mean:*

$$E_q = \langle E \rangle_q = \frac{\sum_i p_i^q E_i}{\sum_i p_i^q} \tag{45}$$

which for $q = 1$ becomes an ordinary linear mean, when $P = \{p_i\}_{i=1}^n$ are normalized to one. When we use this class of means in the Maximum entropy principle, the normalization is enforced by the normalization condition $f_0(P) = 0$, therefore for $q = 1$ we obtain the same results. Nevertheless, by taking $q = 1$ for the results with escort distribution, the energy constraint is actually expressed as

$$\frac{\sum p_i E_i}{\sum p_i} - E \tag{46}$$

can be understood in the same way as considered before in this section, i.e., as a combination of a normalization constraint and energy constraint. In this case the function g has the following form,

$$g(x,y) = \frac{y+E}{x+1} - E. \tag{47}$$

Therefore, we obtain that $G^{(1,0)} = -E$ and $G^{(0,1)} = 1$, which correspond to the previous example for $E' = E$. Therefore, the latent energy mean can be understood in terms of MaxEnt procedure as the shift of the energy spectrum by its average energy.

5. Conclusions

In this paper, we have discussed the calibration invariance of MEP, which means that for a given MaxEnt distribution, there exists a whole class of entropies and constraints that lead to different thermodynamics (Thermodynamic quantities and response coefficients generally have different behavior. For example, from intensive temperature we can obtain temperature that explicitly depends on the size of the system). We have stressed that the MEP procedure consists of two parts, where the first part, consisting of determining the MaxEnt distribution, is rather a mathematical tool, while the second part, making connection between Lagrange multipliers and thermodynamic quantities, is a specific for application of MEP in statistical physics. Indeed, the paper does not cover all possible transformations leading to the same MaxEnt distribution (let us mention, at least, the additive duality of Tsallis entropy, where maximizing S_{2-q} with linear constraint leads to the same result as maximizing S_q with escort constraints [47]). The main lesson of this paper is that in order to fully determine a thermal system in equilibrium, we need to measure not only probability distribution, but also all relevant thermodynamic quantities (as entropy). Moreover, the transformation between Lagrange parameters and its connection to thermodynamic potentials can be useful in situations when one is not certain about the exact form of entropy.

Funding: This research was funded by the Austrian Science Fund (FWF), project I 3073, Austrian Research Promotion agency (FFG), project 882184 and by the Grant Agency of the Czech Republic (GAČR), grant No. 19-16066S.

Acknowledgments: I would like to thank Petr Jizba for helpful discussions.

Conflicts of Interest: The author declares no conflict of interest.

References

1. Jaynes, E.T. Information Theory and Statistical Mechanics. *Phys. Rev.* **1957**, *106*, 620. [CrossRef]
2. Jaynes, E.T. Information Theory and Statistical Mechanics. II. *Phys. Rev.* **1957**, *108*, 171. [CrossRef]
3. Burg, J.P. The relationship between maximum entropy spectra and maximum likelihood spectra. *Geophysics* **1972**, *37*, 375–376. [CrossRef]
4. Rényi, A. *Selected Papers of Alfréd Rényi*; Akademia Kiado: Budapest, Hungary, 1976; Volume 2.
5. Havrda, J.H.; Charvát, F. Quantification Method of Classification Processes. Concept of Structural α-Entropy. *Kybernetika* **1967**, *3*, 30–35.
6. Sharma, B.D.; Mitter, J.; Mohan, M. On Measures of "Useful" Information. *Inf. Control* **1978**, *39*, 323–336. [CrossRef]
7. Tsallis, C. Possible generalization of Boltzmann-Gibbs statistics. *J. Stat. Phys.* **1988**, *52*, 479–487. [CrossRef]
8. Frank, T.; Daffertshofer, A. Exact time-dependent solutions of the Renyi Fokker-Planck equation and the Fokker-Planck equations related to the entropies proposed by Sharma and Mittal. *Physica A* **2000**, *285*, 351–366. [CrossRef]
9. Kaniadakis, G. Statistical mechanics in the context of special relativity. *Phys. Rev. E* **2002**, *66*, 056125. [CrossRef]
10. Jizba, P.; Arimitsu, T. The world according to Rényi: Thermodynamics of multifractal systems. *Ann. Phys.* **2004**, *312*, 17–59. [CrossRef]
11. Hanel, R.; Thurner, S. A comprehensive classification of complex statistical systems and an ab-initio derivation of their entropy and distribution functions. *Europhys. Lett.* **2011**, *93*, 20006. [CrossRef]
12. Thurner, S.; Hanel, R.; Klimek, P. *Introduction to the Theory of Complex Systems*; Oxford University Press: Oxford, UK, 2018.
13. Korbel, J.; Hanel, R.; Thurner, S. Classification of complex systems by their sample-space scaling exponents. *New J. Phys.* **2018**, *20*, 093007. [CrossRef]
14. Tempesta, P.; Jensen, H.J. Universality classes and information-theoretic Measures of complexity via Group entropies. *Sci. Rep.* **2020**, *10*, 1–11. [CrossRef]
15. Ilić, V.M.; Stankovixcx, M.S. Generalized Shannon-Khinchin axioms and uniqueness theorem for pseudo-additive entropies. *Physica A* **2014**, *411*, 138–145. [CrossRef]
16. Ilić, V.M.; Scarfone, A.M.; Wada, T. Equivalence between four versions of thermostatistics based on strongly pseudoadditive entropies. *Phys. Rev. E* **2019**, *100*, 062135. [CrossRef]
17. Czachor, M. Unifying Aspects of Generalized Calculus. *Entropy* **2020**, *22*, 1180. [CrossRef]
18. Beck, C.; Schlögl, F. *Thermodynamics of Chaotic Systems: An Introduction*; Cambridge University Press: Cambridge, UK, 1993.
19. Abe, S. Geometry of escort distributions. *Phys. Rev. E* **2003**, *68*, 031101. [CrossRef] [PubMed]
20. Bercher, J.-F. On escort distributions, q-gaussians and Fisher information. *AIP Conf. Proc.* **2011**, *1305*, 208.
21. Czachor, M.; Naudts, J. Thermostatistics based on Kolmogorov-Nagumo averages: Unifying framework for extensive and nonextensive generalizations. *Phys. Lett. A* **2002**, *298*, 369–374. [CrossRef]
22. Scarfone, A.M.; Matsuzoe, H.; Wada, T. Consistency of the structure of Legendre transform in thermodynamics with the Kolmogorov-Nagumo average. *Phys. Lett. A* **2016**, *380*, 3022–3028. [CrossRef]
23. Bercher, J.-F. Tsallis distribution as a standard maximum entropy solution with 'tail' constraint. *Phys. Lett. A* **2008**, *372*, 5657–5659. [CrossRef]
24. Pressé, S.; Ghosh, K.; Lee, J.; Dill, K.A. Nonadditive Entropies Yield Probability Distributions with Biases not Warranted by the Data. *Phys. Rev. Lett.* **2013**, *111*, 180604. [CrossRef] [PubMed]
25. Oikonomou, T.; Bagci, B. Misusing the entropy maximization in the jungle of generalized entropies. *Phys. Lett. A* **2017**, *381*, 207–211. [CrossRef]
26. Tsallis, C.; Mendes, R.S.; Plastino, A.R. The role of constraints within generalized nonextensive statistics. *Phys. A* **1998**, *286*, 534–554. [CrossRef]
27. Martínez, S.; Nicolás, F.; Peninni, F.; Plastino, A. Tsallis' entropy maximization procedure revisited. *Phys. A* **2000**, *286*, 489–502. [CrossRef]
28. Plastino, A.; Plastino, A.R. On the universality of thermodynamics' Legendre transform structure. *Phys. Lett. A* **1997**, *226*, 257–263. [CrossRef]
29. Rama, S.K. Tsallis Statistics: Averages and a Physical Interpretation of the Lagrange Multiplier β. *Phys. Lett. A* **2000**, *276*, 103–108. [CrossRef]
30. Campisi, M.; Bagci, G.B. Tsallis Ensemble as an Exact Orthode. *Phys. Lett. A* **2007**, *362*, 11–15. [CrossRef]
31. Dixit, P.D.; Wagoner, J.; Weistuch, C.; Pressé, S.; Ghosh, K.; Dill, K.A. Perspective: Maximum caliber is a general variational principle for dynamical systems. *J. Chem. Phys.* **2018**, *148*, 010901. [CrossRef]
32. Lucia, U. Stationary Open Systems: A Brief Review on Contemporary Theories on Irreversibility. *Physica A* **2013**, *392*, 1051–1062. [CrossRef]
33. Palazzo, P. Hierarchical Structure of Generalized Thermodynamic and Informational Entropy. *Entropy* **2018**, *20*, 553. [CrossRef]
34. Shore, J.E.; Johnson, R.W. Axiomatic derivation of the principle of maximum entropy and the principle of minimum cross-entropy. *IEEE Trans. Inf. Theor.* **1980**, *26*, 26–37. [CrossRef]

35. Shore, J.E.; Johnson, R.W. Properties of cross-entropy minimization. *IEEE Trans. Inf. Theor.* **1981**, *27*, 472–482. [CrossRef]
36. Uffink, J. Can the maximum entropy principle be explained as a consistency requirement? *Stud. Hist. Philos. Mod. Phys.* **1995**, *26*, 223–261. [CrossRef]
37. Tsallis, C. Conceptual Inadequacy of the Shore and Johnson Axioms for Wide Classes of Complex Systems. *Entropy* **2015**, *17*, 2853–2861. [CrossRef]
38. Pressé, S.; Ghosh, K.; Lee, J.; Dill, K.A. Reply to C. Tsallis' "Conceptual Inadequacy of the Shore and Johnson Axioms for Wide Classes of Complex Systems". *Entropy* **2015**, *17*, 5043–5046. [CrossRef]
39. Oikonomou, T.; Bagci, G.B. Rényi entropy yields artificial biases not in the data and incorrect updating due to the finite-size data. *Phys. Rev. E* **2019**, *99*, 032134. [CrossRef]
40. Jizba, P.; Korbel, J. Comment on "Rényi entropy yields artificial biases not in the data and incorrect updating due to the finite-size data". *Phys. Rev. E* **2019**, *100*, 026101 [CrossRef]
41. Oikonomou, T.; Bagci, G.B. Reply to "Comment on Rényi entropy yields artificial biases not in the data and incorrect updating due to the finite-size data". *Phys. Rev. E* **2019**, *100*, 026102. [CrossRef]
42. Jizba, P.; Korbel, J. Maximum Entropy Principle in Statistical Inference: Case for Non-Shannonian Entropies. *Phys. Rev. Lett.* **2019**, *122*, 120601. [CrossRef]
43. Jizba, P.; Korbel, J. When Shannon and Khinchin meet Shore and Johnson: Equivalence of information theory and statistical inference axiomatics. *Phys. Rev. E* **2020**, *101*, 042126. [CrossRef]
44. Plastino, A.; Plastino, A.R. Tsallis Entropy and Jaynes' Information Theory Formalism. *Braz. J. Phys.* **1999**, *29*, 50–60. [CrossRef]
45. Naudts, J. *Generalized Thermostatistics*; Springer: London, UK, 2011.
46. Biró, T.S.; Ván, P. Zeroth law compatibility of nonadditive thermodynamics. *Phys. Rev. E* **2011**, *83*, 061147. [CrossRef] [PubMed]
47. Wada, T.; Scarfone, A.M. Connections between Tsallis' formalisms employing the standard linear average energy and ones employing the normalized q-average energy. *Phys. Lett. A* **2005**, *335*, 351–362. [CrossRef]

Article
Unifying Aspects of Generalized Calculus

Marek Czachor

Wydział Fizyki Technicznej i Matematyki Stosowanej, Politechnika Gdańska, 80-233 Gdańsk, Poland; mczachor@pg.edu.pl

Received: 28 September 2020; Accepted: 17 October 2020; Published: 19 October 2020

Abstract: Non-Newtonian calculus naturally unifies various ideas that have occurred over the years in the field of generalized thermostatistics, or in the borderland between classical and quantum information theory. The formalism, being very general, is as simple as the calculus we know from undergraduate courses of mathematics. Its theoretical potential is huge, and yet it remains unknown or unappreciated.

Keywords: non-Newtonian calculus; non-Diophantine arithmetic; Kolmogorov–Nagumo averages; escort probabilities; generalized entropies

1. Introduction

Studies of a calculus based on generalized forms of arithmetic were initiated in the late 1960s by Grossman and Katz, resulting in their little book *Non-Newtonian Calculus* [1–3]. Some twenty years later, the main construction was independently discovered in a different context and pushed in a different direction by Pap [4–6]. After another two decades the same idea, but in its currently most general form, was rediscovered by myself [7–15]. In a wider perspective, non-Newtonian calculus is conceptually related to the works of Rashevsky [16] and Burgin [17–20] on non-Diophantine arithmetics of natural numbers, and to Benioff's attempts [21–25] of basing physics and mathematics on a common fundamental ground. Traces of non-Newtonian and non-Diophantine thinking can be found in the works of Kaniadakis on generalized statistics [26–34]. A relatively complete account of the formalism can be found in the forthcoming monograph [35].

In the paper, we will discuss links between generalized arithmetics; non-Newtonian calculus; generalized entropies; and classical, quantum, and escort probabilities. As we will see, certain constructions such as Rényi entropies or exponential families of probabilities have direct relations to generalized arthmetics and calculi. Some of the constructions one finds in the literature are literally non-Newtonian. Some others only look non-Newtonian, but closer scrutiny reveals formal inconsistencies, at least from a strict non-Newtonian perspective.

Our goal is to introduce non-Newtonian calculus as a sort of unifying principle, simultaneously sketching new theoretical directions and open questions.

2. Non-Diophantine Arithmetic and Non-Newtonian Calculus

The most general form of non-Newtonian calculus deals with functions A defined by the commutative diagram ($f_\mathbb{X}$ and $f_\mathbb{Y}$ are arbitrary bijections)

$$\begin{array}{ccc} \mathbb{X} & \xrightarrow{A} & \mathbb{Y} \\ f_\mathbb{X} \downarrow & & \downarrow f_\mathbb{Y} \\ \mathbb{R} & \xrightarrow{\tilde{A}} & \mathbb{R} \end{array} \qquad (1)$$

The only assumption about the domain $\mathbb{X}$ and the codomain $\mathbb{Y}$ is that they have the same cardinality as the continuum $\mathbb{R}$. The latter guarantees that bijections $f_\mathbb{X}$ and $f_\mathbb{Y}$ exist. The bijections are automatically continuous in the topologies they induce from the open-interval topology of $\mathbb{R}$, even if they are discontinuous in metric topologies of $\mathbb{X}$ and $\mathbb{Y}$ (a typical situation in fractal applications, or in cases where $\mathbb{X}$ or $\mathbb{Y}$ are not subsets of $\mathbb{R}$). In general, one does *not* assume anything else about $f_\mathbb{X}$ and $f_\mathbb{Y}$. In particular, their differentiability in the usual (Newtonian) sense is not assumed. No topological assumptions are made about $\mathbb{X}$ and $\mathbb{Y}$. Of course, the structure of the diagram implies that $\mathbb{X}$ and $\mathbb{Y}$ may be regarded as Banach manifolds with global charts $f_\mathbb{X}$ and $f_\mathbb{Y}$, but one does not make the usual assumptions about changes of charts.

Non-Newtonian calculus begins with (generalized, non-Diophantine) arithmetics in $\mathbb{X}$ and $\mathbb{Y}$, induced from $\mathbb{R}$,

$$x_1 \oplus_\mathbb{X} x_2 = f_\mathbb{X}^{-1}(f_\mathbb{X}(x_1) + f_\mathbb{X}(x_2)), \tag{2}$$
$$x_1 \ominus_\mathbb{X} x_2 = f_\mathbb{X}^{-1}(f_\mathbb{X}(x_1) - f_\mathbb{X}(x_2)), \tag{3}$$
$$x_1 \odot_\mathbb{X} x_2 = f_\mathbb{X}^{-1}(f_\mathbb{X}(x_1) \cdot f_\mathbb{X}(x_2)), \tag{4}$$
$$x_1 \oslash_\mathbb{X} x_2 = f_\mathbb{X}^{-1}(f_\mathbb{X}(x_1) / f_\mathbb{X}(x_2)) \tag{5}$$

(and analogously in $\mathbb{Y}$).

Example 1. *According to one of the axioms of standard quantum mechanics, states of a quantum system belong to a separable Hilbert space. All separable Hilbert spaces are isomorphic, so state spaces of any two quantum systems are isomorphic. Does it mean that all quantum systems are equivalent? No, it only shows that mathematically isomorphic structures can play physically different roles. Similarly, the arithmetic given by (2)–(5) is isomorphic to the standard arithmetic of $\mathbb{R}$, but it does not imply that the two arithmetics are physically equivalent.*

Example 2. *The origin of Einstein's special theory of relativity goes back to the observation that the velocity of a source of light does not influence the velocity of light itself, contradicting our everyday experiences with velocities in trains or football. Relativistic addition of velocities is based on a fundamental unit c and the dimensionless parameter β, related to velocity by $v = \beta c$. $\beta \in \mathbb{X} = (-1, 1)$ while the bijection reads $f_\mathbb{X}(\beta) = \text{arctanh}\,\beta$. The velocities are added or subtracted by means of (2) and (3),*

$$\beta_1 \oplus_\mathbb{X} \beta_2 = \tanh(\text{arctanh}\,\beta_1 + \text{arctanh}\,\beta_2). \tag{6}$$

Interestingly, (4) and (5) are not directly employed in special relativity. The presence of the fundamental unit c is a signature of a general non-Diophantine arithmetic (which typically works with dimensionless numbers). Numbers $\pm 1 \in \mathbb{R}$ play the roles of infinities, $\pm 1_\mathbb{R} = \pm \infty_\mathbb{X}$. The velocity of light is therefore literally infinite in the non-Diophantine sense. The neutral element of multiplication, $1_\mathbb{X} = f_\mathbb{X}^{-1}(1) = \tanh 1 = 0.76$ (i.e., $v = 0.76c$), does not seem to play in relativistic physics any privileged role.

Sometimes, for example in the context of Bell's theorem, one works with mixed arithmetics of the form [13]

$$x_1 \odot_\mathbb{Z}^{\mathbb{XY}} y_2 = f_\mathbb{Z}^{-1}(f_\mathbb{X}(x_1) \cdot f_\mathbb{Y}(y_2)), \quad \odot_\mathbb{Z}^{\mathbb{XY}} : \mathbb{X} \times \mathbb{Y} \to \mathbb{Z}, \quad \text{etc.} \tag{7}$$

Mixed arithmetics naturally occur in Taylor expansions of functions whose domains and codomains involve different arithmetics, and in the chain rule for derivatives (see Example 6).

In order to define calculus one needs limits "to zero", and thus the notion of zero itself. In the arithmetic context a zero is a neutral element of addition, for example, $x \oplus_\mathbb{X} 0_\mathbb{X} = x$ for any $x \in \mathbb{X}$. Obviously, such a zero is arithmetic-dependent. The same concerns a "one", a neutral element of multiplication, fulfilling $x \odot_\mathbb{X} 1_\mathbb{X} = x$ for any $x \in \mathbb{X}$. Once the arithmetic in $\mathbb{X}$ is specified, both neutral elements are uniquely given by the general formula: $r_\mathbb{X} = f_\mathbb{X}^{-1}(r)$ for any $r \in \mathbb{R}$. Therefore, in particular, $0_\mathbb{X} = f_\mathbb{X}^{-1}(0), 1_\mathbb{X} = f_\mathbb{X}^{-1}(1)$. One easily verifies that

$$r_\mathbb{X} \oplus s_\mathbb{X} = (r+s)_\mathbb{X}, \tag{8}$$
$$r_\mathbb{X} \odot s_\mathbb{X} = (rs)_\mathbb{X}, \tag{9}$$

for all $r, s \in \mathbb{R}$, which extends also to mixed arithmetics,

$$r_\mathbb{X} \oplus_\mathbb{X}^{\mathbb{X}\mathbb{Y}} s_\mathbb{Y} = (r+s)_\mathbb{X}, \tag{10}$$
$$r_\mathbb{X} \oplus_\mathbb{Y}^{\mathbb{X}\mathbb{Y}} s_\mathbb{Y} = (r+s)_\mathbb{Y}, \tag{11}$$
$$r_\mathbb{X} \oplus_\mathbb{Z}^{\mathbb{X}\mathbb{Y}} s_\mathbb{Y} = (r+s)_\mathbb{Z}, \text{ etc.} \tag{12}$$

If there is no danger of ambiguity one can simplify the notation by $\oplus_\mathbb{X}^{\mathbb{X}\mathbb{X}} = \oplus_\mathbb{X}$ or $\oplus_\mathbb{X}^{\mathbb{X}\mathbb{Y}} = \oplus_\mathbb{X}^\mathbb{Y}$. Mixed arithmetics can be given an interpretation in terms of communication channels. Mixed multiplication is in many respects analogous to a tensor product [13].

Example 3. *Consider* $\mathbb{X} = \mathbb{R}_+, \mathbb{Y} = -\mathbb{R}_+, f_\mathbb{X}(x) = \ln x, f_\mathbb{X}^{-1}(r) = e^r, f_\mathbb{Y}(x) = \ln(-x), f_\mathbb{Y}^{-1}(r) = -e^r$. *"Two plus two equals four" looks here as follows,*

$$2_\mathbb{X} \oplus_\mathbb{X} 2_\mathbb{X} = f_\mathbb{X}^{-1}(2+2) = 4_\mathbb{X} = e^4, \tag{13}$$
$$2_\mathbb{X} \oplus_\mathbb{X}^\mathbb{Y} 2_\mathbb{Y} = f_\mathbb{X}^{-1}(2+2) = 4_\mathbb{X} = e^4, \tag{14}$$
$$2_\mathbb{Y} \oplus_\mathbb{Y} 2_\mathbb{Y} = f_\mathbb{Y}^{-1}(2+2) = 4_\mathbb{Y} = -e^4, \tag{15}$$
$$2_\mathbb{X} \oplus_\mathbb{Y}^\mathbb{X} 2_\mathbb{Y} = f_\mathbb{Y}^{-1}(2+2) = 4_\mathbb{Y} = -e^4, \tag{16}$$

where $2_\mathbb{X} = f_\mathbb{X}^{-1}(2) = e^2, 2_\mathbb{Y} = f_\mathbb{Y}^{-1}(2) = -e^2$. *From the point of view of communication channels the situation is as follows. There are two parties ("Alice" and "Bob"), each computing by means of her/his own rules. They communicate their results and agree the numbers they have found are the same, namely, "two" and "four". However, for an external observer (an eavesdropper "Eve"), their results are opposite, say e^4 and $-e^4$. Mixed arithmetic plays a role of a "connection" relating different local arithmetics. This is why, in the terminology of Burgin, these types or arithmetics are non-Diophantine (from Diophantus of Alexandria who formalized the standard arithmetic). Similarly to nontrivial manifolds, non-Diophantine arithmetics do not have to admit a single global description (which we nevertheless assume in this paper).*

A limit such as $\lim_{x' \to x} A(x') = A(x)$ is defined by the diagram (1) as follows,

$$\lim_{x' \to x} A(x') = f_\mathbb{Y}^{-1}\left(\lim_{r \to f_\mathbb{X}(x)} \tilde{A}(r) \right) \tag{17}$$

i.e., in terms of an ordinary limit in $\mathbb{R}$. A non-Newtonian derivative is then defined by

$$\frac{DA(x)}{Dx} = \lim_{\delta \to 0} \left(A(x \oplus_\mathbb{X} \delta_\mathbb{X}) \ominus_\mathbb{Y} A(x) \right) \oslash_\mathbb{Y} \delta_\mathbb{Y} = f_\mathbb{Y}^{-1}\left(\frac{d\tilde{A}(f_\mathbb{X}(x))}{df_\mathbb{X}(x)} \right), \tag{18}$$

if the Newtonian derivative $d\tilde{A}(r)/dr$ exists. It is additive,

$$\frac{D[A(x) \oplus_\mathbb{Y} B(x)]}{Dx} = \frac{DA(x)}{Dx} \oplus_\mathbb{Y} \frac{DB(x)}{Dx}, \tag{19}$$

and satisfies the Leibniz rule,

$$\frac{D[A(x) \odot_\mathbb{Y} B(x)]}{Dx} = \left(\frac{DA(x)}{Dx} \odot_\mathbb{Y} B(x) \right) \oplus_\mathbb{Y} \left(A(x) \odot_\mathbb{Y} \frac{DB(x)}{Dx} \right). \tag{20}$$

A general chain rule for compositions of functions involving arbitrary arithmetics in domains and codomains can be derived [12] (see Example 6). It implies, in particular, that the bijections defining the arithmetics are themselves always non-Newtonian differentiable (with respect to the derivatives they define). The resulting derivatives are "trivial",

$$\frac{Df_\mathbb{X}(x)}{Dx} = 1 = \frac{Df_\mathbb{Y}(y)}{Dy}, \quad \frac{Df_\mathbb{X}^{-1}(r)}{Dr} = 1_\mathbb{X}, \quad \frac{Df_\mathbb{Y}^{-1}(r)}{Dr} = 1_\mathbb{Y}. \tag{21}$$

A non-Newtonian integral is defined by the requirement that, under typical assumptions paralleling those from the fundamental theorem of Newtonian calculus, one finds

$$\frac{D}{Dx} \int_y^x A(x') Dx' = A(x), \tag{22}$$

$$\int_y^x \frac{DA(x')}{Dx'} Dx' = A(x) \ominus_\mathbb{Y} A(y), \tag{23}$$

which uniquely implies that

$$\int_y^x A(x') Dx' = f_\mathbb{Y}^{-1}\left(\int_{f_\mathbb{X}(y)}^{f_\mathbb{X}(x)} \tilde{A}(r) dr \right). \tag{24}$$

Here, as before, $\tilde{A}$ is defined by (1) and dr denotes the usual Newtonian (Riemann, Lebesgue, etc.) integration. To have a feel of the potential inherent in this simple formula, let us mention that for a Koch-type fractal (24) turns out to be equivalent to the Hausdorff integral [12,36,37]. In applications, typically the only nontrivial element is to find the explicit form of $f_\mathbb{X}$. It should be stressed that (24) reduces any integral to the one over a subset of $\mathbb{R}$. The fact that such a counterintuitive possibility exists was noticed already by Wiener in his 1933 lectures on Fourier analysis [38].

3. Non-Newtonian Exponential Function and Logarithm

Once we know how to differentiate and integrate, we can turn to differential equations. The so-called exponential family plays a crucial role in thermodynamics, both standard and generalized [39–43]. Many different deformations of the usual e^x can be found in the literature. However, from the non-Newtonian perspective, the exponential function $\text{Exp} : \mathbb{X} \to \mathbb{Y}$ is defined by

$$\frac{D\text{Exp}(x)}{Dx} = \text{Exp}(x), \quad \text{Exp}(0_\mathbb{X}) = 1_\mathbb{Y}. \tag{25}$$

Integrating (25) (in a non-Newtonian way) one finds the unique solution

$$\text{Exp}(x) = f_\mathbb{Y}^{-1}\left(e^{f_\mathbb{X}(x)} \right), \quad \text{Exp}(x_1 \oplus_\mathbb{X} x_2) = \text{Exp}(x_1) \odot_\mathbb{Y} \text{Exp}(x_2). \tag{26}$$

In thermodynamic applications, one often encounters exponents of negative arguments, e^{-x}. In a non-Newtonian context the correct form of a minus is $\ominus_\mathbb{X} x = 0_\mathbb{X} \ominus_\mathbb{X} x = f_\mathbb{X}^{-1}(-f_\mathbb{X}(x))$. The example discussed in the next section will involve $\mathbb{X} = \mathbb{R}$ and $f_\mathbb{X}^{-1}(-r) = -f_\mathbb{X}^{-1}(r)$. In consequence, it will be correct to write $\ominus_\mathbb{X} x = -x$, but in general such a simple rule may be meaningless (because "$-$", as opposed to $\ominus_\mathbb{X}$, may be undefined in $\mathbb{X}$).

Example 4. Let $\mathbb{X} = (\mathbb{R}_+, \oplus, \odot)$, with the arithmetic defined by $f_\mathbb{X} : \mathbb{R}_+ \to \mathbb{R}$, $f_\mathbb{X}(x) = \ln x$, $f_\mathbb{X}^{-1}(r) = e^r$. Then

$$\ominus_\mathbb{X} x = f_\mathbb{X}^{-1}(-f_\mathbb{X}(x)) = e^{-\ln x} = 1/x \in \mathbb{R}_+. \tag{27}$$

The same number can be both positive and negative, depending on the arithmetic.

A (natural) logarithm is the inverse of Exp, namely, Ln : $\mathbb{Y} \to \mathbb{X}$,

$$\text{Ln}(y) = f_{\mathbb{X}}^{-1}\left(\ln f_{\mathbb{Y}}(x)\right), \quad \text{Ln}(y_1 \odot_{\mathbb{Y}} y_2) = \text{Ln}(y_1) \oplus_{\mathbb{X}} \text{Ln}(y_2). \tag{28}$$

Expressions such as Exp $x + $ Ln y are in general meaningless even if $\mathbb{X} \subset \mathbb{R}_+$ and $\mathbb{Y} \subset \mathbb{R}_+$. However, formulas such as

$$(\text{Exp } x) \oplus_{\mathbb{Z}}^{\mathbb{Y}\mathbb{X}} (\text{Ln } y) = f_{\mathbb{Z}}^{-1}\left(e^{f_{\mathbb{X}}(x)} + \ln f_{\mathbb{Y}}(y)\right) \tag{29}$$

make perfect sense. For example, if $p_k \in \mathbb{X}$, then an entropy can be defined as

$$S = \bigoplus_{k} \mathbb{Z} p_k \odot_{\mathbb{Z}}^{\mathbb{X}\mathbb{Y}} \text{Ln}\left(1_{\mathbb{X}} \oslash_{\mathbb{X}} p_k\right) \tag{30}$$

$$= f_{\mathbb{Z}}^{-1}\left[\sum_{k} f_{\mathbb{Z}}\left(p_k \odot_{\mathbb{Z}}^{\mathbb{X}\mathbb{Y}} \text{Ln}(1_{\mathbb{X}} \oslash_{\mathbb{X}} p_k)\right)\right] \tag{31}$$

$$= f_{\mathbb{Z}}^{-1}\left[\sum_{k} f_{\mathbb{X}}(p_k) \ln\left(1/f_{\mathbb{X}}(p_k)\right)\right]. \tag{32}$$

Many intriguing questions occur if one asks about normalization of probabilities. We will come to it later.

Non-Newtonian constructions of Exp and Ln are systematic, general, and flexible. There seems to exist a relation between the arithmetic formalism and the method of monotone embedding discussed in information geometry [44], but the problem requires further studies.

Example 5. *In order to appreciate the difference between Newtonian and non-Newtonian differentiation let us differentiate the function $A(x) = x$, $A : \mathbb{X} \to \mathbb{Y}$, but in two cases. The first one is trivial, $\mathbb{X} = \mathbb{Y} = (\mathbb{R}, +, \cdot)$, with the arithmetic defined by the identity $f_{\mathbb{X}} = f_{\mathbb{Y}} = \text{id}_{\mathbb{R}}$. Then, the non-Newtonian and Newtonian derivatives coincide, so*

$$\frac{DA(x)}{Dx} = \frac{dA(x)}{dx} = 1. \tag{33}$$

The second case involves, as before, the codomain $\mathbb{Y} = (\mathbb{R}, +, \cdot)$, with the arithmetic defined by the identity $f_{\mathbb{Y}} = \text{id}_{\mathbb{R}}$. However, as the domain we choose $\mathbb{X} = (\mathbb{R}_+, \oplus, \odot)$, with the arithmetic defined by $f_{\mathbb{X}} : \mathbb{R}_+ \to \mathbb{R}$, $f_{\mathbb{X}}(x) = \ln x$, $f_{\mathbb{X}}^{-1}(r) = e^r$. Now,

$$\frac{DA(x)}{Dx} = \lim_{\delta \to 0}\left(A(x \oplus_{\mathbb{X}} \delta_{\mathbb{X}}) \ominus_{\mathbb{Y}} A(x)\right) \oslash_{\mathbb{Y}} \delta_{\mathbb{Y}} = \lim_{\delta \to 0}\frac{\left(x \oplus_{\mathbb{X}} f_{\mathbb{X}}^{-1}(\delta)\right) - x}{\delta}$$

$$= \lim_{\delta \to 0}\frac{e^{\ln x + \delta} - x}{\delta} = x = A(x). \tag{34}$$

As, $0_{\mathbb{X}} = f_{\mathbb{X}}^{-1}(0) = e^0 = 1$, we find $A(0_{\mathbb{X}}) = 0_{\mathbb{X}} = 1 = 1_{\mathbb{Y}}$, and conclude that $A(x) = x$, $A : \mathbb{R}_+ \to \mathbb{R}$ belongs to the exponential family. Indeed,

$$A(x_1 \oplus_{\mathbb{X}} x_2) = x_1 \oplus_{\mathbb{X}} x_2 = e^{\ln x_1 + \ln x_2} = x_1 \cdot x_2 = A(x_1) \odot_{\mathbb{Y}} A(x_2). \tag{35}$$

To understand the result, write $A(x) = f_{\mathbb{Y}}^{-1}\left(\tilde{A}(f_{\mathbb{X}}(x))\right) = \tilde{A}(\ln x) = x$, so that $\tilde{A}(r) = e^r$. Then, by the second form of derivative in (18),

$$\frac{DA(x)}{Dx} = f_{\mathbb{Y}}^{-1}\left(\frac{d\tilde{A}(f_{\mathbb{X}}(x))}{df_{\mathbb{X}}(x)}\right) = \frac{de^{f_{\mathbb{X}}(x)}}{df_{\mathbb{X}}(x)} = e^{f_{\mathbb{X}}(x)} = e^{\ln x} = x. \tag{36}$$

The map A does not affect the value of x, but changes its arithmetic properties. It behaves as if it assigned a different meaning to the same word. The example becomes even more intriguing if one realizes that logarithm is known to approximately relate stimulus with sensation in real-life sensory systems (hence the logarithmic scale of decibels and star magnitudes) [35].

Example 6. *Many calculations in thermodynamics reduce to formulas of the form*

$$dU(S,V) = \left(\frac{\partial U}{\partial S}\right)_V dS + \left(\frac{\partial U}{\partial V}\right)_S dV, \tag{37}$$

being equivalent to the derivative $dU(S(t), V(t))/dt$ of a composite function of several variables. The latter has a unique formulation in non-Newtonian calculus: One only needs to specify the arithmetics. For example, let U be a map $U : \mathbb{S} \times \mathbb{V} \to \mathbb{U}$, and let $S : \mathbb{T} \to \mathbb{S}$, $V : \mathbb{T} \to \mathbb{V}$. Then,

$$\frac{DU(S(t), V(t))}{Dt} = \lim_{\delta \to 0} \left(U((S(t \oplus_\mathbb{T} \delta_\mathbb{T}), V(t \oplus_\mathbb{T} \delta_\mathbb{T}))) \ominus_\mathbb{U} U(S(t), V(t)) \right) \oslash_\mathbb{U} \delta_\mathbb{U}. \tag{38}$$

As

$$\lim_{x' \to x} A(x) \oplus_\mathbb{Y} B(x) = \left(\lim_{x' \to x} A(x) \right) \oplus_\mathbb{Y} \left(\lim_{x' \to x} B(x) \right) \tag{39}$$

(see Appendix A), we rewrite (38) as

$$\frac{DU(S(t), V(t))}{Dt} = \lim_{\delta \to 0} \left(U((S(t \oplus_\mathbb{T} \delta_\mathbb{T}), V(t \oplus_\mathbb{T} \delta_\mathbb{T}))) \ominus_\mathbb{U} U((S(t \oplus_\mathbb{T} \delta_\mathbb{T}), V(t))) \right) \oslash_\mathbb{U} \delta_\mathbb{U}$$
$$\oplus_\mathbb{U} \lim_{\delta \to 0} \left(U((S(t \oplus_\mathbb{T} \delta_\mathbb{T}), V(t))) \ominus_\mathbb{U} U(S(t), V(t)) \right) \oslash_\mathbb{U} \delta_\mathbb{U}. \tag{40}$$

Under the usual assumptions about continuity of $\tilde{U} : \mathbb{R} \times \mathbb{R} \to \mathbb{R}$ in

$$\begin{array}{ccc} \mathbb{S} \times \mathbb{V} & \xrightarrow{U} & \mathbb{U} \\ f_\mathbb{S} \downarrow f_\mathbb{V} \downarrow & & \downarrow f_\mathbb{U} \\ \mathbb{R} \times \mathbb{R} & \xrightarrow{\tilde{U}} & \mathbb{R} \end{array}, \tag{41}$$

we reduce (40) to

$$\frac{DU(S(t), V(t))}{Dt} = \lim_{\delta \to 0} \left(U((S(t), V(t \oplus_\mathbb{T} \delta_\mathbb{T}))) \ominus_\mathbb{U} U((S(t), V(t))) \right) \oslash_\mathbb{U} \delta_\mathbb{U}$$
$$\oplus_\mathbb{U} \lim_{\delta \to 0} \left(U((S(t \oplus_\mathbb{T} \delta_\mathbb{T}), V(t))) \ominus_\mathbb{U} U(S(t), V(t)) \right) \oslash_\mathbb{U} \delta_\mathbb{U}, \tag{42}$$

and then to two instances of the non-Newtonian chain rule,

$$\frac{D(B \circ A)(x)}{Dx} = f_\mathbb{Z}^{-1}\left[f_\mathbb{Z}\left(\frac{DB(A(x))}{DA(x)}\right) f_\mathbb{Y}\left(\frac{DA(x)}{Dx}\right) \right] = \frac{DB(A(x))}{DA(x)} \odot_\mathbb{Z}^{\mathbb{ZY}} \frac{DA(x)}{Dx}, \tag{43}$$

valid for the composition

$$\begin{array}{ccccc} \mathbb{X} & \xrightarrow{A} & \mathbb{Y} & \xrightarrow{B} & \mathbb{Z} \\ f_\mathbb{X} \downarrow & & f_\mathbb{Y} \downarrow & & f_\mathbb{Z} \downarrow \\ \mathbb{R} & \xrightarrow{\tilde{A}} & \mathbb{R} & \xrightarrow{\tilde{B}} & \mathbb{R} \end{array} \tag{44}$$

of maps. Finally,

$$\frac{DU(S(t),V(t))}{Dt} = \frac{DU(S(t),V(t))}{DS(t)} \odot_U^{US} \frac{DS(t)}{Dt} \oplus_U \frac{DU(S(t),V(t))}{DV(t)} \odot_U^{UV} \frac{DV(t)}{Dt}. \tag{45}$$

Effectively,

$$DU(S,V) = \left(\frac{DU}{DS}\right)_V \odot_U^{US} DS \oplus_U \left(\frac{DU}{DV}\right)_S \odot_U^{UV} DV, \tag{46}$$

is the non-Newtonian formula for a differential.

The next section shows that the above mentioned subtleties with arithmetics of domains and codomains have straightforward implications for generalized thermostatistics.

4. Kaniadakis κ-Calculus Versus Non-Newtonian Calculus

Kaniadakis, in a series of papers [26–34], developed a generalized form of arithmetic and calculus, with numerous applications to statistical physics, and beyond. In the present section, we will clarify links between his formalism and non-Newtonian calculus. As we will see, some of the results have a straightforward non-Newtonian interpretation, but not all.

Assume $\mathbb{X} = \mathbb{R}$, with the bijection $f_{\mathbb{X}} \equiv f_\kappa : \mathbb{R} \to \mathbb{R}$ given explicitly by

$$f_\kappa(x) = \frac{1}{\kappa}\operatorname{arcsinh}\kappa x, \tag{47}$$

$$f_\kappa^{-1}(x) = \frac{1}{\kappa}\sinh\kappa x. \tag{48}$$

Kaniadakis' κ-calculus begins with the arithmetic,

$$x \overset{\kappa}{\oplus} y = f_\kappa^{-1}(f_\kappa(x) + f_\kappa(y)), \tag{49}$$

$$x \overset{\kappa}{\ominus} y = f_\kappa^{-1}(f_\kappa(x) - f_\kappa(y)), \tag{50}$$

$$x \overset{\kappa}{\odot} y = f_\kappa^{-1}(f_\kappa(x) \cdot f_\kappa(y)), \tag{51}$$

$$x \overset{\kappa}{\oslash} y = f_\kappa^{-1}(f_\kappa(x)/f_\kappa(y)). \tag{52}$$

As $f_0(x) = x$, the case $\kappa = 0$ corresponds to the usual field $\mathbb{R}_0 = (\mathbb{R}, +, \cdot)$, which we will shortly denote by $\mathbb{R}$. The neutral element of addition, $0_\kappa = f_\kappa^{-1}(0) = 0$, is the same for all κs. The neutral element of κ-multiplication is nontrivial, $1_\kappa = f_\kappa^{-1}(1) \neq 1$. The fields $\mathbb{R}_\kappa = (\mathbb{R}, \overset{\kappa}{\oplus}, \overset{\kappa}{\odot})$ are isomorphic to one another due to their isomorphism with $\mathbb{R}_0$,

$$f_\kappa(x \overset{\kappa}{\oplus} y) = f_\kappa(x) + f_\kappa(y), \tag{53}$$

$$f_\kappa(x \overset{\kappa}{\odot} y) = f_\kappa(x) \cdot f_\kappa(y). \tag{54}$$

Kaniadakis defines his κ-derivative of a real function $A(x)$ as

$$\frac{dA(x)}{d_\kappa x} = \lim_{\delta \to 0} \frac{A(x+\delta) - A(x)}{(x+\delta) \overset{\kappa}{\ominus} x} = \frac{dA(x)}{dx}\bigg/\frac{df_\kappa(x)}{dx} = \frac{dA(x)}{dx}\sqrt{1+\kappa^2 x^2}. \tag{55}$$

We will now specify in which sense the κ-derivative is non-Newtonian. First consider a function A,

Its non-Newtonian derivative

$$\begin{CD} \mathbb{R}_{\kappa_1} @>A>> \mathbb{R}_{\kappa_2} \\ @Vf_{\kappa_1}VV @VVf_{\kappa_2}V \\ \mathbb{R} @>\tilde{A}>> \mathbb{R} \end{CD} \qquad (56)$$

Its non-Newtonian derivative

$$\frac{DA(x)}{Dx} = \lim_{\delta \to 0} \left(A(x \stackrel{\kappa_1}{\oplus} \delta_{\kappa_1}) \stackrel{\kappa_2}{\ominus} A(x) \right) \stackrel{\kappa_2}{\oslash} \delta_{\kappa_2}, \qquad (57)$$

if compared with (55), suggests $\kappa_2 = 0$. Setting $\kappa_1 = \kappa$, $\kappa_2 = 0$, we find

$$\frac{DA(x)}{Dx} = \lim_{\delta \to 0} \frac{A(x \stackrel{\kappa}{\oplus} \delta_\kappa) - A(x)}{\delta} = \lim_{\delta \to 0} \frac{A[x \stackrel{\kappa}{\oplus} f_\kappa^{-1}(\delta)] - A(x)}{\delta} = \lim_{\delta \to 0} \frac{A(x \stackrel{\kappa}{\oplus} \delta) - A(x)}{\delta}, \qquad (58)$$

as $f_\kappa^{-1}(\delta) \approx \delta$ for $\delta \approx 0$. Denoting $x \stackrel{\kappa}{\oplus} \delta = x + \delta'$ we find $\delta = (x + \delta') \stackrel{\kappa}{\ominus} x$, and

$$\frac{DA(x)}{Dx} = \lim_{\delta' \to 0} \frac{A(x + \delta') - A(x)}{(x + \delta') \stackrel{\kappa}{\ominus} x}, \qquad (59)$$

in agreement with the Kaniadakis formula. However, as a by-product of the calculation we have proved that κ-calculus is applicable only to functions mapping $\mathbb{R}_\kappa$ into $\mathbb{R}$. Kaniadakis exponential function satisfies

$$\frac{D\mathrm{Exp}(x)}{Dx} = \mathrm{Exp}(x), \quad \mathrm{Exp}(0) = 1, \qquad (60)$$

with $0 = 0_\kappa$, $1 = 1_0$. Accordingly,

$$\mathrm{Exp}(x) = f_\mathbb{Y}^{-1}\left(e^{f_\mathbb{X}(x)}\right) = e^{f_\kappa(x)} = e^{\frac{1}{\kappa} \operatorname{arcsinh} \kappa x}, \qquad (61)$$

which is indeed the Kaniadakis result. Recalling that $f_\mathbb{Y}(x) = x$, we find the explicit form of the logarithm, $\mathrm{Ln} : \mathbb{R} \to \mathbb{R}_\kappa$,

$$\mathrm{Ln}(y) = f_\mathbb{X}^{-1}\left(\ln f_\mathbb{Y}(y)\right) = \frac{1}{\kappa} \sinh(\kappa \ln y), \qquad (62)$$

which again agrees with the Kaniadakis definition.

Yet, the readers must be hereby warned that it is *not* allowed to apply the Kaniadakis definition of derivative to $\mathrm{Ln}\, x$. The correct non-Newtonian form is

$$\frac{D\mathrm{Ln}(y)}{Dy} = \lim_{\delta \to 0} \left(\mathrm{Ln}(y+\delta) \stackrel{\kappa}{\ominus} \mathrm{Ln}(y)\right) \stackrel{\kappa}{\oslash} \delta_\kappa = f_\mathbb{X}^{-1}(1/f_\mathbb{Y}(y)) = \frac{1}{\kappa} \sinh(\kappa/y), \qquad (63)$$

because Ln maps $\mathbb{R}$ into $\mathbb{R}_\kappa$. Kaniadakis is aware of the subtlety and thus introduces also another derivative, meant for differentiation of inverse functions,

$$\frac{d_\kappa A(y)}{dy} = \lim_{u \to y} \frac{A(y) \stackrel{\kappa}{\ominus} A(u)}{y - u} = \lim_{\delta \to 0} \frac{A(y+\delta) \stackrel{\kappa}{\ominus} A(y)}{\delta}, \qquad (64)$$

a definition which, from the non-Newtonian standpoint, must be nevertheless regarded as incorrect ('/' should be replaced by $\stackrel{\kappa}{\oslash}$ typical of the codomain $\mathbb{R}_\kappa$). As a result,

$$\frac{d_\kappa \mathrm{Ln}(y)}{dy} = \frac{1}{y} \neq \frac{D\mathrm{Ln}(y)}{Dy} = \frac{1}{\kappa}\sinh\frac{\kappa}{y}. \qquad (65)$$

This is probably why (64), as opposed to (55), has not found too many applications.

Let us finally check what would have happened if instead of (61) one considered the exponential function mapping $\mathbb{R}_\kappa$ into itself, $f_\mathbb{Y} = f_\mathbb{X} = f_\kappa$,

$$\mathrm{Exp}(x) = f_\mathbb{Y}^{-1}\left(e^{f_\mathbb{X}(x)}\right) = f_\kappa^{-1}\left(e^{f_\kappa(x)}\right) = \frac{1}{\kappa}\sinh\left(\kappa\, e^{\frac{1}{\kappa}\operatorname{arcsinh}\kappa x}\right). \tag{66}$$

As in thermodynamic applications one typically encounters Exp of a negative argument, one expects that physical differences between $\mathrm{Exp}: \mathbb{R}_\kappa \to \mathbb{R}_\kappa$ and $\mathrm{Exp}: \mathbb{R}_\kappa \to \mathbb{R}$ should not be essential. Moreover, indeed, Figure 1 shows that both exponents lead to identical asymptotic tails.

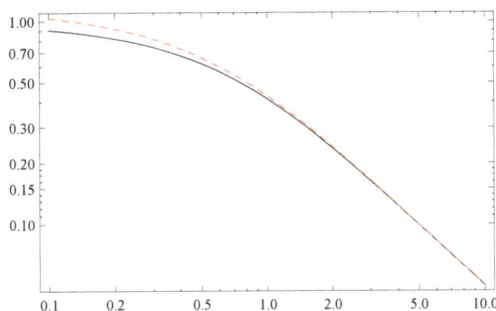

Figure 1. Log-log plots of $\mathrm{Exp}(-x)$ for $\kappa_1 = 1$, $\kappa_2 = 0$ (black), and $\kappa_1 = \kappa_2 = 1$ (red). The tails are identical.

5. A Cosmological Aspect of the Kaniadakis Arithmetic

Kaniadakis explored possible relativistic implications of his formalism. In particular, he noted that fluxes of cosmic rays depend on energy in a way that seems to indicate $\kappa > 0$. It is therefore intriguing that essentially the same arithmetic was recently shown [14] to have links with the problem of accelerated expansion of the Universe, one of the greatest puzzles of contemporary physics.

Cosmological expansion is well described by the Friedman equation,

$$\frac{da(t)}{dt} = \sqrt{\Omega_\Lambda a(t)^2 + \frac{\Omega_M}{a(t)}}, \quad a(t) > 0, \tag{67}$$

for a dimensionless scale factor $a(t)$ evolving in a dimensionless time t (in units of the Hubble time $t_H \approx 13.58 \times 10^9$ yr). The observable parameters are $\Omega_M = 0.3$, $\Omega_\Lambda = 0.7$ [45,46]. $\Omega_\Lambda \neq 0$ is typically interpreted as an indication of dark energy. Equation (67) is solved by

$$a(t) = \left(\sqrt{\frac{\Omega_M}{\Omega_\Lambda}}\sinh\frac{3\sqrt{\Omega_\Lambda}t}{2}\right)^{2/3}, \quad t > 0. \tag{68}$$

Now assume that

$$\begin{array}{ccc} \mathbb{X} & \xrightarrow{a} & \mathbb{R} \\ f_\mathbb{X} \downarrow & & \downarrow f_\mathbb{R} = \mathrm{id}_\mathbb{R} \\ \mathbb{R} & \xrightarrow{\tilde{a}} & \mathbb{R} \end{array}, \tag{69}$$

whereas the Friedman equation involves no Ω_Λ,

$$\frac{Da(t)}{Dt} = \sqrt{\frac{\Omega}{a(t)}}, \quad a(t) > 0, \tag{70}$$

for some Ω. Its solution by non-Newtonian techniques reads

$$a(t) = \left(\frac{3}{2}\sqrt{\Omega} f_{\mathbb{X}}(t)\right)^{2/3}, \tag{71}$$

so, comparing (71) with (68), we find

$$f_{\mathbb{X}}(t) = \frac{2}{3\sqrt{0.7}}\sqrt{\frac{\Omega_M}{\Omega}}\sinh\frac{3\sqrt{0.7}}{2}t = \sqrt{\frac{\Omega_M}{\Omega}} f_\kappa^{-1}(t), \quad \text{for } \kappa = 1.255. \tag{72}$$

Accelerated expansion of the Universe looks like a combined effect of non-Euclidean geometry and non-Diophantine arithmetic. The resulting dynamics is non-Newtonian in both meanings of this term.

The presence of the inverse bijection f_κ^{-1} and $\kappa > 1$ raises a number of interesting questions. It is related to the fundamental duality between Diophantine and non-Diophantine arithmetics. Namely, any equation of the form, say

$$x_1 \oplus x_2 = f^{-1}(f(x_1) + f(x_2)), \tag{73}$$

can be inverted by $f(x) = y$ into

$$y_1 + y_2 = f(f^{-1}(y_1) \oplus f^{-1}(y_2)), \tag{74}$$

suggesting that it is $\oplus$ and not $+$ which is the Diophantine arithmetic operation. Having two isomorphic arithmetics we, in general, do not have any criterion telling us which of the two is "normal", and which is "generalized".

6. Kolmogorov–Nagumo Averages and Non-Diophantine/Non-Newtonian Probability

Another non-Diophantine/non-Newtonian aspect that can be identified in the context of information theory and thermodynamics is implicitly present in the works of Kolmogorow, Nagumo, and Rényi. Let us recall that a Kolmogorov–Nagumo average is defined as [47–54]

$$\langle a \rangle_f = f^{-1}\left(\sum_k p_k f(a_k)\right). \tag{75}$$

Rewriting (75) as

$$\langle a \rangle_f = f^{-1}\left(\sum_k f(p'_k) f(a_k)\right) = \bigoplus_k p'_k \odot a_k, \tag{76}$$

where $p'_k = f^{-1}(p_k)$, one interprets the average as the one typical of a non-Diophantine-arithmetic-valued probability. Apparently, neither Kolmogorov nor Nagumo nor Rényi had interpreted their results from this arithmetic point of view [7].

The lack of arithmetic perspective is especially visible in the works of Rényi [49] who, while deriving his α-entropies, began with a general Kolmogorov–Nagumo average. Trying to derive a meaningful class of fs he demanded that

$$\langle a + c \rangle_f = \langle a \rangle_f + c \tag{77}$$

be valid for any constant random variable c, and this led him to the exponential family $f_\alpha(x) = 2^{(1-\alpha)x}$ (up to a general affine transformation $f \mapsto Af + B$, which does not affect Kolmogorov–Nagumo averages). In physical applications, it is more convenient to work with natural logarithms, so let us replace f_α by $f_q(x) = e^{(1-q)x}$, $f_q^{-1}(x) = \frac{1}{1-q}\ln x$, $q \in \mathbb{R}$. With this particular choice of f one finds

$$\langle a \rangle_{f_q} = \frac{1}{1-q}\ln\left(\sum_k p_k e^{(1-q)a_k}\right). \tag{78}$$

As is well known, the standard linear average is the limiting case $\lim_{q \to 1}\langle a \rangle_{f_q} = \sum_k p_k a_k$, that includes the entropy of Shannon, $S = \sum_k p_k \ln(1/p_k) = S_1$, as the limit $q \to 1$ of the Rényi entropy

$$S_q = \frac{1}{1-q}\ln\left(\sum_k p_k e^{(1-q)\ln(1/p_k)}\right) = \frac{1}{1-q}\ln\sum_k p_k^q. \tag{79}$$

Still, notice that $\langle a \oplus b \rangle_f = \langle a \rangle_f \oplus \langle b \rangle_f$ for any f, so had Rényi been thinking in arithmetic categories, he would not have arrived at his f_α. Yet, f_α is an interesting special case. For example,

$$p'_k = f_q^{-1}(p_k) = \frac{1}{q-1}\ln(1/p_k). \tag{80}$$

The random variable $a_k = \log_b(1/p_k)$ is, according to Shannon [49,55], the amount of information obtained by observing an event whose probability is p_k. The choice of b defines units of information. Therefore, Rényi's non-Diophantine probability p'_k is the amount of information encoded in p_k.

7. Escort Probabilities and Quantum Mechanical Hidden Variables

Non-Diophantine arithmetics have several properties that make them analogous to sets of values of incompatible random variables in quantum mechanics. Generalized arithmetics and non-Newtonian calculi have nontrivial consequences for the problem of hidden variables and completeness of quantum mechanics.

Example 7. *Pauli matrices σ_1 and σ_2 represent random variables whose values are $s_1 = \pm 1$ and $s_2 = \pm 1$, respectively. However, it is not allowed to assume that $\sigma_1 + \sigma_2$ represents a random variable whose possible values are $s_1 + s_2 = 0, \pm 2$, even though an average of $\sigma_1 + \sigma_2$ ia a sum of independent averages of σ_1 and σ_2. In non-Diophantine arithmetic one encounters a similar problem. In general it makes no sense to perform additions of the form $x_\mathbb{X} + y_\mathbb{Y}$ even if $x_\mathbb{X} \in \mathbb{R}$ and $y_\mathbb{Y} \in \mathbb{R}$. One should not be surprised if non-Diophantine probabilities turn out to be analogous to quantum probabilities, at least in some respects.*

Normalization of probability implies

$$1_\mathbb{X} = f^{-1}(1) = f^{-1}\left(\sum_k p_k\right) = f^{-1}\left(\sum_k f(p'_k)\right) = \bigoplus_k p'_k. \tag{81}$$

In principle, $1_\mathbb{X} \neq 1$. An interesting and highly nontrivial case occurs if both p_k and $p'_k = f^{-1}(p_k)$ are probabilities in the ordinary sense, i.e., in addition to (81) one finds $1_\mathbb{X} = 1$, $0 \leq p'_k \leq 1$, and $\sum_k p'_k = 1$. What can be then said about f? We can formalize the question as follows.

Problem 1. *Find a characterization of those functions $g : [0,1] \to [0,1]$ that satisfy*

$$\sum_k g(p_k) = 1, \quad \text{for any choice of probabilities } p_k. \tag{82}$$

In analogy to the generalized thermostatistics literature we can term $p'_k = g(p_k)$ the escort probabilities [56–58]. Notice that we are *not* in interested in the trivial solution, often employed in the context of Tsallis and Rényi entropies, where p_k is replaced by p_k^q and then *renormalized*,

$$P_k = \frac{p_k^q}{\sum_j p_j^q} = g_k(p_1, \ldots, p_n, \ldots) \tag{83}$$

as $g_k(p_1, \ldots, p_n, \ldots) \neq g(p_k)$ for a single function g of one variable. As we will shortly see, the solution of (82) turns out to have straightforward implications for the quantum mechanical problem of hidden variables, and relations between classical and quantum probabilities.

The most nontrivial result is found for binary probabilities, $p_1 + p_2 = 1$.

Lemma 1. $g(p_1) + g(p_2) = 1$ *for all* $p_1 + p_2 = 1$ *if and only if*

$$g(p) = \frac{1}{2} + h\left(p - \frac{1}{2}\right) \tag{84}$$

where $h(-x) = -h(x)$.

Proof. See Appendix B. □

The lemma has profound consequences for foundations of quantum mechanics, as it allows to circumvent Bell's theorem by non-Newtonian hidden variables. For more details the readers are referred to [13,15], but here just a few examples.

Example 8. *The trivial case* $g(p) = p$ *implies* $h(x) = x$, *where* $0 \leq p \leq 1$ *and* $-1/2 \leq x \leq 1/2$.

Example 9. *Consider* $g(p) = \sin^2 \frac{\pi}{2} p$. *Then,*

$$h(x) = g\left(x + \frac{1}{2}\right) - \frac{1}{2} = \frac{1}{2} \sin \pi x. \tag{85}$$

Let us cross-check,

$$g(p) + g(1-p) = \sin^2 \frac{\pi}{2} p + \sin^2 \frac{\pi}{2}(1-p) = \sin^2 \frac{\pi}{2} p + \cos^2 \frac{\pi}{2} p = 1. \tag{86}$$

Now let $p = (\pi - \theta)/\pi$ be the probability of finding a point belonging to the overlap of two half-circles rotated by θ. Then,

$$g(p) = \sin^2 \frac{\pi}{2} \frac{\pi - \theta}{\pi} = \cos^2 \frac{\theta}{2} \tag{87}$$

is the quantum-mechanical law describing the conditional probability for two successive measurements of spin-1/2 in two Stern–Gerlach devices placed one after another, with relative angle θ. Escort probability has become a quantum probability.

Example 10. *Let us continue the analysis of Example* 9. *Function* $g : [0,1] \to [0,1]$, $g(p) = \sin^2 \frac{\pi}{2} p$, *is one-to-one. It can be continued to the bijection* $g : \mathbb{R} \to \mathbb{R}$ *by the periodic repetition,*

$$g(x) = n + \sin^2 \frac{\pi}{2}(x - n), \quad n \leq x \leq n+1, \quad n \in \mathbb{Z}. \tag{88}$$

Now let $f = g^{-1}$. (88) leads to a non-Diophantine arithmetic and non-Newtonian calculus. Let $\theta = \alpha - \beta$, $0 \leq \theta \leq \pi$, be an angle between two vectors representing directions of Stern-Gerlach devices. Quantum conditional probability (87) can be represented in a non-Newtonian hidden-variable form,

$$\cos^2 \frac{\alpha - \beta}{2} = \sin^2 \frac{\pi}{2} \frac{\pi - (\alpha - \beta)}{\pi} = f^{-1}\left(\frac{1}{\pi} \int_\alpha^{\pi+\beta} dr\right) = f^{-1}\left(\int_{f(\alpha')}^{f(\pi' \oplus \beta')} \tilde{\rho}(r) dr\right)$$
$$= \int_{\alpha'}^{\pi' \oplus \beta'} \rho(\lambda) D\lambda, \tag{89}$$

where $x' = f^{-1}(x)$. Here, ρ is a conditional probability density of non-Newtonian hidden-variables (the half-circle is a result of conditioning by the first measurement).

Non-Newtonian calculus shifts the discussion on relations between classical and quantum probability, or classical and quantum information, into unexplored areas.

Example 11. *In typical Bell-type experiments one deals with four probabilities, corresponding to four combinations $(\pm, \pm)$, $(\pm, \mp)$ of pairs of binary results. The corresponding non-Newtonian model is obtained by rescaling $g(p_k) \mapsto p g(p_k/p)$, with $p = 1/2$. The rescaled bijection satisfies $g(p_1) + g(p_2) = p$ for any $p_1 + p_2 = p$. Explicitly,*

$$g(p_{++}) + g(p_{+-}) + g(p_{-+}) + g(p_{--}) = 1 = p_{++} + p_{+-} + p_{-+} + p_{--}. \tag{90}$$

The resulting hidden-variable model is local, but standard Bell's inequality cannot be proved [15]. Why? Mainly because the non-Newtonian integral is not a linear map with respect to the ordinary Diophantine addition and multiplication (unless f is linear), whereas the latter is always assumed in proofs of Bell-type inequalities.

A generalization to arbitrary probabilities, $p_1 + \cdots + p_n = 1$, leads to an affine deformation of arithmetic, an analogue of Benioff number scaling [21–25]. Affine transformations do not affect Kolmogorov–Nagumo averages.

Lemma 2. *Consider probabilities $p_1, \ldots, p_n$, $n \geq 3$. $g(p_k)$ are probabilities for any choice of p_k if and only if $g(p_k) = \frac{1-a+2ap_k}{n+(2-n)a}$, $-1 \leq a \leq 1$.*

Proof. See Appendix C. □

The bijection g implied by Lemma 2 depends on n. In infinitely dimensional systems, that is when n can be arbitrary, the only option is $a = 1$ and thus $g(p) = p$ is the only acceptable solution. However, in spin systems there exits an alternative interpretation of this property: The dimension n grows with spin in such a way that $g_n(p) \to p$ with $n \to \infty$ is a correspondence principle meaning that very large spins are practically classical. The transition non-Diophantine $\to$ Diophantine, non-Newtonian $\to$ Newtonian becomes an analogue of non-classical $\to$ classical.

Example 12. *Limitations imposed by Lemma 2 can be nevertheless circumvented in various ways. For example, let $g(1) = 1$ for a solution g from Lemma 1, so that $1_\mathbb{X} = 1$. Obviously,*

$$1 = 1_\mathbb{X} \odot \cdots \odot 1_\mathbb{X} = 1 \odot \cdots \odot 1 = 1 \cdot \ldots \cdot 1. \tag{91}$$

Replacing each of the 1s by an appropriate sum of binary conditional probabilities

$$1 = g(p_{k_1 \ldots k_n 1}) + g(p_{k_1 \ldots k_n 2}) = g(p_{k_1 \ldots k_n 1}) \oplus g(p_{k_1 \ldots k_n 2}) \tag{92}$$

we can generate various conditional classical or quantum probabilities typical of a generalized Bernoulli-type process, representing several classical or quantum filters placed one after another.

8. Non-Newtonian Maximum Entropy Principle

Let us finally discuss the implications of our non-Newtonian form (32) of entropy for maximum entropy principles. Assume probabilities belong to $\mathbb{X}$. Define the Massieu function [43] by

$$\Phi = S \ominus_\mathbb{Z} \alpha_\mathbb{Z} \odot_\mathbb{Z} N \ominus_\mathbb{Z} \beta_\mathbb{Z} \odot_\mathbb{Z} H, \tag{93}$$

$$N = \bigoplus_k {}^\mathbb{X}_\mathbb{Z} p_k = f_\mathbb{Z}^{-1}\left(\sum_k f_\mathbb{X}(p_k)\right), \tag{94}$$

$$H = \bigoplus_k z p_k \odot_\mathbb{Z}^{\mathbb{X}\mathbb{E}} E_k = f_\mathbb{Z}^{-1}\left(\sum_k f_\mathbb{X}(p_k) f_\mathbb{E}(E_k)\right), \tag{95}$$

where $E_k \in \mathbb{E}$, and $\alpha_\mathbb{Z} = f_\mathbb{Z}^{-1}(\alpha)$, $\beta_\mathbb{Z} = f_\mathbb{Z}^{-1}(\beta)$ are Lagrange multipliers. Explicitly,

$$\Phi = f_\mathbb{Z}^{-1}\left[\sum_k f_\mathbb{X}(p_k)\ln(1/f_\mathbb{X}(p_k)) - \alpha \sum_k f_\mathbb{X}(p_k) - \beta \sum_k f_\mathbb{X}(p_k) f_\mathbb{E}(E_k)\right]. \tag{96}$$

Vanishing of the derivative of Φ,

$$\frac{D\Phi}{Dp_l} = 0_\mathbb{Z}, \tag{97}$$

is equivalent to the standard formula for probabilities $f_\mathbb{X}(p_k)$ (see the second form of non-Newtonian derivative in (18)),

$$\frac{d}{df_\mathbb{X}(p_l)}\left(\sum_k f_\mathbb{X}(p_k)\ln(1/f_\mathbb{X}(p_k)) - \alpha \sum_k f_\mathbb{X}(p_k) - \beta \sum_k f_\mathbb{X}(p_k) f_\mathbb{E}(E_k)\right) = 0. \tag{98}$$

Accordingly, the solution reads

$$p_k = f_\mathbb{X}^{-1}\left(e^{-\beta f_\mathbb{E}(E_k)}/\tilde{Z}(\beta)\right) = \mathrm{Exp}(\ominus_\mathbb{E}\beta_\mathbb{E} \odot_\mathbb{E} E_k) \oslash_\mathbb{X} Z_\mathbb{X}(\beta), \tag{99}$$

$$Z_\mathbb{X}(\beta) = f_\mathbb{X}^{-1}(\tilde{Z}(\beta)) = f_\mathbb{X}^{-1}\left(\tilde{Z}(f_\mathbb{E}(\beta_\mathbb{E}))\right) = Z(\beta_\mathbb{E}), \tag{100}$$

and involves the exponential function $\mathrm{Exp} : \mathbb{E} \to \mathbb{X}$ we have encountered before. The normalization,

$$1_\mathbb{X} = \bigoplus_k {}_\mathbb{X} p_k = f_\mathbb{X}^{-1}\left(\sum_k e^{-\beta f_\mathbb{E}(E_k)}/\tilde{Z}(\beta)\right) = f_\mathbb{X}^{-1}(1), \tag{101}$$

implies the usual relation $\tilde{Z}(\beta) = \sum_k e^{-\beta f_\mathbb{E}(E_k)}$.

Equivalently, directly at the level of $\mathbb{X}$,

$$Z_\mathbb{X}(\beta) = f_\mathbb{X}^{-1}\left(\sum_k e^{-\beta f_\mathbb{E}(E_k)}\right) = f_\mathbb{X}^{-1}\left(\sum_k f_\mathbb{X} \circ f_\mathbb{X}^{-1}\left(e^{f_\mathbb{E}(\ominus_\mathbb{E}\beta_\mathbb{E}\odot_\mathbb{E} E_k)}\right)\right)$$

$$= f_\mathbb{X}^{-1}\left(\sum_k f_\mathbb{X}\left(\mathrm{Exp}(\ominus_\mathbb{E}\beta_\mathbb{E} \odot_\mathbb{E} E_k)\right)\right) = \bigoplus_k {}_\mathbb{X}\mathrm{Exp}(\ominus_\mathbb{E}\beta_\mathbb{E} \odot_\mathbb{E} E_k) = Z(\beta_\mathbb{E}). \tag{102}$$

All the standard tricks one finds in thermodynamics textbooks will work here. For example,

$$H = f_\mathbb{Z}^{-1}\left(\sum_k f_\mathbb{X}(p_k)f_\mathbb{E}(E_k)\right) = f_\mathbb{Z}^{-1}\left(\sum_k e^{-\beta f_\mathbb{E}(E_k)}f_\mathbb{E}(E_k)/\tilde{Z}(\beta)\right) = f_\mathbb{Z}^{-1}\left(-\frac{d\ln\tilde{Z}(\beta)}{d\beta}\right)$$
$$= \ominus_\mathbb{Z} f_\mathbb{Z}^{-1}\left(\frac{d\ln\tilde{Z}(\beta)}{d\beta}\right) = \ominus_\mathbb{Z} f_\mathbb{Z}^{-1}\left(\frac{d\ln\tilde{Z}(\beta)}{df_\mathbb{E}(\beta_\mathbb{E})}\right) = \ominus_\mathbb{Z} f_\mathbb{Z}^{-1}\left(\frac{d\tilde{A}(f_\mathbb{E}(\beta_\mathbb{E}))}{df_\mathbb{E}(\beta_\mathbb{E})}\right), \quad (103)$$

for some function

$$\begin{array}{ccc} \mathbb{E} & \xrightarrow{A} & \mathbb{Z} \\ f_\mathbb{E}\downarrow & & \downarrow f_\mathbb{Z} \\ \mathbb{R} & \xrightarrow{\tilde{A}} & \mathbb{R} \end{array} \quad (104)$$

we yet have to determine. Clearly,

$$\ln\tilde{Z}(\beta) = \tilde{A}(f_\mathbb{E}(\beta_\mathbb{E})) = \tilde{A}(\beta), \quad (105)$$
$$A(x) = f_\mathbb{Z}^{-1}\left(\tilde{A}(f_\mathbb{E}(x))\right) = f_\mathbb{Z}^{-1}\left(\ln\left(\tilde{Z}(f_\mathbb{E}(x))\right)\right) = f_\mathbb{Z}^{-1}\left(\ln\left(f_\mathbb{X}\circ f_\mathbb{X}^{-1}[\tilde{Z}(f_\mathbb{E}(x))]\right)\right)$$
$$= f_\mathbb{Z}^{-1}\left(\ln f_\mathbb{X}(Z(x))\right) = \text{Ln } Z(x), \quad (106)$$

where $Z: \mathbb{E} \to \mathbb{X}$, $\text{Ln}: \mathbb{X} \to \mathbb{Z}$. Ultimately,

$$H = \ominus_\mathbb{Z}\frac{D(\text{Ln}\circ Z)(\beta_\mathbb{E})}{D\beta_\mathbb{E}}. \quad (107)$$

9. Final Remarks

Non-Newtonian calculus, and the non-Diophantine arithmetics behind it, are as simple as the undergraduate arithmetic and calculus we were taught at schools. Their conceptual potential is immense but they remain largely unexplored and unappreciated. Apparently, physicists in general do not feel any need of going beyond standard Diophantine arithmetic operations, in spite of the fact that the two greatest revolutions of the 20th century physics were, in their essence, arithmetic (i.e., relativistic addition of velocities and quantum mechanical addition of probabilities). It is thus intriguing that two of the most controversial issues of modern science—dark energy and Bell's theorem—reveal new aspects when reformulated in generalized arithmetic terms.

One should not be surprised that those who study generalizations of Boltzmann–Gibbs statistics are naturally more inclined to accept non-aprioric rules of physical arithmetic. Anyway, the very concept of non-extensivity, the core of many studies on generalized entropies, is implicitly linked with generalized forms of addition, multiplication, and differentiation [54,59–61].

Funding: This research received no external funding.

Acknowledgments: I am indebted to Jan Naudts for comments.

Conflicts of Interest: The author declares no conflict of interest.

Appendix A. Proof of (39)

Let us treat this as an exercise in non-Newtonian calculus. Begin with the three diagrams

$$\begin{array}{ccc} \mathbb{X} & \xrightarrow{A} & \mathbb{Y} \\ f_\mathbb{X}\downarrow & & \downarrow f_\mathbb{Y} \\ \mathbb{R} & \xrightarrow{\tilde{A}} & \mathbb{R} \end{array}, \quad \begin{array}{ccc} \mathbb{X} & \xrightarrow{B} & \mathbb{Y} \\ f_\mathbb{X}\downarrow & & \downarrow f_\mathbb{Y} \\ \mathbb{R} & \xrightarrow{\tilde{B}} & \mathbb{R} \end{array}, \quad \begin{array}{ccc} \mathbb{X} & \xrightarrow{A\oplus_\mathbb{Y} B} & \mathbb{Y} \\ f_\mathbb{X}\downarrow & & \downarrow f_\mathbb{Y} \\ \mathbb{R} & \xrightarrow{\tilde{A}+\tilde{B}} & \mathbb{R} \end{array}. \quad (A1)$$

Indeed,

$$A \oplus_\mathbb{Y} B(x) = A(x) \oplus_\mathbb{Y} B(x) = f_\mathbb{Y}^{-1}(f_\mathbb{Y} \circ A(x) + f_\mathbb{Y} \circ B(x)) = f_\mathbb{Y}^{-1}(\tilde{A} \circ f_\mathbb{X}(x) + \tilde{B} \circ f_\mathbb{X}(x))$$
$$= f_\mathbb{Y}^{-1}((\tilde{A} + \tilde{B}) \circ f_\mathbb{X}(x)) \quad (A2)$$

By the definition (17) of the limit,

$$\lim_{x' \to x} A(x') \oplus_\mathbb{Y} B(x') = \lim_{x' \to x} (A \oplus_\mathbb{Y} B)(x') = f_\mathbb{Y}^{-1}\left(\lim_{r \to f_\mathbb{X}(x)} (A \widetilde{\oplus_\mathbb{Y}} B)(r)\right)$$
$$= f_\mathbb{Y}^{-1}\left(f_\mathbb{Y} \circ f_\mathbb{Y}^{-1}\left(\lim_{r \to f_\mathbb{X}(x)} \tilde{A}(r)\right) + f_\mathbb{Y} \circ f_\mathbb{Y}^{-1}\left(\lim_{r \to f_\mathbb{X}(x)} \tilde{B}(r)\right)\right)$$
$$= f_\mathbb{Y}^{-1}\left(\lim_{r \to f_\mathbb{X}(x)} \tilde{A}(r)\right) \oplus_\mathbb{Y} f_\mathbb{Y}^{-1}\left(\lim_{r \to f_\mathbb{X}(x)} \tilde{B}(r)\right)$$
$$= \left(\lim_{x' \to x} A(x')\right) \oplus_\mathbb{Y} \left(\lim_{x' \to x} B(x')\right). \quad (A3)$$

Appendix B. Proof of Lemma 1

Ref. (84) may be regarded as a definition of h. If $h(-x) = -h(x)$ then

$$g(1-p) + g(p) = \frac{1}{2} + h\left(1 - p - \frac{1}{2}\right) + \frac{1}{2} + h\left(p - \frac{1}{2}\right) \quad (A4)$$
$$= 1 + h\left(\frac{1}{2} - p\right) + h\left(p - \frac{1}{2}\right) \quad (A5)$$
$$= 1 - h\left(p - \frac{1}{2}\right) + h\left(p - \frac{1}{2}\right) = 1 \quad (A6)$$

Now let $g(1-p) + g(p) = 1$. Then

$$1 = g(1-p) + g(p) \quad (A7)$$
$$= \frac{1}{2} + h\left(1 - p - \frac{1}{2}\right) + \frac{1}{2} + h\left(p - \frac{1}{2}\right) \quad (A8)$$
$$= 1 + h\left(\frac{1}{2} - p\right) + h\left(p - \frac{1}{2}\right). \quad (A9)$$

Denoting $x = p - 1/2$ we find $h(-x) = -h(x)$.

Appendix C. Proof of Lemma 2

$g(p_1) + \cdots + g(p_n) = 1$ must hold for any choice of probabilities. Setting $p_1 = p$, $p_2 = 1 - p$, we find

$$g(p) + g(1-p) + (n-2)g(0) = 1, \quad (A10)$$

If $g(0) = 0$ then, by Lemma 1, $g(p) = 1/2 + h(p - 1/2)$, with antisymmetric h. Returning to arbitrary p_k, we get

$$1 = \frac{n}{2} + \sum_{k=1}^{n-1} h\left(p_k - \frac{1}{2}\right) + h\left(1 - \sum_{k=1}^{n-1} p_k - \frac{1}{2}\right). \quad (A11)$$

By antisymmetry of h,

$$1 - \frac{n}{2} - \sum_{k=2}^{n-1} h\left(p_k - \frac{1}{2}\right) = h\left(p_1 - \frac{1}{2}\right) - h\left(p_1 - \frac{1}{2} + \sum_{k=2}^{n-1} p_k\right), \quad (A12)$$

which implies that the right-hand side of (A12) is independent of p_1 for any choice of $p_2, \ldots, p_{n-1}$. In other words, the difference $h(x) - h(x+p)$ is independent of x for any $0 \leq p \leq 1/2 - x$, so $h(x) = ax$. $g(0) = 0$ implies $h(1/2) = 1/2$, $a = 1$, and $g(p) = p$ for any p.

Now let $g(0) > 0$. Normalization

$$g(1) + (n-1)g(0) = 1 \tag{A13}$$

combined with (A10), imply

$$g(p) + g(1-p) = g(0) + g(1) > 0. \tag{A14}$$

Accordingly, $G(p) = g(p)/(g(0) + g(1))$ satisfies $G(p) + G(1-p) = 1$, so that

$$G(p) = \frac{1}{2} + H\left(p - \frac{1}{2}\right), \tag{A15}$$

where $H(-x) = -H(x)$. Returning to

$$g(p) = (g(0) + g(1))\left[\frac{1}{2} + H\left(p - \frac{1}{2}\right)\right], \tag{A16}$$

we find

$$\frac{1}{g(0) + g(1)} = \frac{n}{2} + \sum_{k=1}^{n-1} H\left(p_k - \frac{1}{2}\right) + H\left(1 - \sum_{k=1}^{n-1} p_k - \frac{1}{2}\right). \tag{A17}$$

and $H(x) = ax$ by the same argument as before. Now,

$$g(p) = (g(0) + g(1))\frac{1 - a + 2ap}{2} \tag{A18}$$

Summing over all the probabilities,

$$1 = \sum_{k=1}^{n} g(p_k) = (g(0) + g(1))\frac{n - an + 2a}{2}, \tag{A19}$$

we get

$$g(p) = \frac{1 - a + 2ap}{n + (2-n)a}, \tag{A20}$$

$$g(0) = \frac{1 - a}{n + (2-n)a}, \tag{A21}$$

$$g(1) = \frac{1 + a}{n + (2-n)a}. \tag{A22}$$

For $a = 1$ we reconstruct the case $g(0) = 0$, $g(p) = p$. $g(0) > 0$ and $g(1) \geq 0$ imply either

$$1 - a > 0, \quad 1 + a \geq 0, \quad n + (2-n)a > 0, \tag{A23}$$

or

$$1 - a < 0, \quad 1 + a \leq 0, \quad n + (2-n)a < 0, \tag{A24}$$

but (A24) is inconsistent. The first two inequalities of (A23) imply $-1 \leq a < 1$, but then $n + (2-n)a > 0$ is fulfilled automatically for $n \geq 3$. Non-negativity of $g(p)$ requires $0 \leq 1 - a + 2ap$ for all $0 \leq p \leq 1$. For positive a the affine function $p \mapsto 1 - a + 2ap$ is minimal at $p = 0$, implying $0 < a \leq 1$. For negative a the map $p \mapsto 1 - a + 2ap$ is minimal at $p = 1$, so $-1 \leq a < 0$. Finally, $-1 \leq a \leq 1$ covers all the cases. The case $a = 0$ implies $g(p_k) = 1/n$, which is possible, but uninteresting for non-Newtonian applications as such a g is not one-to-one.

References

1. Grossman, M.; Katz, R. *Non-Newtonian Calculus*; Lee Press: Pigeon Cove, MA, USA, 1972.
2. Grossman, M. *The First Nonlinear System of Differential and Integral Calculus*; Mathco: Rockport, ME, USA, 1979.
3. Grossman, M. *Bigeometric Calculus: A System with Scale-Free Derivative*; Archimedes Foundation: Rockport, ME, USA, 1983.
4. Pap, E. g-calculus. *Zb. Rad. Prirod. Fak. Ser. Mat.* **1993**, *23*, 145–156.
5. Pap, E. Generalized real analysis and its applications. *Int. J. Approx. Reason.* **2008**, *47*, 368–386. [CrossRef]
6. Grabisch, M.; Marichal, J.-L.; Mesiar, R.; Pap, E. *Aggregation Functions*; Cambridge University Press: Cambridge, UK, 2009.
7. Czachor, M. Relativity of arithmetic as a fundamental symmetry of physics. *Quantum Stud. Math. Found.* **2016**, *3*, 123–133. [CrossRef]
8. Aerts, D.; Czachor, M.; Kuna, M. Crystallization of space: Space-time fractals from fractal arithmetic. *Chaos Solit. Frac.* **2016**, *83*, 201–211. [CrossRef]
9. Aerts, D.; Czachor, M.; Kuna, M. Fourier transforms on Cantor sets: A study in non-Diophantine arithmetic and calculus. *Chaos Solit. Frac.* **2016**, *91*, 461–468. [CrossRef]
10. Czachor, M. If gravity is geometry, is dark energy just arithmetic? *Int. J. Theor. Phys.* **2017**, *56*, 1364–1381. [CrossRef]
11. Aerts, D.; Czachor, M.; Kuna, M. Simple fractal calculus from fractal arithmetic. *Rep. Math. Phys.* **2018**, *81*, 357–370. [CrossRef]
12. Czachor, M. Waves along fractal coastlines: From fractal arithmetic to wave equations. *Acta Phys. Polon. B* **2019**, *50*, 813–831. [CrossRef]
13. Czachor, M. A loophole of all 'loophole-free' Bell-type theorems. *Found. Sci.* **2020**. [CrossRef]
14. Czachor, M. Non-Newtonian mathematics instead of non-Newtonian physics: Dark matter and dark energy from a mismatch of arithmetics. *Found. Sci.* **2020**. [CrossRef]
15. Czachor, M. Arithmetic loophole in Bell's theorem: An overlooked threat for entangled-state quantum cryptography. *arXiv* **2020**, arXiv:2004.04097.
16. Rashevsky, P.K. On the dogma of the natural numbers. *Uspekhi Mat. Nauk.* **1973**, *28*, 243–246. (In Russian) [CrossRef]
17. Burgin, M.S. Nonclassical models of the natural numbers. *Uspekhi Mat. Nauk* **1977**, *32*, 209–210. (In Russian)
18. Burgin, M. *Non-Diophantine Arithmetics, or Is It Possible that 2 + 2 Is Not Equal to 4?* Ukrainian Academy of Information Sciences: Kiev, Ukraine, 1997. (In Russian)
19. Burgin, M. Introduction to projective arithmetics. *arXiv* **2010**, arXiv:1010.3287.
20. Burgin, M.; Meissner, G. $1 + 1 = 3$: Synergy arithmetics in economics. *Appl. Math.* **2017**, *8*, 133–134. [CrossRef]
21. Benioff, P. Towards a coherent theory of physics and mathematics. *Found. Phys.* **2002**, *32*, 989–1029. [CrossRef]
22. Benioff, P. Towards a coherent theory of physics and mathematics: The theory-experiment connection. *Found. Phys.* **2005**, *35*, 1825–1856. [CrossRef]
23. Benioff, P. Fiber bundle description of number scaling in gauge theory and geometry. *Quantum Stud. Math. Found.* **2015**, *2*, 289–313. [CrossRef]
24. Benioff, P. Space and time dependent scaling of numbers in mathematical structures: Effects on physical and geometric quantities. *Quantum Inf. Process.* **2016**, *15*, 1081–1102. [CrossRef]
25. Benioff, P. Effects of a scalar scaling field on quantum mechanics. *Quantum Inf. Process.* **2016**, *15*, 3005–3034. [CrossRef]
26. Kaniadakis, G. Nonlinear kinetics underlying generalized statistics. *Physica A* **2001**, *296*, 405. [CrossRef]
27. Kaniadakis, G. H-theorem and generalized entropies within the framework of nonlinear kinetics. *Phys. Lett. A* **2001**, *288*, 283. [CrossRef]
28. Kaniadakis, G. Statistical mechanics in the context of special relativity. *Phys. Rev. E* **2002**, *66*, 056125. [CrossRef]

29. Kaniadakis, G.; Scarfone, A.M. A new one parameter deformation of the exponential function. *Physica A* **2002**, *305*, 69. [CrossRef]
30. Kaniadakis, G.; Lissia, M.; Scarfone, A.M. Two-parameter deformations of logarithm, exponential, and entropy: A consistent framework for generalized statistical mechanics. *Phys. Rev. E* **2005**, *71*, 046128. [CrossRef]
31. Kaniadakis, G. Statistical mechanics in the context of special relativity (II). *Phys. Rev. E* **2005**, *72*, 036108. [CrossRef]
32. Biró, T.S.; Kaniadakis, G. Two generalizations of the Boltzmann equation. *Eur. Phys. J. B* **2006**, *50*, 3. [CrossRef]
33. Kaniadakis, G. Towards a relativistic statistical theory. *Physica A* **2006**, *365*, 17–23. [CrossRef]
34. Kaniadakis, G. Theoretical foundations and mathematical formalism of the power-law tailed statistical distributions. *Entropy* **2013**, *15*, 3983–4010. [CrossRef]
35. Burgin, M.; Czachor, M. *Non-Diophantine Arithmetics in Mathematics, Physics, and Psychology*; World Scientific: Singapore, 2020.
36. Epstein, M.; Śniatycki, J. Fractal mechanics. *Physica D* **2006**, *220*, 54–68. [CrossRef]
37. Epstein, M.; Śniatycki, J. The Koch curve as a smooth manifold. *Chaos Solitons Fractals* **2008**, *38*, 334–338. [CrossRef]
38. Wiener, N. *The Fourier Integral and Certain of Its Applications*; Cambridge University Press: Cambridge, UK, 1933.
39. Tsallis, C. What are the numbers that experiments provide? *Quim. Nova* **1994**, *17*, 468.
40. Naudts, J. Deformed exponentials and logarithms in generalized thermostatistics. *Physica A* **2002**, *316*, 323–334. [CrossRef]
41. Ay, N.; Jost, J.; Le, H.V.; Schwachhöfer, L. *Information Geometry*; Springer: Berlin/Heidelberg, Germany, 2017.
42. Naudts, J. Generalised exponential families and associated entropy functions. *Entropy* **2008**, *10*, 131–149. [CrossRef]
43. Naudts, J. *Generalized Thermostatistics*; Springer: London, UK, 2011.
44. Zhang, J.; Naudts, J. Information geometry under monotone embedding. Part I: Divergence functios. In *Geometric Science of Information*; Lecture Notes in Computer Science; Nielsen, F., Barbaresco, F., Eds.; Springer: Berlin/Heidelberg, Germany, 2017; Volume 10589, pp. 205–214.
45. Reiss, A.G. Observational evidence from supernovae for an accelerating Universe and a cosmological constant. *Astron. J.* **1998**, *116*, 1009–1039. [CrossRef]
46. Perlmutter, S. Measurements of Ω and Λ from 42 high-redshift supernovae. *Appl. J.* **1999**, *517*, 565–586. [CrossRef]
47. Kolmogorov, A.N. Sur la notion de la moyenne. *Atti Acad. Naz. Lincei Rend.* **1930**, *12*, 388–391; Reprinted in *Selected Works of A. N. Kolmogorov*; Mathematics and Mechanics; Tikhomirov, V.M., Ed.; Kluwer: Dordrecht, The Netherlands, 1991; Volume 1.
48. Nagumo, M. Über eine Klasse der Mittelwerte. *Jpn. J. Math.* **1930**, *7*, 71–79; Reprinted in *Mitio Nagumo Collected Papers*; Yamaguti, M., Nirenberg, L., Mizohata, S., Sibuya, Y., Eds.; Springer: Tokyo, Japan, 1993. [CrossRef]
49. Rényi, A. Some fundamental questions of information theory. *MTA III Oszt. Közl.* **1960**, *10*, 251–282. Reprinted in *Selected Papers of Alfréd Rényi*; Turán, P., Ed.; Akadémiai Kiadó: Budapest, Hungary, 1976.
50. Jizba, P.; Arimitsu, T. Observability of Rényi's entropy. *Phys. Rev. E* **2004**, *69*, 026128. [CrossRef]
51. Jizba, P.; Arimitsu, T. The world according to Rényi: Thermodynamics of fractal systems. *AIP Conf. Proc.* **2001**, *597*, 341. [CrossRef]
52. Czachor, M.; Naudts, J. Thermostatistics based on Kolmogorov-Nagumo averages: Unifying framework for extensive and nonextensive generalizations. *Phys. Lett. A* **2002**, *298*, 369–374. [CrossRef]
53. Massi, M. On the extended Kolmogorov–Nagumo information-entropy theory, the $q \to 1/q$ duality and its possible implications for a non-extensive two-dimensional Ising model. *Physica A* **2007**, *377*, 67–78. [CrossRef]
54. Jizba, P.; Korbel, J. When Shannon and Khinchin meet Shore and Johnson: Equivalence of information theory and statistical inference axiomatics. *Phys. Rev. E* **2020**, *101*, 042126. [CrossRef] [PubMed]
55. Shannon, C.E. A mathematical theory of communication. *Bell Syst. Tech. J.* **1948**, *27*, 379–423, 623–653. [CrossRef]

56. Tsallis, C.; Mendes, R.; Plastino, A. The role of constraints within generalized nonextensive statistics. *Physica A* **1998**, *261*, 543–554. [CrossRef]
57. Naudts, J. Estimators, escort probabilities, and phi-exponential families in statistical physics. *J. Ineq. Pure Appl. Math.* **2004**, *5*, 102.
58. Naudts, J. Escort operators and generalized quantum information measures. *Open Syst. Inf. Dyn.* **2005**, *12*, 13–22. [CrossRef]
59. Touchette, H. When is a quantity additive, and when is it extensive? *Physica A* **2002**, *305*, 84–88. [CrossRef]
60. Nivanen, L.; Le Mehaute, A.; Wang, Q.A. Generalized algebra within a nonextensive statistics. *Rep. Math. Phys.* **2003**, *52*, 437–444. [CrossRef]
61. Borges, E.P. A possible deformed algebra and calculus inspired in nonextensive thermostatistics. *Physica A* **2004**, *340*, 95–101. [CrossRef]

Publisher's Note: MDPI stays neutral with regard to jurisdictional claims in published maps and institutional affiliations.

© 2020 by the author. Licensee MDPI, Basel, Switzerland. This article is an open access article distributed under the terms and conditions of the Creative Commons Attribution (CC BY) license (http://creativecommons.org/licenses/by/4.0/).

Article

Dynamic and Renormalization-Group Extensions of the Landau Theory of Critical Phenomena

Miroslav Grmela [1],*, Václav Klika [2] and Michal Pavelka [3]

1. École Polytechnique de Montréal, C.P.6079 suc. Centre-Ville, Montréal, QC H3C 3A7, Canada
2. Department of Mathematics—FNSPE, Czech Technical University in Prague, Trojanova 13, 12000 Prague, Czech Republic; vaclav.klika@fjfi.cvut.cz
3. Mathematical Institute, Faculty of Mathematics, Charles University, Sokolovská 83, 18675 Prague, Czech Republic; pavelka@karlin.mff.cuni.cz
* Correspondence: miroslav.grmela@polymtl.ca

Received: 14 August 2020; Accepted: 31 August 2020; Published: 2 September 2020

Abstract: We place the Landau theory of critical phenomena into the larger context of multiscale thermodynamics. The thermodynamic potentials, with which the Landau theory begins, arise as Lyapunov like functions in the investigation of the relations among different levels of description. By seeing the renormalization-group approach to critical phenomena as inseparability of levels in the critical point, we can adopt the renormalization-group viewpoint into the Landau theory and by doing it bring its predictions closer to results of experimental observations.

Keywords: entropy; critical phenomena; renormalization; multiscale thermodynamics; GENERIC

1. Introduction

Our point of departure is the Landau theory of critical phenomena [1]. We formulate it in two steps highlighting its relation and role in multiscale thermodynamics. The first step is the *2-level formulation of equilibrium thermodynamics*. The first level is the *equilibrium level* with the number of moles N and the energy E (both per unit volume), serving as state variables. The second level is an *upper level* with the variable x serving as the state variable. For example, x could be temperature and chemical potential together with an order parameter (as it is in the original formulation of the Landau theory) or it could also be one particle distribution function or other state variables used in mesoscopic theories of macroscopic systems. Equilibrium thermodynamics enters the 2-level formulation in the upper reducing thermodynamic relation (consisting of three real valued functions $S^\uparrow(x), N^\uparrow(x), E^\uparrow(x)$) and in the maximum entropy principle (MaxEnt principle) transforming it (details are in Section 2) to the equilibrium reduced thermodynamic relation $S = S(E,N), E = E, N = N$. Very often the mesoscopic state variables x are fields that enter the specification of $S^\uparrow(x), N^\uparrow(x), E^\uparrow(x)$ as mean fields. The 2-level formulation is therefore often called a *mean-field approach to thermodynamics*.

The second step in the Landau theory is the specification of the upper reducing thermodynamic relation (i.e., specification of the three potentials $S^\uparrow(x), N^\uparrow(x), E^\uparrow(x)$). The particularity of the physics of the macroscopic system under investigation is expressed in these three potentials. Specifications of $S^\uparrow(x), N^\uparrow(x), E^\uparrow(x)$ thus requires commitment to a specific system. For example, in the Gibbs equilibrium statistical mechanics the upper level is the microscopic level with x being the n-particle distribution function ($n \sim 10^{23}$), $S^\uparrow(x)$ is the universal Gibbs entropy, the energy $E^\uparrow(x)$ is the average microscopic energy (expressing the particularity of the system under investigation), and $N^\uparrow(x)$ is the universal potential expressing the number of moles; see, for example [2].

Landau has noted that in the critical region all three potentials $S^\uparrow(x), N^\uparrow(x), E^\uparrow(x)$ tend to be universal. The criticality overrides the particularity of the physical nature of the systems under investigation. Having the universal potentials $S^\uparrow(x), N^\uparrow(x), E^\uparrow(x)$, MaxEnt principle transforms them into a universal equilibrium critical behavior.

The universality of the upper reducing thermodynamic relation in the critical region is based on two observations, one is physical, the other is mathematical. The observation of the physical nature addresses the appearance of criticality on the equilibrium and on the upper levels. While the criticality is more visible, both in experimental observations and in its mathematical representation, on the equilibrium level, it manifests itself also on upper levels. For example, observations of fluctuations in the results of lower-level experimental observations are observations reaching beyond the lower level towards upper levels. Fluctuations appear to be indeed more pronounced in the critical region. The observation of the mathematical nature addresses the universality of potential functions in the critical region investigated in the catastrophe theory [3].

The above two-step formulation of the Landau theory extends naturally to dynamic critical phenomena. In the first step we replace the 2-level formulation of thermodynamics with the 2-level formulation of rate-thermodynamics. The equilibrium level is replaced by a mesoscopic level that still takes into account fewer details than the upper level but it is a level on which the time evolution takes place. We recall that no time evolution takes place on the equilibrium level. In fact, we choose to use the vector field governing the lower time evolution as the state variable on the lower level. The rate-thermodynamics on the upper level is expressed in the upper reducing rate-thermodynamic relation $(\Sigma^\uparrow(x), Y^\uparrow(x))$, where $\Sigma^\uparrow(x)$ is the rate entropy, $Y(x)$ is the lower level vector field expressed in terms of x. The MaxRent principle (Maximum Rate Entropy principle—see details in Section 3), replacing the MaxEnt principle in equilibrium thermodynamics, transforms then the upper reducing rate-thermodynamic relation to the lower reduced rate-thermodynamic relation $(\Sigma(Y), Y)$, where $\Sigma(Y)$ is the lower rate entropy and Y is the lower vector field.

Regarding the comparison of predictions of Landau's theory with results of experimental observations, the agreement is only qualitative. The problem is in the multiscale nature of critical phenomena. The closer is the critical point, the closer is the upper level to the lower level and in the critical point itself all levels become inseparable. This observation is then taken as a basis for the renormalization-group theory of critical phenomena [4]. In Section 4 we extend the Landau theory to 3-level formulation, which then provides a setting for the renormalization-group theory of critical phenomena seen as an extension of the Landau theory.

2. Landau's Theory of Static Critical Phenomena

Level of description is an autonomous collection of results of certain type of experimental observations (different for different levels) together with a model that allows to organize them, to reproduce them, and to make predictions. The model, based on the insight inspired by the experimental data and by investigating relations to nearby levels involving less or more details, offers also an understanding of the physics involved. For instance, the *equilibrium level* with the energy E, number of moles N, and volume V serving as state variables and the *microscopic level* with position and momenta of $\sim 10^{23}$ particles composing the macroscopic system serving as state variable are examples of two different autonomous levels of description. The latter is more microscopic (it takes into account more details) than the former. We call the latter level an *upper level* and the former the *lower level*. In this section the lower level will always be the equilibrium level. The state variable on the upper level is denoted by the symbol x. For example, in the Landau theory x is usually the equilibrium temperature and an appropriately chosen order parameter. On the level of kinetic theory $x = f(r, v)$ or on the level of hydrodynamics $x = (\rho(r), e(r), u(r))$, where $f(r, v)$ is one particle distribution function, r is the position vector and v momentum of one particle; $\rho(r)$ is the field of mass density, $e(r)$ the field of internal energy, $u(r)$ the field of momentum. The state space on the equilibrium level is denoted by the symbol $\mathbb{M}$, i.e., $(E, N, V) \in \mathbb{M}$, the state space on the upper level is denoted by the symbol $M^\uparrow$, i.e., $x \in M^\uparrow$.

Every level is autonomous and differs from other levels in the amount of details (that are taken into account in both experimental observations and the mathematical formulation) and in the range of applicability. However self-contained are the levels, their mathematical formulation is closely related to their relationship to other levels. From investigating relations to upper levels (i.e., to levels involving more details) comes a structure that we shall call *reduced structure* and from investigating relations to lower levels (i.e., to levels involving fewer details) comes the *reducing structure*. Both structures equip the state space with a geometry and a vector field. The geometry is a mathematical formulation of thermodynamics. Every level has thus reduced and reducing thermodynamics and reduced and reducing vector fields. Below, we limit ourselves only to the reducing thermodynamics on the upper level and the reduced thermodynamics on the lower level (which is in this section the equilibrium level).

We emphasize that the term "reduction" has in this paper the same meaning as "emergence". Some details on the upper level are lost in the reduction from an upper level to a lower level but at the same time an emerging overall pattern is gained. The process of reduction, as well as processes conductive to an emergence of overall features (pattern-recognition processes), involve both a loss and a gain. The terms "upper" and "lower" levels that we use in this paper have a different meaning than they have in, say, social sciences. The lower level is inferior from the upper level in the amount of details but superior in the ability to see overall patterns.

2.1. 2-Level Equilibrium Thermodynamics

Among many questions about the origin of both reduced and reducing structures and about their relations, we shall discuss only the one that is directly relevant to the Landau theory. We shall investigate the passage from the reducing thermodynamics on the upper level to the reduced thermodynamics on the equilibrium level. A few comments about the placement of the investigation of this passage in the larger context of multiscale thermodynamics are discussed at the end of this section.

The upper reducing thermodynamic relation

$$S^{\uparrow}(x), E^{\uparrow}(x), N^{\uparrow}(x) \tag{1}$$

is one of several possible forms of the mathematical formulation of the upper reducing thermodynamics. The quantities introduced in (1) are the upper energy per unit volume $E^{\uparrow}: M^{\uparrow} \to \mathbb{R}$, the upper number of moles per unit volume $N^{\uparrow}: M^{\uparrow} \to \mathbb{R}$, and the upper reducing entropy per unit volume $S^{\uparrow}: M^{\uparrow} \to \mathbb{R}$ and are assumed to be sufficiently regular.

The reduced equilibrium thermodynamic relation

$$S(E, N), E, N \tag{2}$$

is obtained from (1) by the following reducing Legendre transformation. We make this transformation in four steps.

Step 1: We introduce upper reducing thermodynamic potential (Note that x can be any state variable, e.g., distribution function, hydrodynamic fields, electromagnetic fields, see [5])

$$\Phi^{\uparrow}(x; E^*, N^*) = -S^{\uparrow}(x) + E^* E^{\uparrow}(x) + N^* N^{\uparrow}(x) \tag{3}$$

where (E^*, N^*) are conjugate equilibrium state variables. In the standard equilibrium thermodynamic notation $E^* = \frac{1}{T}$, where T is the equilibrium temperature, and $N^* = -\frac{\mu}{T}$, where μ is the equilibrium chemical potential.

Step 2: We solve the equation

$$\Phi^{\uparrow}_x = 0 \tag{4}$$

We use hereafter the notation: $\Phi^{\uparrow}_x = \frac{\partial \Phi^{\uparrow}}{\partial x}$, where $\frac{\partial}{\partial x}$ is an appropriate functional derivative if x is a function (i.e., an element of an infinite dimensional space). Let $\hat{x}(E^*, N^*)$ be the solution to (4).

Step 3: We introduce
$$S^*(E^*, N^*) = \Phi^\uparrow(\hat{x}(E^*, N^*); E^*, N^*) \tag{5}$$
called a reduced conjugate entropy.

Step 4: Finally, we pass from $S^*(E^*, N^*)$ to $S(E, N)$ by the Legendre transformation (i.e., we introduce the equilibrium thermodynamic potential $\Phi^*(E^*, N^*; E, N) = -S^*(E^*, N^*) + EE^* + NN^*$, solve $\Phi^*_E = 0$, $\Phi^*_N = 0$, and arrive at $S(E, N) = \Phi^*(\hat{E}^*(E, N), \hat{N}^*(E, N); E, N)$, where $(\hat{E}^*(E, N), \hat{N}^*(E, N))$ are solutions to $\Phi^*_E = 0$, $\Phi^*_N = 0$.

The reducing Legendre transformation (1) → (2) can also be seen as maximization of the upper reducing entropy $S^\uparrow(x)$ subjected to constraints $E^\uparrow(x), N^\uparrow(x)$ [5,6]. The conjugate equilibrium state variables E^*, N^* play the role of Lagrange multipliers. This viewpoint then gives the passage (1) → (2) the name Maximum Entropy principle (MaxEnt principle)

Summing up, the MaxEnt passage from the upper level to the equilibrium level, via the upper reducing thermodynamic relation (1), is the following sequence of two mappings

$$(S^\uparrow(x), E^\uparrow(x), N^\uparrow(x)) \mapsto (S^*(E^*, N^*), E^*, N^*). \mapsto (S(E, N), E, N) \tag{6}$$

The second mapping in (6) is the standard Legendre transformation. The first mapping is the upper reducing Legendre transformation expressing the MaxEnt principle.

In the particular case when $x = (E, N)$ and $N^\uparrow(E, N) = N$, $E^\uparrow(E, N) = E$, there is no reduction in (6) and both arrows in (6) are (one-to-one) standard Legendre transformations:

$$(S(E, N), E, N) \mapsto (S^*(E^*, N^*), E^*, N^*) \mapsto (S(E, N), E, N) \tag{7}$$

To conclude this section we turn to questions like where the potentials $S^\uparrow(x), E^\uparrow(x), N^\uparrow(x)$ come from and why the upper entropy $S^\uparrow(x)$ is maximized subjected to constraints $E^\uparrow(x)$ and $N^\uparrow(x)$. These questions are answered simply by the existence of the autonomous upper level and the existence of the autonomous equilibrium level. The autonomous existence implies that there exists a way to prepare macroscopic systems for their investigations on the equilibrium level and that the time evolution describing the preparation process can be formulated on the upper level as a reducing time evolution. It is in this reducing time evolution where the potentials $S^\uparrow(x), E^\uparrow(x), N^\uparrow(x)$ make their first appearance. The entropy $S^\uparrow(x)$ generates the reducing time evolution. Maximization of $S^\uparrow(x)$ reflects the property of its solutions expressing mathematically the approach to the equilibrium level. We note that in the context of the classical formulation of equilibrium thermodynamics the existence of the preparation process for the equilibrium level (the existence of equilibrium states) is a subject of the zero axiom (see [2]). The 2-level formulation of equilibrium thermodynamics can be thus seen as a way to bring the zero axiom to an active participation in the formulation of equilibrium thermodynamics.

2.2. 2-Level Equilibrium Thermodynamics in the Critical Region

Equilibrium-level experimental observations of phase transitions are mathematically expressed in various types of singularities of the equilibrium reduced thermodynamic relation $S(E, N)$. We call the subspace of the equilibrium level state space at which the Hessian (matrix of second derivatives) of $S(E, N)$ has a nontrivial nullspace as a critical submanifold. Its neighborhood is called a critical region.

Experimental observations made on upper levels (i.e., levels involving more details than the equilibrium level) show that the critical behavior seen on the equilibrium level is also seen on the upper levels. For instance, it is well established that fluctuations in the results of the equilibrium-level measurements (that is an example of measurements that involve more details than the equilibrium-level measurements) become very pronounced in the critical region. The mathematical manifestation of the criticality in the upper reducing thermodynamic relation $S^\uparrow(x), E^\uparrow(x), N^\uparrow(x)$ is however different from its mathematical manifestation on the equilibrium level. The potentials $S^\uparrow(x), E^\uparrow(x), N^\uparrow(x)$ remain completely smooth, but Equation (4) has two or more solutions.

2.3. Van der Waals Theory

We now illustrate the Landau theory on the van der Waals theory of a gas composed of particles interacting via long range attractive and short range repulsive forces. The macroscopic physical system investigated in this illustration is a gas composed of particles interacting via long range attractive forces and short range repulsive forces (van der Waals gas). The latter forces are treated as constraints and their influence enters the entropy rather than energy. The mathematical model of this system on the equilibrium level is the well known classical van der Waals model, on the level of kinetic theory the van Kampen model [7] (see also [8]) and its dynamical extension ([9]).

The upper level is the level of kinetic theory with the one particle distribution function

$$x = f(r, v) \tag{8}$$

serving as a single state variable, r is the position coordinate and v the momentum of one particle. In this example we specify explicitly the upper reducing thermodynamic relation $S^\uparrow(x), E^\uparrow(x), N^\uparrow(x)$ by using arguments developed mainly in the Gibbs equilibrium statistical mechanics. Having $S^\uparrow(x), E^\uparrow(x), N^\uparrow(x)$, we identify the critical point and subsequently restrict $S^\uparrow(x), E^\uparrow(x), N^\uparrow(x)$ to the critical region. The resulting potentials take the form of the Landau critical thermodynamic potentials. This illustration has already been presented in [8], we can therefore omit details.

Following van Kampen [7], the upper reducing thermodynamic relation (1) representing on the level of kinetic theory the van der Waals gas is

$$
\begin{aligned}
E^\uparrow(f) &= \int dr \int dv \left(\frac{v^2}{2} f(r, v) + \frac{1}{2} \int dr_1 \int dv_1 V_{pot}(|r - r_1|) f(r, v) f(r_1, v_1) \right) \\
N^\uparrow(f) &= \int dr \int dv f(r, v) \\
S^\uparrow(f) &= \int dr \int dv \left(-f(r, v) \ln f(r, v) - f(r, v) \frac{\partial \theta}{\partial n(r)} \right)
\end{aligned}
\tag{9}
$$

where we put the volume of the region in which the van der Waals gas is confined equal to one, the mass of one particle is also equal to one, $V_{pot}(|r - r_1|)$ is the potential energy,

$$
\begin{aligned}
n(r) &= \int dv f(r, v) \\
\theta(n(r)) &= \frac{1 - Bn(r)}{B} (\ln(1 - Bn(r)) - 1)
\end{aligned}
\tag{10}
$$

and where $B \in \mathbb{R}$ is a small (proportional to the volume of one particle) parameter. Note that $n(r) = \int dv f(r, v)$ is the local particle density. In $E^\uparrow$, the first term is the kinetic energy, the second the potential energy. In $S^\uparrow$, the first term is the Boltzmann entropy, the second term is the contribution to the entropy due to the excluded volume constraint.

By making the transformations in (6), we arrive at the classical well known van der Waals thermodynamic relation (see details in [5] (pp. 43–44) and [8]).

Now we proceed to investigate the critical region. First note that van Kampen [7] showed that the critical points corresponding to the end points of the critical curve in van der Waals gas are spatially homogeneous. This knowledge can be translated within this multiscale framework formulation into a restriction of the MaxEnt distribution function and the reduction to the level with state variable n (see the Appendix A and [8] for details). In short, we begin by looking for solutions of (4) only among $n(r)$ that are independent of r. With this restriction, we arrive at

$$\Phi^\uparrow(n; \alpha, \beta) = n \ln n + n \frac{d\theta}{dn} - \frac{1}{2} \beta V_{pot} n^2 - \left(\alpha - \frac{3}{2} \ln \frac{\beta}{2\pi} \right) n \tag{11}$$

where we use the shorthand notation $\beta = \frac{1}{T}$, $\alpha = -\frac{\mu}{T}$ and $V_{pot} = \int d\mathbf{r}_1 V_{pot}(|\mathbf{r}-\mathbf{r}_1|)$.

The critical point is

1. A MaxEnt value: an extremum of reducing thermodynamic potential that governs the evolution and hence this extremum corresponds to an equilibrium state;
2. A point where loss of convexity occurs: the extremum (equilibrium point) is ambiguous, multiple or continuum of extrema are plausible;
3. Critical point of the whole system (including the inverse temperature $\beta = \frac{1}{T}$), i.e., the lowest temperature (T) and chemical potential (α) for which a critical point given by the above two points still exists: an extremal point, i.e., the third derivative with respect to n has to vanish as follows from Taylor expansion and the fact that the two above requirements can be translated into vanishing first two derivatives.

In short, the critical point is a point (n_c, α_c, β_c) where the potential has a stationary point, just loses convexity and has a minimum. By taking the Taylor expansion about the critical point,

$$\Phi^{\uparrow}(n, \alpha_c, \beta_c) = \Phi^{\uparrow}(n_c, \alpha_c, \beta_c) + \frac{\partial \Phi^{\uparrow}}{\partial n} \delta n + \frac{1}{2!} \frac{\partial^2 \Phi^{\uparrow}}{\partial n^2} (\delta n)^2 \qquad (12)$$
$$+ \frac{1}{3!} \frac{\partial^3 \Phi^{\uparrow}}{\partial n^3} (\delta n)^3 + \frac{1}{4!} \frac{\partial^4 \Phi^{\uparrow}}{\partial n^4} (\delta n)^4 + \mathcal{O}(\delta n)^5,$$

the requirements on the critical point lead to equations

$$\Phi^{\uparrow}_n = 0;\ \Phi^{\uparrow}_{nn} = 0;\ \Phi^{\uparrow}_{nnn} = 0;\ \Phi^{\uparrow}_{nnnn} > 0. \qquad (13)$$

Hence the critical point of van der Waals gas is given by

$$n_c = \frac{1}{3B};\ \beta_c = \frac{27B}{4V_{pot}}$$
$$\alpha_c = \frac{1}{2}\ln(3B) + \frac{3}{2}\ln\frac{B}{V_{pot}} + \frac{3}{4} + 4\ln\frac{3}{2} - \frac{3}{2}\ln(2\pi). \qquad (14)$$

See Appendix A for more details.

We now make an explicit choice of the order parameter

$$\zeta = n - n_c,$$

where n_c is the critical value of n, and define

$$\Phi^{\uparrow}_{crit}(n; \omega_1, \omega_2, \omega_3) = \Phi(n; \alpha, \beta) - \Phi(n_c; \alpha, \beta). \qquad (15)$$

With the explicit knowledge of the reducing fundamental thermodynamic potential (11) we know that in a neighborhood of the critical point we arrive at

$$\Phi^{\uparrow}_{crit}(\zeta; \omega_1, \omega_2, \omega_3) = \omega_1 \zeta + \frac{1}{2}\omega_2 \zeta^2 + \frac{1}{24}\omega_3 \zeta^4, \qquad (16)$$

where

$$\omega_1 = a_1(\alpha - \alpha_c) + a_2(\beta - \beta_c) \qquad (17)$$
$$\omega_2 = a_3(\beta - \beta_c)$$

as the coefficient of α in $\Phi^{\uparrow}$ is linear in n while the coefficient of β is quadratic in n. Note that the cubic term is missing in the expansion as its coefficient is independent of α, β and hence $\Phi^{\uparrow}_{nnn}(n_c; \alpha, \beta) = \Phi^{\uparrow}_{nnn}(n_c; \alpha_c, \beta_c) = 0$ due to (13). This form of the thermodynamic relation in the critical region $\Phi^{\uparrow}_{crit}$ can be put in a more general framework due to Landau [10] (Chapter XIV). Expressions for a_1, a_2, a_3, ω_3 involve the parameters B, V_{pot}, serving as the material parameters in the van der Waals theory.

From (16) we obtain

$$S^{\downarrow*}_{crit}(\alpha, \beta) = \Phi^{\uparrow}_{crit}(\hat{\xi}(\alpha, \beta); \alpha, \beta) \tag{18}$$

where $\hat{\xi}(\alpha, \beta)$ is a solution to

$$\left(\Phi^{\uparrow}_{crit}\right)_{\xi} = 0. \tag{19}$$

The lower reduced entropy (18) provides complete information about the behavior (the behavior seen in equilibrium-thermodynamic observations) of the van der Waals gas in a small neighborhood of the critical point. In particular, we obtain the critical exponents arising in the dependence of $S^{\downarrow*}_{crit}(\alpha, \beta)$ on α and β.

A simple way to see that $S^{\downarrow*}_{crit}(\alpha, \beta)$ is a generalized homogeneous function and thus to identify the critical exponents is to use (16) with $\omega_{1,2,3}$ now being the variables instead of α, β. We replace ξ in (16) with $\lambda^{-1/4}\xi$. We obtain

$$\Phi^{\uparrow}_{crit}(\lambda^{-1/4}\xi; \omega_1, \omega_2, \omega_3) = \lambda^{-1}\left((\omega_1 \lambda^{3/4})\xi + (\omega_2 \lambda^{1/2})\xi^2 + \omega_3 \xi\right)$$
$$= \lambda^{-1}\Phi^{\uparrow}_{crit}(\xi; (\omega_1 \lambda^{3/4}), (\omega_2 \lambda^{1/2}), \omega_3)$$

and consequently, noting ω_3 is unaffected by the rescaling,

$$S^{\downarrow*}_{crit}(\omega_1, \omega_2) = \lambda^{-1} S^{\downarrow*}_{crit}(\lambda^{3/4}\omega_1, \lambda^{1/2}\omega_2). \tag{20}$$

Finally, one could invert (ω_1, ω_2) from (17) to get a (generalized) scaling for $S^{\downarrow*}_{crit}(\alpha, \beta)$.

Still another view of this relation can serve as an introduction to the renormalization-group theory of critical phenomena discussed below in Section 4. We start again with (16) and write it in the form $\Phi^{\uparrow}_{crit}(\xi; \omega)$, where $\omega = (\omega_1, \omega_2, \omega_3)$ is given in (17). Our aim is to introduce a *renormalization time evolution* (i.e., renormalization group of transformations generated by a vector field) of ω and of $\Phi^{\uparrow}_{crit}$ such that:

$$\Phi^{\uparrow}_{crit}(\xi; \omega(\tau), \tau) = \Phi^{\uparrow}_{crit}(\xi; \omega) \; \forall \; \tau > 0$$

with the initial conditions

$$\begin{aligned} \omega(0) &= \omega \\ \Phi^{\uparrow}_{crit}(\xi; \omega, 0) &= \Phi^{\uparrow}_{crit}(\xi; \omega) \end{aligned} \tag{21}$$

and the constraint

$$\omega_3(\tau) = \omega_3 \; \forall \; \tau > 0 \tag{22}$$

The renormalization time is denoted by the symbol τ. We emphasize that the renormalization time τ has nothing to do with the real time t. The renormalization time evolution will become the basis for a new definition of critical points discussed in Section 4. From the physical point of view, the constraint expresses the requirement that the material parameter B entering the repulsive short range forces in (10) remains unchanged in the renormalization process.

We begin with

$$(\Phi^{\uparrow}_{crit})_{\tau} = -\chi \Phi^{\uparrow}_{crit} \tag{23}$$

with $\chi > 0$ being at this point an unspecified parameter and with the initial condition given by the second line in (21). It can be easily verified that [8]

$$\dot{\omega} = \mathcal{R}(\chi, \omega) \tag{24}$$

with the initial condition given by the first line in (21) and

$$\mathcal{R}(\chi,\omega) = \begin{pmatrix} \chi - 1 & 0 & 0 \\ 0 & \chi - 2 & 0 \\ 0 & 0 & \chi - 4 \end{pmatrix} \omega^T. \tag{25}$$

We see now that with $\chi = 4$ we satisfy both (21) and the constraint (23).

The fixed point of the renormalization time evolution is the critical point and the eigenvalues of the vector field linearized about the fixed point are the critical exponents.

This statement, which has arisen as a simple observation in the particular context discussed above, is in fact a definition of the critical points and the critical exponents in the renormalization-group theory of critical phenomena (see Section 4). In the case of (25) the linearization is, of course, unnecessary since the vector field is already linear.

Finally we compare the classical analysis of the van der Waals gas with the analysis based on the Landau theory. The starting point of the classical analysis is the physical insight that led us to the upper reducing thermodynamic relation (9). By restricting it to the critical region we have arrived at the Landau expression (16). The starting point of the Landau theory is the expression (16). The quantity ξ, called in the Landau theory an order parameter, does not need to have a specific physical interpretation, nor the coefficients a_1, a_2, a_3 are specified in the Landau theory.

The extra information about the critical phenomena that the classical van der Waals theory provides (but only for the van der Waals gas) is thus: (i) the location of the critical point in the state space $M^\uparrow$, (ii) physical interpretation of the order parameter, (iii) a detailed knowledge of the critical behavior beyond a small neighborhood of the critical point. On the other hand, the advantage of the Landau theory is its universal applicability. In Section 4 we make a comment about the renormalization group theory, the objective of which is to bring the critical exponents implied by the van der Waals theory (and thus also the Landau theory) closer to those seen in experiments.

Before leaving the van der Waals theory, we mention that the static version of the theory recalled above has been upgraded to the dynamical theory in [9]. The kinetic equation of which solutions make the maximization of the entropy $S^\uparrow(f)$ subjected to constraints $E^\uparrow(f), N^\uparrow(f)$ (see (9)) is the Enskog Vlasov kinetic equation.

3. Landau's Theory of Dynamic Critical Phenomena

In the 2-level formulation of the equilibrium thermodynamics we replace the equilibrium level with a lower level that still takes into account fewer details than the upper level, but it is a mesoscopic level on which the time evolution, called a lower time evolution, takes place. We recall that no time evolution takes place on the equilibrium level that served us as the lower level in the preceding section. We again assume that both the upper and the lower levels are well established (well tested with experimental observations) autonomous levels. This then means that by investigating solutions to the upper time evolution equations we have to be able to split the upper time evolution into a reducing time evolution describing the preparation process for using the lower level and a reduced time evolution that is the lower time evolution. The investigation leading to the split is essentially a pattern recognition process in solutions to the upper governing equations.

There are two types of the reducing and the reduced time evolutions. The reducing time evolution can be either the time evolution taking place in $M^\uparrow$ and approaching an invariant (or in most cases a quasi-invariant) manifold $\mathcal{M}^\downarrow \subset M^\uparrow$ that represents in $M^\uparrow$ the state space $M^\downarrow$ used on the lower level or it can be the time evolution of vector fields $Y^\uparrow(x) \in \mathfrak{X}(M^\uparrow)$ taking the vector field generating the upper time evolution to the vector field generating the lower time evolution. The former viewpoint is discussed for example in [11–14]. In this paper we follow the second route, discussed in [15], since on this route we can directly transpose the 2-level equilibrium thermodynamics introduced in the previous section to 2-level rate-thermodynamics. We use "rate" to point out that the state space is the space of vector fields.

The upper reducing rate-thermodynamic relation

$$\Sigma^\uparrow(x), Y^\uparrow(x) \tag{26}$$

replaces the upper reducing thermodynamic relation (1). The passage from the upper reducing rate-thermodynamic relation (26) to the lower reduced rate-thermodynamic relation

$$\Sigma(Y), Y \tag{27}$$

remains the same as the passage from the upper reducing thermodynamic relation (1) to the lower reduced thermodynamic relation (2) in 2-level thermodynamics, discussed in Section 2. We introduce an upper reducing rate-thermodynamic potential

$$\Psi^\uparrow(x; Y^*) = -\Sigma^\uparrow(x) + \langle Y^*, Y^\uparrow(x) \rangle \tag{28}$$

where Y^* are conjugate lower vector fields. The sequence of mappings

$$(\Sigma^\uparrow(x), Y^\uparrow(x)) \mapsto (\Sigma^{\downarrow*}(Y^*(y)), Y^*(y)) \mapsto (\Sigma^\downarrow(Y(y)), Y(y)) \tag{29}$$

corresponds in MaxRent to the sequence of mappings (6) in MaxEnt. The lower vector field is $Y(y) = \Sigma^{\downarrow*}_{Y^*(y)}$.

How do we specify the upper reducing relations (1) or (26)? The following three routes can be taken.

(i) Both relations (1) and (26) arise from a detail investigation of the upper time evolution. Since both the upper and the lower levels are well established, the upper level has to reduce, by following a certain preparation process in which time evolution is described by the reducing time evolution, to the lower level. The reducing time evolution then introduces the upper reducing thermodynamics relation (1) or upper reducing rate-thermodynamic relation (26). In this paper we do not introduce and discuss explicitly the reducing time evolution, neither in the equilibrium thermodynamics nor in the rate-thermodynamics. It is important to recall that the upper reducing thermodynamic relation (1) representing an ideal gas on the level of kinetic theory has been originally obtained by Boltzmann from analyzing solutions (Boltzmann's H-theorem) of the Boltzmann kinetic equation describing the reducing time evolution. In the Boltzmann analysis the kinetic equation is primary, and the Boltzmann entropy arises as a result.

(ii) In the critical region the upper reducing thermodynamic potentials (3) and (28) are determined by mathematical results arising in the catastrophe theory [3].

(iii) The association between specific physical systems and the upper reducing thermodynamic relations (1) in equilibrium thermodynamics can also be investigated by physical arguments developed mainly in the Gibbs equilibrium statistical mechanics (as we did in the illustration in Section 2.3).

Before proceeding to the illustration we make a few remarks.

Our investigation in the preceding section was limited to equilibrium. We have considered only systems that are allowed to reach equilibrium states. Behavior of macroscopic systems that are prevented from reaching the equilibrium states (either by external or internal forces) cannot be described on the equilibrium level but can be described on a lower level. For instance the experimentally observed behavior of a Rayleigh–Bénard system (a thin horizontal layer of a fluid heated from below) is well described on the level of hydrodynamics (in Boussinesq equations) [16,17]. The lower reduced thermodynamic relation that we are getting on the lower level from relating it to an upper level provides thus thermodynamics also for such externally or internally forced systems [18].

The equilibrium can be reached either directly (*upper level*→ *equilibrium level*) or indirectly (*upper level*→ *lower level*→*equilibrium level*). We require that the equilibrium thermodynamic relations obtained by following both routes are identical for consistency in the multilevel framework. This requirement implies the following relation between quantities entering the equilibrium and rate-thermodynamic relations

$$\dot{S}^\uparrow(y) = [\langle Y^*, \Sigma_{Y^*}^{\downarrow *}\rangle]_{Y^*=S_y^\uparrow(y)} \tag{30}$$

By y we denote the state variables on the lower level, $\dot{S}^\uparrow(y)$ is the upper entropy generating the approach from the lower level to the equilibrium level. The relation (30) makes precise the connection between the rate entropy $\Sigma^\uparrow(x)$ on the upper level and the entropy production $\dot{S}^\uparrow(y)$ on the lower level.

3.1. Illustration: Immiscible Fluids

Dispersions of two immiscible fluids (fluid A and fluid B) have two essentially different morphologies. One in which the fluid A is dispersed in the form of droplets in the fluid B that forms a continuous phase. The second is the inverse, fluid B is dispersed and fluid A is continuous. The transition between these two morphologies is a dynamic critical point called phase inversion. Dispersions under consideration are subjected to externally imposed flows.

As it was in the case of the van der Waals gas in Section 2.3, we want to identify the critical point (in particular the critical concentrations) and to investigate the behavior in the critical region (in particular the flow behavior of the dispersion). As we saw in the case of the investigation of static critical phenomena in Section 2.3, both questions are answered if we know explicitly the upper reducing thermodynamic potential. In the case of phase inversion it would be the explicit knowledge of the upper reducing rate-thermodynamic potential (28). In general, the problem of finding thermodynamic potentials corresponding to specific physical systems is more difficult in rate-thermodynamics than in thermodynamics. For example in the specification of the van der Waals upper reducing thermodynamic potential (11) we have used the insight offered by Gibbs investigations in which the upper level is the Microscopic level and the lower level the equilibrium level. No such powerful source of insights seems to be available in rate-thermodynamics. Nevertheless, we know that the rate-thermodynamic potentials exist. This is because the upper and the lower levels exist as autonomous levels and consequently the upper level approaches the lower level. In the case of dispersions the lower level is the level of hydrodynamics and the upper level can be, for instance, the Microscopic level or it could also be the level of kinetic theory. Just the knowledge of the existence of the upper and lower levels gives us the right to use the Landau theory (which will address the behavior in the critical region) and also certain arguments that are based on partial knowledge rate-thermodynamic potentials that will address the problem of identifying the point of phase inversion.

We turn first to the latter investigation. In the absence of a complete knowledge of the rate-thermodynamic potentials, we can attempt to identify them separately on both sides of the phase inversion. At the point of phase inversion the two potentials must be equal. Their equality is then an equation determining the point of phase inversion. With the surface energy playing the role of the potentials, this analysis has been made in [19,20] and with the rate-thermodynamic potential (28) in [21]. We have seen in (30) that the upper reducing rate-thermodynamic potential is related to but not identical to the entropy production. In [21] the entropy production on both sides of the phase inversion is specified and then put (as an approximation) on the place of rate-thermodynamic potentials.

The Landau theory has been applied to the problem of phase inversion in [22]. The order parameter is an unspecified characterization of the morphology of the dispersion. It can be for instance an average (oriented) curvature of the interface separating the two fluids.

3.2. Illustration: Shear Banding

It was experimentally observed in [23] that the Taylor–Couette flow of a special shear banding fluid exhibits unusual behavior. In the experiments either the force required to rotate the outer cylinder (shear stress) or speed of the rotation (shear rate) can be controlled, the other being measured. It turns out that when varying shear stress, shear rate behaves continuously while when varying shear rate, shear stress exhibits a jump. Such behavior, also called dissipative phase transition, can be captured by a non-convex dissipation potential giving relation between shear stress and shear rate [24].

In rate thermodynamics (also called CR-thermodynamics [21]) the roles of state and conjugate variables are played by thermodynamic forces and fluxes (or vice versa), and the role of entropy is played by dissipation potential. In order to see a phase transition in the rate thermodynamics (as in the above-introduced experiment), one thus needs to be equipped with a non-convex dissipation potential. Such potential was proposed in [24] for a dissipative phase transition in complex fluids,

$$\Xi = 0.01 x^2 + \frac{1}{2} - \frac{1}{2(1+x^2)}. \tag{31}$$

Note that the potential is written in a non-dimensional form and that x represents the norm of the deviatoric stress tensor. The dissipation potential is clearly non-convex, as is apparent from Figure 1.

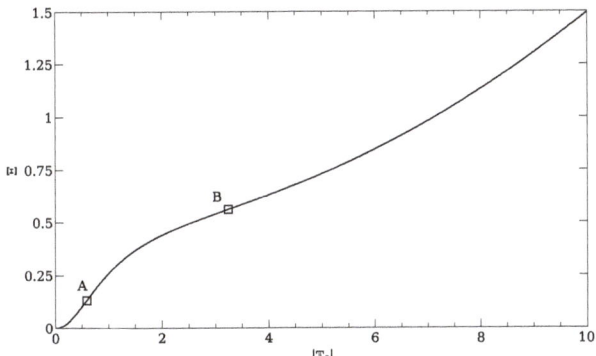

Figure 1. Dissipation potential (31). The points where convexity of lost are highlighted. This potential was found in [24] and is based on a work by Le Roux and Rajagopal [25]. The rate thermodynamic analysis of the qualitative implied by the potential was then confirmed by numerical simulations in [26].

In order to obtain the stress tensor, one has to perform the Legendre transformation

$$\frac{\partial \Phi}{\partial x} = 0 \quad \text{for} \quad \Phi(x, \gamma) = -\Xi(x) + \gamma x, \tag{32}$$

where γ represents the shear rate.

Let us now find the critical points (there are two) of potential Φ. The loss of convexity at the critical points is expressed by the equation

$$\Phi_{xx} = 0, \tag{33}$$

the solutions of which are the critical stresses $x_{c1} = 3.24294$ and $x_{c2} = 0.591376$. The critical shear rates are then obtained by solving the equations

$$\Phi_x(x_{c1}, \gamma_{c1}) = 0 \quad \text{and} \quad \Phi_x(x_{c2}, \gamma_{c2}) = 0, \tag{34}$$

giving $\gamma_{c1} = 0.0893092$ and $\gamma_{c2} = 0.336446$. The first and second derivatives at the critical point vanish. Note, however, that the third derivative does not vanish and the critical points are thus not minima of the potential Ψ and the rate thermodynamics is thus not stable in the critical points.

We can further expand the potential Ψ in power series in the critical points,

$$\Psi(\gamma_{c1}, \xi) \approx -0.272127 - 0.0035087\xi^3 + 0.0011077\xi^4 \quad \text{for} \quad \xi = x - x_{c1} \tag{35}$$

$$\Psi(\gamma_{c2}, \xi) \approx 0.0659143 + 0.231744\xi^3 - 0.210484\xi^4 \quad \text{for} \quad \xi = x - x_{c2}. \tag{36}$$

Shifting the potential around the critical point by a constant value so that their value at the critical point is zero, the general expansion around the critical point (also in the direction of the parameter γ) then reads

$$\Psi(\gamma_{c1}, \xi) \approx \omega_1 (\gamma - \gamma_{c1})\xi^2 \quad \text{for} \quad \xi = x - x_{c1} \tag{37}$$

$$\Psi(\gamma_{c2}, \xi) \approx \omega_2 (\gamma - \gamma_{c2})\xi^2 \quad \text{for} \quad \xi = x - x_{c2}. \tag{38}$$

The first derivative disappears, since we have one parameter γ that can be used to keep it zero, but the coefficients in front of the second derivatives only disappear in the critical points.

In summary, the geometric analysis of critical phenomena can be carried out also in the realm of rate thermodynamics, where the thermodynamic forces and fluxes play the role of state and conjugate variables. The universal behavior near the critical points is observed similarly as in the classical theory.

4. Renormalization-Group Theory of Critical Phenomena in the Setting of Landau's Theory

The universality of the upper reducing thermodynamic relations in the critical region is based on mathematical arguments [3]. The mathematical universality then implies the universality of physical behavior that can be observed experimentally. Is the experimentally observed critical behavior indeed universal? The answer is well known. Predictions of the Landau theory agree with results of experimental observations only qualitatively. How can we explain it?

The problem is in the autonomy of levels in the critical region. The closer the critical point, the more difficult it is to separate the levels. This general observation is often illustrated on the example of the observation of fluctuations. We recall that fluctuations seen in results of experimental observations made on level $\mathcal{L}$ are in fact observations that reach beyond the $\mathcal{L}$-level towards observations belonging to a level involving more details. This means that large fluctuations seen on the level $\mathcal{L}$ indicate that the level $\mathcal{L}$ ceases to be autonomous. Some of the details ignored on the level $\mathcal{L}$ cannot be ignored anymore in order to keep the level $\mathcal{L}$ autonomous. In the critical point itself the levels become inseparable. This feature of criticality is then taken as the basis for the renormalization-group theory of critical phenomena.

Our objective in this section is to formulate the renormalization-group theory of critical phenomena as an extension of the Landau theory. For the sake of simplicity we make below the extension only for the Landau theory of static critical phenomena. Its dynamical version will be the subject of a future paper.

The first step in the extension is a replacement of the upper level with more upper levels. In the enlarged family of upper levels we keep the original upper level and add a one parameter (the parameter is denoted by the symbol τ) family of new levels. These new levels involve more details than the original upper level. We call them UPPER levels. We construct them by taking a sharper view of the macroscopic system under investigation. We separate the particles composing it into two classes. Originally, the particles are indistinguishable (for instance all particles are white), now the particles are either white or red. The parameter $\tau \in \mathbb{R}$ labels the extra degrees of freedom arising on UPPER levels due to sharper view of particles. Both passages *upper level* $\to$ *UPPER levels* and *UPPER levels* $\to$ *upper level* are thus known. The former is made by taking glasses allowing to recognize colors, the latter is made by becoming colourblind.

Besides the straightforward passage *UPPER levels* $\to$ *upper level* made simply by colour-blindness, there is another way to make the same passage. The extra degrees of freedom that arise on *UPPER levels* due to the sharper viewpoint are MaxEnt eliminated. In other words, we pass from *UPPER levels* to *upper level* in the same way as we passed from the upper level to the equilibrium level in the preceding section. The MaxEnt reduction of *UPPER levels* to the *upper level* will be termed *MaxEnt-reductions*, see [5].

A comparison of the upper reduced thermodynamic potential with the upper MaxEnt-reduced thermodynamic potential, both restricted to the critical region, is then the essence of the renormalization-group viewpoint of critical phenomena. Let the coefficients in the critical polynomials be ω for the upper reduced thermodynamic potential and $\Omega(\tau)$ in the one parameter family of the upper MaxEnt-reduced thermodynamic potential. The difference between two UPPER levels, one corresponding to τ_1 and the other to $\tau_2 \neq \tau_1$ are manifested mathematically in $\Omega(\tau_1) \neq \Omega(\tau_2)$. The inseparability of UPPER levels in the critical point is mathematically expressed in $\Omega(\tau)$ becoming independent of τ. Let Ω_{crit} be such a fixed point. Eigenvalues of the linearized renormalization dynamics (i.e., dynamics in which τ plays the role of the renormalization time—see the end of Section 2.3) are then the renormalized critical exponents.

The main features of this viewpoint of the renormalization-group theory of critical phenomena have already appeared in [8,27]. Also the illustration of the formulation presented below has been largely developed in [8]. In the original formulation of the renormalization-group theory [4] the upper level is the Microscopic level used as the upper level in the Gibbs equilibrium statistical mechanics. The state variable x in the Gibbs theory is the n-particle distribution function ($n \sim 10^{23}$ is the number of particles composing the macroscopic system under investigation). The upper reducing thermodynamic relation consists of the Gibbs entropy, the average microscopic energy, and normalization of the distribution function. The family of UPPER levels is constructed by extending the system in all directions by a scale factor τ. The MaxEnt-reduced levels are obtained by seeing the extension from the upper level to UPPER levels as replacement of every point with a "box" and the MaxEnt-passage from UPPER levels to the upper level as a MaxEnt reduction (with the Microscopic fundamental thermodynamic relation) of all boxes back to points.

The main difference between the original formulation of the renormalization group theory is thus the choice of the upper level. In the original formulation it is the Microscopic level. The Microscopic thermodynamic relation consists of the universal Gibbs entropy and an energy (Hamiltonian) in which the individual nature of the macroscopic system under investigation is expressed. The Ginzburg–Landau form of the energy is often used [28]. In our formulation the upper level is a general mesoscopic level and the upper reducing thermodynamic relation is its universal form (Landau polynomials arising the catastrophe theory) in the critical region. The main advantage of our formulation is thus its universal applicability and adaptability to dynamic critical phenomena.

Illustration

In order to illustrate the renormalization-group theory of critical phenomena that is cast into the setting of the Landau theory we turn to the van der Waals theory recalled in Section 2.3. We keep the same equilibrium level and the same upper level. In addition we introduce an UPPER level with the state variables

$$x = (f(\mathbf{r}, \mathbf{v}), g(\mathbf{r}, \mathbf{v})) \tag{39}$$

and the upper reducing thermodynamic relation

$$\begin{aligned}
\mathbb{E}^\uparrow(f,g) &= \int d\mathbf{r}\int d\mathbf{v}\left(\frac{v^2}{2}f(\mathbf{r},\mathbf{v}) + \frac{v^2}{2}g(\mathbf{r},\mathbf{v})\right) \\
&\quad + \frac{1}{2}\int d\mathbf{r}\int d\mathbf{v}\int d\mathbf{r}_1\int d\mathbf{v}_1 V_{pot}(|\mathbf{r}-\mathbf{r}_1|) \\
&\quad \times (f(\mathbf{r},\mathbf{v})f(\mathbf{r}_1,\mathbf{v}_1) + g(\mathbf{r},\mathbf{v})g(\mathbf{r}_1,\mathbf{v}_1) + 2f(\mathbf{r},\mathbf{v})g(\mathbf{r}_1,\mathbf{v}_1)) \\
\mathbb{N}^\uparrow(f,g) &= \int d\mathbf{r}\int d\mathbf{v}(f(\mathbf{r},\mathbf{v}) + g(\mathbf{r},\mathbf{v})) \\
\mathbb{S}^\uparrow(f,g) &= \int d\mathbf{r}\int d\mathbf{v}\left(-f(\mathbf{r},\mathbf{v})\ln f(\mathbf{r},\mathbf{v}) - g(\mathbf{r},\mathbf{v})\ln g(\mathbf{r},\mathbf{v})\right. \\
&\quad \left. - f(\mathbf{r},\mathbf{v})\frac{\partial\theta}{\partial n(\mathbf{r})} - g(\mathbf{r},\mathbf{v})\frac{\partial\theta}{\partial m(\mathbf{r})}\right) \quad (40)
\end{aligned}$$

where $n(\mathbf{r}) = \int d\mathbf{v} f(\mathbf{r},\mathbf{v})$, $m(\mathbf{r}) = \int d\mathbf{v} g(\mathbf{r},\mathbf{v})$, $\theta(n,m) = \theta(n+m)$. The UPPER level represents a more detailed view of the van der Waals gas in the sense that the gas particles are no longer indistinguishable. They are divided into two groups. One group is composed of the same particles as on the upper level. We can call them now f-particles. Their states are characterized by the one particle distribution function $f(\mathbf{r},\mathbf{v})$. The second group is composed of g-particles, the state variable is the one particle distribution function $g(\mathbf{r},\mathbf{v})$. The f-particles and g-particles remain identical. In particular, interactions among the f-particles, among the g-particles, and among f-particles and g-particles are exactly the same as on the interactions of the f-particles on the upper level. The upper level differs from the UPPER level only in our ability to distinguish the f-particles from g-particles (for instance by having a different colour). The UPPER level thus indeed takes into account more details than the upper level. We are able to distinguish two colours.

We follow now the analysis that we made on the upper level in Section 2.3. We restrict ourselves to n and m that are independent of $\mathbf{r}$ and introduce the UPPER reducing thermodynamic potential

$$\begin{aligned}
\Psi^\uparrow(n,m;\beta,A^{(n)},A^{(m)}) &= n\ln n + m\ln m + (n+m)\theta'(n+m) - \frac{1}{2}V_{pot}(n+m)^2 \\
&\quad - (\ln A^{(n)})n - (\ln A^{(m)})m \quad (41)
\end{aligned}$$

where we use the symbol Ψ instead of Φ to distinguish the UPPER level from the upper level and $\ln A = \alpha - \frac{3}{2}\ln\frac{\beta}{2\pi}$.

Next, we transform the UPPER level into a one parameter family of UPPER levels. We introduce first a one parameter family of the potentials $\Psi^\uparrow$ by inserting into (41) $A^{(n)} = e^{-\tau}A$; $A^{(m)} = (1-e^{-\tau})A$, where $\tau \in \mathbb{R}; \tau > 0$ is the parameter. The UPPER reducing thermodynamic potential (41) turns into the one parameter family

$$\begin{aligned}
\Psi^\uparrow(n,m;\beta,A,\tau) &= n\ln n + m\ln m + (n+m)\theta'(n+m) - \frac{1}{2}V_{pot}(n+m)^2 \\
&\quad - \ln(e^{-\tau}A)n - \ln((1-e^{-\tau})A)m \quad (42)
\end{aligned}$$

where $\theta'(n) = \frac{d\theta}{dn}$. We note that solution to $\Psi^\uparrow_n = 0$ is $e^{-\tau}n$, solution to $\Psi^\uparrow_m = 0$ is $(1-e^{-\tau})n$, and n is a solution to $\Phi^\uparrow_n = 0$ with $\Phi^\uparrow$ given in (11). This means that if we are colour blind, then n_c of the potentials (42) and (11) are the same. Also the reduced equilibrium thermodynamic relation implied by (42) and (11) are the same.

Now we pass from the UPPER level back to the upper level, see Figure 2. We can follow two routes:

Route 1:

On the first route we simply ignore the g-particles.

$$[\Psi^\uparrow(n,m;\beta,A,\tau)]_{m=0} = \Phi^\uparrow(n,A,\tau) \qquad (43)$$

where $\Phi^\uparrow$ is given in (11)

Route 2:

On the second route we eliminate the presence of g-particles with MaxEnt. In this way we arrive at

$$[\Psi^\uparrow(n,m;\beta,A,\tau)]_{\Psi^\uparrow_m=0} \qquad (44)$$

We have transformed f-particles into f-quasi-particles, i.e., f-particles that are modified by taking into account the presence of g-particles. The passage from particles to quasi-particles is a pattern recognition process. The more are the quasi-particles different from the original particles the more pronounced is the pattern.

Following our terminology and notation (we recall that the thermodynamic potential on the upper level is denoted by the symbol Φ) we denote the potential (43) by the symbol $\Phi^\downarrow$ and call it upper reduced potential. The potential (44) is denoted $\Phi^{(\downarrow ME)}$ and called upper MaxEnt-reduced potential in order to point out its provenance.

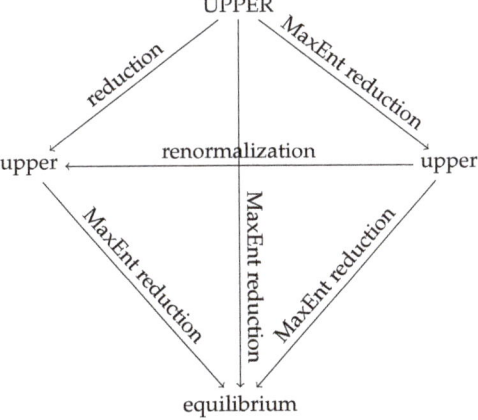

Figure 2. Diagram of the levels of description.

We expect that the two upper thermodynamic potentials (43) and (44) are different. The former is just the original upper reducing potentials (42) without the g-particle. The latter is also the upper reducing potential (42) without the g-particles but with their presence felt through MaxEnt. The passage from the UPPER potentials (42) to the upper MaxEnt-reduced potential (44) can be seen as a process of recognizing a pattern on the UPPER level. The recognized pattern is then expressed in terms of the upper-level state variables. In the absence of such patterns on the UPPER level the upper MaxEnt-reduced potential (44) will be the same as the upper reduced potential (43). This is expected to happen when the inclusion of more details on the UPPER level does not reveal anything new (new with respect to what is seen on the upper level). The experimentally observed inseparability of levels in the critical region then suggests to define the critical point as the point at which the potentials (43) and (44) are identical.

In order to be able to compare the critical part of the upper MaxEnt-reduced potential (44) with the critical part of the upper reduced potential (43), we cast them into the form of the Landau polynomials.

If we choose (23) for the renormalization time evolution of $\Psi_{crit}^{(\downarrow ME)}$, then the upper MaxEnt-reduced thermodynamic potential takes the form

$$\Phi_{crit}^{(\downarrow ME)}(n,\beta;A,B,\tau) = e^{-\chi\tau}\left(\Omega_1(\Omega,\chi,\tau)\zeta + \Omega_2(\Omega,\chi,\tau)\zeta^2 + \Omega_3(\Omega,\chi,\tau)\zeta^4\right)$$

where $\Omega = (\Omega_1, \Omega_2, \Omega_3)$ is a solution of (compare with (24))

$$\dot{\Omega} = \mathcal{R}^{(ME)}(\Omega, \chi). \tag{45}$$

It remains to show the relation of Ω, ζ and $\mathcal{R}^{(ME)}$ to (n, β, A, B) appearing in the upper potential (43) and to investigate the renormalization time evolution governed by (45).

Regarding the former task, we only indicate the route and refer to [8] for details. We recall that solution to $\Psi_n^{\uparrow} = 0 = \Psi_m^{\uparrow}$ is $n = e^{-\tau} n^{\Phi^{\uparrow}}$, $m = (1 - e^{-\tau}) n^{\Phi^{\uparrow}}$, where $n^{\Phi^{\uparrow}}$ is a solution to $\Phi_n^{\uparrow} = 0$ with $\Phi^{\uparrow}$ given in (11). This means that the reduced equilibrium thermodynamic relation implied by (42) and (11) are the same.

Next, we note that

$$\Phi^{(\downarrow ME)}(n;\beta,A) = \Phi^{\uparrow}(n;\beta,e^{-\tau}A) - \tau n e^{-\Phi_n^{\uparrow}(n;\beta,A)} + O(\tau^2) \tag{46}$$

This relation follows from

$$\Psi_m^{\uparrow} = 0 \Rightarrow \ln\left(\frac{m}{n}\right) = -\Phi_n^{\uparrow}(n;\beta,A) + \ln\left(\frac{A^{(m)}}{A}\right)$$

and from $\Psi^{\uparrow}(n,m;\beta,A^{(n)},A^{(m)}) = \Phi^{\uparrow}(n;\beta,A^{(n)}) - m + O(\tau^2)$. The remaining details can be found in [8].

Regarding the latter task, solutions to the renormalization time evolution governed by (45) have been investigated in [8]. Three fixed points have been identified and the largest eigenvalue of the linearized (45) about one of them equals 0.8. The largest eigenvalue in (25) (which is the classical critical exponent corresponding to the approach to the coexistence line in a transverse direction) is 0.75. The value of this type of critical exponent measured in experiments is indeed close to 0.8 [8,29]. The physical significance of this agreement remains to be investigated. Our main objective in this section was to illustrate the renormalization-group theory of critical phenomena in the setting of the Landau theory of critical phenomena.

5. Concluding Remarks

Thermodynamics is a theory of relations among theories of macroscopic systems formulated on different autonomous levels of description. Our main objective in this paper is to show that this multiscale viewpoint of thermodynamics unifies investigations of static and dynamic critical phenomena. We emphasize that the reduction of an upper level to a lower level (a level involving fewer details than the upper level) represents a loss of details but a gain of emerging features arising as patterns in the phase portrait of the upper level. Applicability of multi-level thermodynamics is ubiquitous, ranging from chemical engineering, rheology, electrodynamics of matter, kinetic theory to machine learning [30].

Classical equilibrium thermodynamics arises in investigations of relations between an upper level (e.g., the microscopic level or the level of kinetic theory) and the equilibrium level, on which no time evolution takes place. The upper reducing entropy, which generates the approach to the equilibrium level, becomes the equilibrium entropy when the approach is completed. When the approached lower level involves the time evolution, the result of the reduction can either be seen as a reduction in the upper state space (approach to a quasi-invariant submanifold representing the lower state space in the upper state space) or as an approach of the upper vector field to the lower vector field).

The thermodynamics that arises by following the latter viewpoint is termed rate-thermodynamics. The mathematical formulations of thermodynamics and rate-thermodynamics are essentially identical. The thermodynamic potentials are however replaced with rate-thermodynamic potentials. In the particular case of externally unforced systems, when the lower level is allowed to reach the equilibrium level, the upper rate-entropy is closely related to the production of the entropy generating the approach of the lower level to the equilibrium level.

The multiscale thermodynamics acquires two new features in the critical region. First it is the universality of the thermodynamic and rate-thermodynamic potentials and the inseparability of levels. The former is a consequence of the mathematical representation of criticality (catastrophe theory) and the latter the consequence of the physical nature of the criticality. The former feature has been noted by Lev Davidovich Landau and is a basis of his theory of critical phenomena. The latter feature is a basis of the renormalization-group theory of critical phenomena. We show that the multiscale thermodynamics provides a unified setting for the Landau theory of static critical phenomena, for its extension to the dynamic critical phenomena and for the renormalization-group theory of critical phenomena.

Author Contributions: M.G.: conceptualization and writing the original draft. V.K. and M.P.: writing the original draft. All authors have read and agreed to the published version of the manuscript.

Funding: V.K. and M.P. were supported by the Czech Grant Agency, project number 20-22092S. M.P. was supported by Charles University Research Program No. UNCE/SCI/023.

Conflicts of Interest: The authors declare no conflict of interest.

Appendix A. Details of the Calculation of the van der Waals Transition

Let us repeat the calculation of the critical points from Section 2.3 in more detail. Firstly, the entropy on the level of kinetic theory is rather

$$S^{\uparrow}(f) = -k_B \int d\mathbf{r} \int d\mathbf{p} f (\ln(h^3 f) - 1) + k_B \int d\mathbf{r} n(f) \ln(1 - bn(f)) \tag{A1}$$

in order to be in the SI units at to yield the Sackur–Tetrode relation in the case of $b = 0$, see [5].

Solution to the equation $\Phi_f^{\uparrow} = 0$ is sought in the form

$$\tilde{f} = \frac{1}{(2\pi m k_B T)^{3/2}} n(r) e^{-\frac{p^2}{2mk_B T}}, \tag{A2}$$

where m is the mass of one particle. In the case of $n = $ const this leads to an algebraic equation for n.

The equation $\Phi_{nnn} = 0$ (now taking derivative with respect to n due to the Ansatz (A2)) becomes

$$0 = \frac{3k_B b^2}{(1-nb)^2} + 2k_B \frac{nb^3}{(1-nb)^3} - \frac{k_B}{n^2}, \tag{A3}$$

solution of which is $n_c = 1/3b$.

The equation $\Phi_{nn} = 0$ reads

$$0 = \frac{2k_B b}{1-nb} + k_B \frac{nb^2}{(1-nb)^2} - 2aE^* + \frac{k_B}{n}, \tag{A4}$$

where $2a = -\int d\mathbf{r}' V(\mathbf{r} - \mathbf{r}) > 0$. Using the value of n_c, we obtain that

$$\frac{1}{k_B T_c} = E_c^* = \frac{27}{8} \frac{k_B b}{a}. \tag{A5}$$

The equation $\Phi_n = 0$,

$$0 = k_B \ln \lambda^3 - k_B \ln(1 - nb) + k_B \frac{nb}{1 - nb} - 2aE^* n + k_B \ln n + N^*, \quad (A6)$$

where $\lambda = h/\sqrt{2\pi m k_B T}$ is the thermal de Broglie wavelength, leads, using n_c and E_c^*, to $N^* = k_B \ln(b/\lambda^3) k_B \ln(2/3) - 5k_B/4$.

Note also that evaluating the thermodynamic potential itself at the solution (A2) leads to the van der Waals equation of state,

$$-\frac{PV}{T} = \Phi(\tilde{f}) = -k_B \frac{Vn}{1 - nb} + n^2 V \frac{a}{T}, \quad (A7)$$

where a solution $n(E^*, N^*)$ of Equation (A6) should be substituted for n in order to obtain the lower conjugate entropy $S^{\downarrow *}(E^*, N^*)$.

References

1. Landau, L.D. On the theory of phase transitions. *Zhur Eksp. Teor. Fiz.* **1937**, *7*, 19–32. [CrossRef]
2. Callen, H. *Thermodynamics: An Introduction to the Physical Theories of Equilibrium Thermostatics and Irreversible Thermodynamics*; Wiley: Hoboken, NJ, USA, 1960.
3. Arnold, V. *Catastrophe Theory*; Springer: Berlin/Heidelberg, Germany, 1986.
4. Wilson, K. Renormalization Group and Critical Phenomena. I. Renormalization Group and the Kadanoff Scaling Picture. *Phys. Rev. B* **1971**, *4*, 3174–3184. [CrossRef]
5. Pavelka, M.; Klika, V.; Grmela, M. *Multiscale Thermo-Dynamics*; De Gruyter: Berlin, Germany, 2018.
6. Grmela, M.; Klika, V.; Pavelka, M. Reductions and extensions in mesoscopic dynamics. *Phys. Rev. E* **2015**, *92*, 032111. [CrossRef] [PubMed]
7. Kampen, N. Condensation of a classical gas with long-range attraction. *Phys. Rev.* **1964**, *135*, A362. [CrossRef]
8. Grmela, M. Renormalization of the van der Waals theory of critical phenomena. *Phys. Rev. A* **1976**, *14*, 1781–1789. [CrossRef]
9. Grmela, M. Kinetic approach to phase transitions. *J. Stat. Phys.* **1971**, *3*, 347–364. [CrossRef]
10. Landau, L.; Lifschitz, E. *Statistical Physics*; Number pt. 1 in Course of Theoretical Physics; Pergamon Press, New York, NY, USA, 1969.
11. Gorban, A.; Karlin, I. *Invariant Manifolds for Physical and Chemical Kinetics*; Lecture Notes in Physics; Springer: Berlin/Heidelberg, Germany, 2005.
12. Klika, V.; Pavelka, M.; Vágner, P.; Grmela, M. Dynamic Maximum Entropy Reduction. *Entropy* **2019**, *21*, 715. [CrossRef]
13. Pavelka, M.; Klika, V.; Grmela, M. Ehrenfest regularization of Hamiltonian systems. *Phys. D Nonlinear Phenom.* **2019**, *399*, 193–210. [CrossRef]
14. Pavelka, M.; Klika, V.; Grmela, M. Generalization of the dynamical lack-of-fit reduction. *J. Stat. Phys.* **2020**, *181*, 19–52. [CrossRef]
15. Grmela, M.; Klika, V.; Pavelka, M. Gradient and GENERIC evolution towards reduced dynamics. *Phil. Trans. R. Soc. A* **2020**, *378*, 20190472. [CrossRef]
16. Boussinesq, J. *Théorie de l'Écoulement Tourbillonnant Et Tumultueux Des Liquides Dans Les Lits Rectilignes a Grande Section*; Gauthier-Villars et Fils, Paris, France, 1897.
17. Lebon, G.; Jou, D.; Vázquez, J. *Understanding Non-Equilibrium Thermodynamics: Foundations, Applications, Frontiers*; Springer: London, UK, 2008.
18. Bulíček, M.; Málek, J.; Průša, V. Thermodynamics and Stability of Non-Equilibrium Steady States in Open Systems. *Entropy* **2019**, *21*, 704. [CrossRef]
19. Yeo, L.Y.; Matar, O.K.; de Ortiz, E.S.P.; Hewitt, G.F. Simulation studies of phase inversion in agitated vessels using a Monte Carlo technique. *Chem. Eng. Sci.* **2002**, *57*, 1069. [CrossRef]
20. Brauner, N.; Ullmann, A. Modeling of phase inversion phenomenon in two-phase pipe flows. *Int. J. Multiph. Flow* **2002**, *28*, 1177. [CrossRef]

21. Grmela, M. Externally Driven Macroscopic Systems: Dynamics Versus Thermodynamics. *J. Stat. Phys.* **2017**, *166*, 282–316. [CrossRef]
22. Piela, K.; Oomsa, G.; Sengers, J. Phenomenological description of phase inversion. *Phys. Rev. E* **2009**, *79*, 021403. [CrossRef] [PubMed]
23. Boltenhagen, P.; Hu, Y.; Matthys, E.; Pine, D. Observation of bulk phase separation and coexistence in a sheared micellar solution. *Phys. Rev. Lett.* **1997**, *79*, 2359–2362. [CrossRef]
24. Janečka, A.; Pavelka, M. Non-convex dissipation potentials in multiscale non-equilibrium thermodynamics. *Contin. Mech. Thermodyn.* **2018**, *30*, 917–941. [CrossRef]
25. Le Roux, C.; Rajagopal, K. Shear flows of a new class of power-law fluids. *Appl. Math.* **2013**, *58*, 153–177. [CrossRef]
26. Janečka, A.; Málek, J.; Průša, V.; Tierra, G. Numerical scheme for simulation of transient flows of non-Newtonian fluids characterised by a non-monotone relation between the symmetric part of the velocity gradient and the Cauchy stress tensor. *Acta Mech.* **2019**, *230*, 729–747. [CrossRef]
27. Green, M.S. Exact Renormalization in the Statistical Mechanics of Fluids. *Phys. Rev. A* **1973**, *8*, 1998. [CrossRef]
28. Ginzburg, V.; Landau, L. On the theory of superconductivity. *Zhur Eksp. Theor. Fiz.* **1950**, *20*, 1064–1082.
29. Waldram, J. *The Theory of Thermodynamics*; Cambridge University Press: Cambridge, UK, 1985.
30. González, D.; Chinesta, F.; Cueto, E. Learning corrections for hyperelastic models from data. *Front. Mater.* **2019**, *6*, 14. [CrossRef]

© 2020 by the authors. Licensee MDPI, Basel, Switzerland. This article is an open access article distributed under the terms and conditions of the Creative Commons Attribution (CC BY) license (http://creativecommons.org/licenses/by/4.0/).

Article

Entropy-Based Solutions for Ecological Inference Problems: A Composite Estimator

Rosa Bernardini Papalia [1] and Esteban Fernandez Vazquez [2,*]

[1] Department of Statistical Sciences, University of Bologna, 40126 Bologna, Italy; rossella.bernardini@unibo.it
[2] REGIOlab and Department of Applied Economics, University of Oviedo, 33003 Oviedo, Spain
* Correspondence: evazquez@uniovi.es

Received: 18 June 2020; Accepted: 13 July 2020; Published: 17 July 2020

Abstract: Information-based estimation techniques are becoming more popular in the field of Ecological Inference. Within this branch of estimation techniques, two alternative approaches can be pointed out. The first one is the Generalized Maximum Entropy (GME) approach based on a matrix adjustment problem where the only observable information is given by the margins of the target matrix. An alternative approach is based on a distributionally weighted regression (DWR) equation. These two approaches have been studied so far as completely different streams, even when there are clear connections between them. In this paper we present these connections explicitly. More specifically, we show that under certain conditions the generalized cross-entropy (GCE) solution for a matrix adjustment problem and the GME estimator of a DWR equation differ only in terms of the a priori information considered. Then, we move a step forward and propose a composite estimator that combines the two priors considered in both approaches. Finally, we present a numerical experiment and an empirical application based on Spanish data for the 2010 year.

Keywords: ecological inference; generalized cross entropy; distributional weighted regression; matrix adjustment

1. Introduction

Ecological inference (EI) is the process of drawing conclusions about individual-level behavior from aggregate (historically called "ecological") data, when no individual data are available. Situations where the only available data are aggregated at a level other than the level of interest are quite common in many application fields. This is the typical setting for Ecological Inference [1–3], Cross-level Inference [4,5], Small Area Estimation [6], or disaggregation methods [7]. The basic idea is that, in order to study the behavior of the individuals (or sub-groups of individuals), a microeconomic analysis ought to be carried out using fairly localized individual data, and data which are aggregated by areal units may be used in order to investigate the behavior of the individuals comprising those units. In this paper, we specifically refer to the process of drawing conclusions about individual-level behavior from aggregate data, when no individual data are available or when individual data are incomplete. In this inferential context, one problem is that many different possible relationships at the individual (or subgroup) level can generate the same observations at the aggregate (or group) level [8]. In the absence of individual (or subgroup) level measurements (in the form of survey data), such information needs to be inferred. Estimates of the disaggregated values for the variable of interest can be inferred from aggregate data by using appropriate statistical techniques. However, in many situations, given that the micro-data of interest are not available, the accuracy of any predicted value cannot be verified. This research focuses on the estimation on disaggregated indicators by subclasses. Assume that we have an indicator, y_i, that is observable across the different areas $i = 1, \ldots, T$. Our objective is to disaggregate it into an indicator y_{ij} for the $j = 1, \ldots, K$ different sub-categories (or sub-areas) that conform each class (or area) i. The information available for this inference exercise, together with the indicator y_i, is another disaggregated indicator x_{ij} that is related to the target indicator y_{ij}. This paper approaches this estimation problem in an attempt to unify two estimation strategies and it is organized as follows. Section 2 explains the main features of the matrix-adjustment following the ideas of the Generalized Cross Entropy (GCE) estimation introduced in [9], whereas in Section 3 the basis of the Distributionally Weighted Regression (DWR) estimation are explained. Section 4 studies these two strategies under a common approach and propose a composite prior estimator in line with the Data Weighted Prior (DWP) proposed in [10,11]. The comparative performance of the three techniques is evaluated by means of a numerical experiment in Section 5. Finally, Section 6 presents the main conclusions of the paper.

2. Matrix-Adjustment and Distributionally Weighted Regression Problems

Within the family of IT estimators, [10] proposed a general solution for the estimation problem described in the introduction basing on the minimization of the divergence between the target variable and some prior information. Following this approach, each indicator y_{ij} is assumed as a discrete random variable that can take M different values. Defining a supporting vector (for the sake of simplicity assumed as common for all the y_{ij}) $z' = [z_1, z_2, \ldots, z_M]$ that contains the M possible realizations of the targets with unknown probabilities $p'_{ij} = [p_{ij1}, p_{ij2}, \ldots, p_{ijM}]$, y_{ij} can be written as:

$$y_{ij} = \sum_{m=1}^{M} p_{ijm} z_m \qquad (1)$$

Alternatively, this idea can be generalized in order to include an error term and define each y_{ij} as:

$$y_{ij} = \sum_{m=1}^{M} p_{ijm} z_m + \varepsilon_{ij} \qquad (2)$$

In such a case, we assume that the y_{ij} elements are given from two sources: a signal that keeps the resemblance with the priors x_{ij}, plus a noise term (ε_{ij}). The noise components can be included in order to account for potential spatial heterogeneity and our uncertainty about the target variable. Basically, we represent uncertainty about the realizations of the errors treating each element ε_{ij} as a discrete random variable with $L \geq 2$ possible outcomes contained in a convex set $v' = \{v_1, \ldots, v_L\}$, which for the sake of simplicity will be assumed as common for all the ε_{ij}. We also assume that these possible realizations are symmetric around zero ($-v_1 = v_L$). The traditional way of fixing the upper and lower limits of this set is to apply the three-sigma rule [12]. Under these conditions, each ε_{ij} can be defined as:

$$\varepsilon_{ij} = \sum_{l=1}^{L} w_{ijl} v_l; \; \forall i = 1, \ldots, T; \; j = 1, \ldots, K \qquad (3)$$

where w_{ijl} is the unknown probability of the outcome v_l for the cell ij. Now, the y_{ij} elements can be written as:

$$y_{ij} = \sum_{m=1}^{M} p_{ijm} z_m + \sum_{l=1}^{L} w_{ijl} v_l \qquad (4)$$

The solution to the estimation problem is given by the minimization of the Kullback-Leibler divergence between the posteriors distributions p's and the a priori probabilities $q'_{ij} = [q_{ij1}, q_{ij2}, \ldots, q_{ijM}]$. The q's reflect the information we have on the indicators x_{ij}, which are somehow related to our target y_{ij}, being defined by the expression:

$$x_{ij} = \sum_{m=1}^{M} q_{ijm} z_m \tag{5}$$

The solution to the estimation problems is given by minimizing the KL divergence between the p's and the q's. If we do not have an informative prior, the a priori distributions are specified as uniform $\left(q_{ij} = \frac{1}{M};\, \forall m = 1,\ldots, M\right)$, which leads to the GME solution. The uniform distribution is usually set as the natural prior W^0 for the error terms. Specifically, the constrained minimization problem can be written as:

$$\underset{p,W}{Min} D(p, W\|q, W^0) = \sum_{m=1}^{M}\sum_{i=1}^{T}\sum_{j=1}^{K} p_{ijm} ln\left(\frac{p_{ijm}}{q_{ijm}}\right) + \sum_{l=1}^{L}\sum_{i=1}^{T}\sum_{j=1}^{K} w_{ijl} ln\left(\frac{w_{ijl}}{w^0_{ijl}}\right) \tag{6}$$

subject to:

$$y_{i\cdot} = \sum_{j=1}^{K}\left(\sum_{m=1}^{M} p_{ijm} z_m + \sum_{l=1}^{L} w_{ijl} v_l\right) C_{\cdot j};\, i = 1,\ldots, T \tag{7}$$

$$\sum_{m=1}^{M} p_{ijm} = \sum_{l=1}^{L} w_{ijl} = 1;\, 1\, \forall i = 1,\ldots, T;\, j = 1,\ldots, K \tag{8}$$

Restrictions (8) are just normalization constrains, whereas Equation (7) reflects the observable information that we have on the relationship between the aggregates $y_{i\cdot}$ and the indicators y_{ij} through the observable K-dimensional vector $C_{\cdot j}$. Denoting as $\hat{y}^0_{ij}$ to the solution in absence of this information, this is given by the indicator x_{ij}; i.e., $\hat{y}^0_{ij} = x_{ij} = \sum_{m=1}^{M} q_{ijm} z_m$.

Following Golan et al., (1994), the aggregate vectors $y_{i\cdot}$ and $C_{\cdot j}$ are, respectively, row and column margins in a matrix of inter-industry flows. However, the availability of sample (observable) and out-of-sample (unobservable) information could be different in our estimation problem, because in the inter-industry problem it is natural to have known $K + T$ data, but in other estimation problems we only have aggregate information across the dimension of T through $y_{i\cdot}$. For example, if we want to disaggregate the income per capita in each area i ($y_{i\cdot}$) into the income per capita of its sub-populations (men and women, population classified by education levels, etc.) being observable the weight of each sub-population on the total population, but not the overall income per capita of each sub-group.

Sometimes the aggregate $C_{\cdot j}$ is not observable and it is replaced by the observation of the weights given to the sub-category j in each area i (θ_{ij}) that defines the indicator $y_{i\cdot}$ as the weighted sum:

$$y_{i\cdot} = \sum_{j=1}^{K} y_{ij}\theta_{ij};\, i = 1,\ldots, T \tag{9}$$

Additionally, the relation between the target indicators y_{ij} and the prior information x_{ij} will be made explicit by means of a functional relationship like:

$$y_{ij} = \alpha_i + \beta_{ij} x_{ij} + \varepsilon_{ij} \tag{10}$$

and, consequently:

$$y_{i\cdot} = \sum_{j=1}^{K}(\alpha_i + \beta_{ij} x_{ij} + \varepsilon_{ij})\theta_{ij};\, i = 1,\ldots, T \tag{11}$$

Equations (10) and (11) contain the starting point of the traditional approach to spatial disaggregation based on some Distributionally Weighted Regression (DWR) of the type proposed in [13,14]. In Equation (10), the unobservable y_{ij} are defined as a linear function of x_{ij}, allowing for slope heterogeneity (note that the β_{ij} can be different for each area and sub-class) and an specific area indicator α_i plus an error term ε_{ij}. For the estimation of model Equation (10), the same IT-based strategy is followed, by defining for the M possible realizations of each parameter, the support vector $b' = [b_1, b_2, \ldots, b_M]$ (again common for parameters α_i and β_{ij}) with unknown probabilities p^α, p^β to be recovered. The noise components ε_{ij} are treated in the same ways as in Equation (5).

Once the respective supporting vectors and the a priori probability distributions are set, the DWR estimation can be made in the terms of the following GCE program:

$$\operatorname*{Min}_{p^\alpha, p^\beta, W} D(p^\alpha, p^\beta, W \| q^\alpha, q^\beta, W^0) = \sum_{m=1}^{M} \sum_{i=1}^{T} p^\alpha_{mi} \ln\left(\frac{p^\alpha_{mi}}{q^\alpha_{mi}}\right) + \sum_{m=1}^{M} \sum_{i=1}^{T} \sum_{j=1}^{K} p^\beta_{mij} \ln\left(\frac{p^\beta_{mij}}{q^\beta_{mij}}\right) + \sum_{l=1}^{L} \sum_{i=1}^{T} \sum_{j=1}^{K} w_{ijl} \ln\left(\frac{w_{ijl}}{w^0_{ijl}}\right) \quad (12)$$

subject to:

$$y_{i\cdot} = \sum_{j=1}^{K} \left(\sum_{m=1}^{M} p^\alpha_{mi} b^\alpha_m + \sum_{m=1}^{M} p^\beta_{mij} b^\beta_m x_{ij} + \sum_{l=1}^{L} w_{ijl} v_l \right) \theta_{ij}; \; i = 1, \ldots, T \quad (13)$$

$$\sum_{m=1}^{M} p^\alpha_{mi} = \sum_{m=1}^{M} p^\beta_{mij} = \sum_{l=1}^{L} w_{ijl} = 1; \forall i = 1, \ldots, T; \; j = 1, \ldots, K \quad (14)$$

Both for the parameters and the errors, the supporting vectors usually contain values symmetrically centered on zero. If all the a priori distributions (q^α, q^β, W^0) are specified as uniform, then the GCE solution reduces to the GME one.

3. Unifying the Two Approaches: A Composite Prior Estimator

In this section, we will unify the two previous approaches under a common framework showing that the matrix adjustment problem introduced in [9] is simply a case of a DWR equation (if the available observable information is the same) with not necessarily uniform distributions for q^α and q^β. We let out of the discussion the a priori distribution of the errors W^0 because the uniform solution is the most intuitive. We will base our explanation on the most common case of supporting vectors with $M \geq 2$ values distributed symmetrically around zero.

Note that the GME solution to the DWR problem departs from the specification of a priori distributions that assume that the parameters can take any value as long as they remain in the bounds set in the supports. In contrast, in the solution offered in [9] for the inter-industry flows estimation, no area-specific (row-specific in terms of the problem discussed there) effect was considered and the prior expectation on y_{ij} is given by the corresponding cell x_{ij}. These assumptions can be formulated in terms of the a priori distributions used in the DWR approach, which means that both approaches can be treated as particular cases of a general estimation problem.

The a priori distribution q^α can be defined in order to consider the assumption of avoiding any area-specific parameter α_i from Equation (10). As opposed to the GME's solution to the DWR estimation where they are specified as uniform ($q^{\alpha u}$), now we specify an alternative non-uniform distribution ($q^{\alpha n}$) with a point mass at $b^\alpha_m = 0$. Similarly, the a priori distribution q^β should reflect that the uninformative estimation of y_{ij} is the regressor x_{ij}. This non-uniform distribution ($q^{\beta n}$), consequently, should be specified as fulfilling the condition $\hat{y}^0_{ij} = x_{ij}$, or alternatively:

$$\sum_{m=1}^{M} b^\beta_m q^{\beta n}_{ijm} = 1; \; i = 1, \ldots, T; \; j = 1, \ldots, K \quad (15)$$

Appendix A illustrates how specifying such an a priori distribution for the simplest case with $M = 2$ values in the supporting vectors. Having made explicit that, under the same information availability, the two approaches only differ on the a priori distributions specified, it is possible to apply a composite prior estimator that considers both possibilities in the same fashion as in in [10,11]. This estimator is very flexible in the assumptions made on the a priori distributions, given that it allows for including both uniform and non-uniform priors. The estimator it is called Data Weighted Prior (DWP) because it is the information observed which weighs the two alternative priors considered. Furthermore, the authors of [10] prove that its estimates present relatively lower variance than those estimated from a GCE program.

Specifically, the DWP program can be written for our problem as:

$$\min_{p^\alpha, p^\beta, p^\gamma, W} D(p^\alpha, p^\beta, p^\gamma, W \| q^\alpha, q^\beta, q^\gamma, W^0) =$$

$$(1 - \gamma_i^\alpha) \sum_{m=1}^{M} \sum_{i=1}^{T} p_{mi}^{\alpha u} \ln\left(\frac{p_{mi}^{\alpha u}}{q_{mi}^{\alpha u}}\right) + (1 - \gamma_{ij}^\beta) \sum_{m=1}^{M} \sum_{i=1}^{T} \sum_{j=1}^{K} p_{mij}^{\beta u} \ln\left(\frac{p_{mij}^{\beta u}}{q_{mij}^{\beta u}}\right) +$$

$$\gamma_i^\alpha \sum_{m=1}^{M} \sum_{i=1}^{T} p_{mi}^{\alpha n} \ln\left(\frac{p_{mi}^{\alpha n}}{q_{mi}^{\alpha n}}\right) + \gamma_{ij}^\beta \sum_{m=1}^{M} \sum_{i=1}^{T} \sum_{j=1}^{K} p_{mij}^{\beta n} \ln\left(\frac{p_{mij}^{\beta n}}{q_{mij}^{\beta n}}\right) + \qquad (16)$$

$$\sum_{h=1}^{H} \sum_{i=1}^{T} p_{hi}^{\gamma\alpha} \ln\left(\frac{p_{hi}^{\gamma\alpha}}{q_{hi}^{\gamma\alpha}}\right) + \sum_{h=1}^{H} \sum_{i=1}^{T} \sum_{j=1}^{K} p_{hij}^{\gamma\beta} \ln\left(\frac{p_{hij}^{\gamma\beta}}{q_{hij}^{\gamma\beta}}\right) +$$

$$\sum_{l=1}^{L} \sum_{i=1}^{T} \sum_{j=1}^{K} w_{ijl} \ln\left(\frac{w_{ijl}}{w_{ijl}^0}\right)$$

subject to:

$$y_{i\cdot} = \sum_{j=1}^{K} \left(\sum_{m=1}^{M} p_{mi}^\alpha b_m^\alpha + \sum_{m=1}^{M} p_{mij}^\beta b_m^\beta x_{ij} + \sum_{l=1}^{L} w_{ijl} v_l \right) \theta_{ij}; \; i = 1, \ldots, T \qquad (17)$$

$$\sum_{m=1}^{M} p_{mi}^\alpha = \sum_{m=1}^{M} p_{mij}^\beta = \sum_{h=1}^{H} p_{hi}^{\gamma\alpha} = \sum_{h=1}^{H} p_{hij}^{\gamma\beta} = \sum_{l=1}^{L} w_{ijl} = 1; \qquad (18)$$
$$i = 1, \ldots, T; \; j = 1, \ldots, K$$

The γ parameters are estimated simultaneously with the rest of coefficients of the model. Each γ measures the weight given to the uniform prior q^u for each parameter and it is defined as $\gamma = \sum_{h=1}^{H} b_h^\gamma p_h^\gamma$, where $b_1^\gamma = 0$ and $b_H^\gamma = 1$ are, respectively, the lower and upper bound defined in the supporting vectors with H values for these parameters ($b' = (0, \ldots, 1) \rightarrow 0 \leq \gamma \leq 1$). The a priori probability distributions are always uniform ($q_h^\gamma = \frac{1}{H}$) and the same is applied for the errors ($w_{ijl}^0 = \frac{1}{J}$).

To understand the logic of this estimator, an explanation on the objective function of the previous minimization program is required. Note that Equation (16) is divided in four terms. The last term measures the Kullback divergence between the posterior and the prior probabilities for the noise component of the model. The first term quantifies this divergence between the recovered probabilities and the uniform priors for each coefficient, being this divergence weighted by the corresponding $(1 - \gamma)$. Next, the second element of (16) measures the divergence with the non-uniform priors and it is weighted by γ. The third element in (16) relates to the Kullback divergence of the weighting parameters γ. Equation (16) is minimized subject to the set of constraints present in Equations (16)–(18). Again, the restrictions in (18) ensure that the posterior probability distributions of the estimates and the errors are compatible with the observations, and Equation (18) are just normalization constraints.

4. A Numerical Experiment

The numerical simulation compares the performance of the estimation strategies explained previously to estimate a set of latent indicators ($T \times K$). The target will be the unknown elements y_{ij} (output per worker, income per capita, etc.) that measure the amount of certain variable z_{ij} per unit of other auxiliary variable l_{ij}. The values of the later are drawn from a normal distribution as $l_{ij} \sim N(20, 2)$, which define the weights as $\theta_{ij} = l_{ij}/l_i$. We also simulate an observable disaggregated indicator x_{ij} drawn as $x_{ij} \sim N(10, 1)$ related to our unobservable target y_{ij}.

In the context of simulation, we assume that the indicator y_{ij} is generated as a convex combination from two possible schemes:

$$y_{ij} = \delta\left[\alpha_i + \beta_{ij}x_{ij} + \varepsilon_{ij}\right] + (1-\delta)\left[\eta_{ij}x_{ij} + \varepsilon_{ij}\right]; \quad i = 1,..,T; \; j = 1,..,K. \tag{19}$$

This equation contains two sets of slope parameters, namely β_{ij} and η_{ij}, which relate the regressor x_{ij} with the target y_{ij}. Furthermore, a fixed area effect α_i is also included. These parameters have been arbitrarily set as:

$$\begin{aligned} \alpha_i &\sim N(5,1) \\ \beta_{ij} &\sim N(0,0.1) \\ \eta_{ij} &\sim N(1,0.1) \end{aligned} \tag{20}$$

and they are kept constant along the simulations. The error term ε_{ij} is drawn as $\varepsilon_{ij} \sim N(0,0.1)$ and it is generated in each new trial of the experiment.

The first part of the equation ($\alpha_i + \beta_{ij}x_{ij} + \varepsilon_{ij}$) shows that y_{ij} can be generated from a process like the one depicted in (16): a linear function of x_{ij} with slope heterogeneity plus a specific area effect (see 11). The second term ($\eta_{ij}x_{ij} + \varepsilon_{ij}$) does not include any specific area indicator and assumes that y_{ij} is exclusively affected by x_{ij} (see 2). Equation (19) includes the scalar δ bounded between 0 and 1 that weighs the two possible sources that generate the variable. If we make $\delta \to 1$, the first possible mechanism takes over and the contrary happens when we make $\delta \to 0$. Note that if we set $\delta = 1$ we are imposing a data-generating process in line with the assumptions made in the GME program depicted in Equations (12)–(14) for the DWR estimation. On the contrary, if we set $\delta = 0$, this is a scenario compatible with the assumptions of non-uniform priors for the parameters that reflected the belief of absence of area-specific effects and a slope parameter close to 1 (labeled as GCE when the simulation results are shown). Any other value of δ between these two extreme cases shows a data-generating process that is not fully incorporated in the priors of either alternative. It is in this type of intermediate situation with the composite prior estimator (labeled as DWP in the simulation results) described in Equations (16)–(18) can be useful, because both priors are considered and we let the data speak for themselves and favor the most realistic one.

The unobservable indicators generated in (20) will be estimated by the three estimation strategies described in the paper (DWR, GCE and DWP estimators) with equal amounts of observable information (the aggregates $y_{i\cdot} = \sum_{j=1}^{K} y_{ij}\theta_{ij}$). We have specified a common supporting vector for all the parameters with $M = 3$ points at $b' = (-10, 0, 10)$. Similarly, a three-point ($H = 3$) support vector with values 0, 0.5 and 1 has been set for the weighting parameters γ. For the error terms, the support with $L = 3$ values has been chosen, applying the three-sigma rule with uniform a priori weights.

In the experiment, we compare the performance of the three approaches under different scenarios. Three different dimensions (T × K) of the matrix with the target indicators y_{ij} have been considered and for each case we set arbitrarily six different values of scalar δ: 0.0; 0.2; 0.4; 0.6; 0.8 and 1.0. In each one of these 18 scenarios, we have carried out 200 trials and computed the mean of the absolute deviation in percentage between our estimates and the real y_{ij}. Table 1 shows the results:

Table 1. Results of the numerical experiment (1000 replications): deviation figures.

		Matrix 1 (20 × 4)	Matrix 2 (50 × 4)	Matrix 3 (100 × 4)
	DWR	13.126 (0.049) [1.544]	13.642 (0.126) [1.622]	14.837 (0.040) [1.767]
$\gamma = 0.00$	GCE	11.420 (0.126) [1.275]	10.047 (0.054) [1.232]	11.633 (0.038) [1.382]
	DWP	11.546 (0.087) [1.321]	11.267 (0.091) [1.352]	12.645 (0.002) [1.494]

Table 1. Cont.

		Matrix 1 (20 × 4)	Matrix 2 (50 × 4)	Matrix 3 (100 × 4)
$\gamma = 0.20$	DWR	13.697 (0.053) [1.429]	13.667 (0.116) [1.462]	14.791 (0.035) [1.595]
	GCE	12.623 (0.131) [1.249]	10.639 (0.044) [1.276]	11.996 (0.044) [1.297]
	DWP	12.393 (0.091) [1.248]	11.420 (0.081) [1.233]	12.654 (0.004) [1.356]
$\gamma = 0.40$	DWR	15.307 (0.057) [1.382]	14.357 (0.107) [1.357]	15.381 (0.029) [1.479]
	GCE	14.788 (0.136) [1.282]	12.213 (0.035) [1.288]	13.306 (0.049) [1.278]
	DWP	14.247 (0.095) [1.243]	12.258 (0.072) [1.175]	13.406 (0.009) [1.282]
$\gamma = 0.60$	DWR	18.565 (0.062) [1.399]	15.922 (0.097) [1.307]	17.152 (0.024) [1.429]
	GCE	18.603 (0.141) [1.373]	15.264 (0.025) [1.288]	16.098 (0.055) [1.330]
	DWP	17.666 (0.100) [1.300]	14.222 (0.062) [1.182]	15.465 (0.015) [1.278]
$\gamma = 0.80$	DWR	25.047 (0.067) [1.466]	19.109 (0.088) [1.313]	20.898 (0.018) [1.439]
	GCE	26.062 (0.145) [1.519]	20.652 (0.016) [1.409]	21.302 (0.060) [1.449]
	DWP	24.405 (0.105) [1.409]	18.271 (0.053) [1.255]	19.764 (0.020) [1.341]
$\gamma = 1.00$	DWR	42.350 (0.071) [1.578]	26.659 (0.079) [1.374]	31.769 (0.013) [1.506]
	GCE	46.481 (0.149) [1.711]	31.563 (0.006) [1.595]	34.915 (0.066) [1.625]
	DWP	42.903 (0.109) [1.564]	27.362 (0.043) [1.385]	31.829 (0.026) [1.466]

Values on each cell report the mean absolute deviation (in %) between the real generated target values and the estimated ones. Values in parentheses show the average bias, on absolute terms (ABIAS), and the figures in brackets show the root of the mean squared errors of the estimates (RMSE).

Independently of the estimation approach, the numbers on Table 1 show some common patterns to the three of them. The deviations increase with the value of the scalar δ given that high values of this scalar give more weight to the part of the data-generating process that includes an area-specific effect, which makes the y_{ij} indicators more difficult to predict. The errors seem more stable regarding the different sizes of the target matrices.

If we pay attention to the comparative performance among the three approaches evaluated in the experiment, the results indicate (not surprisingly) that, for low values of the scalar δ, it seems preferable considering that the GCE approach does not introduce any area-specific effect and considers the regressor x_{ij} as the best prediction in absence of observable information. The longer the value of this scalar, the better the relative performance of the GME-DWR approach (based on a priori uniform distributions).

The rule of thumb would be, consequently, to use the former when we suspect that no area-specific effect is present (if the second term in Equation (19) dominates) and to favor the latter otherwise (if the first term is more important). In empirical estimation problems, is virtually impossible to know beforehand which one of the two terms is more important. It is in these situations when the use of the composite prior estimator can be helpful. The DWP approach generally outperforms the competing estimators for intermediate values of δ (ranging from 0.4 to 0.8). These medium values indicate some degree of uncertainty about the type of process that generates the data to be estimated. Moreover, the DWP approach can be seen as a conservative solution: even when one of the two parts of the process is clearly dominant ($\delta = 0$ or $\delta = 1$), the composite prior does not perform much worse than the best of the three options. The losses in terms of prediction, however, can be larger if we choose one single-prior estimator when the other is the best option (see the first and last rows of Table 1).

5. An Empirical Application: Obtaining Disaggregated Information on Wages

In order to illustrate the performance of the proposed estimator, it will be applied to solve an empirical problem of disaggregating data of average wages for Spain. The most detailed information about non-agricultural wages in Spain is published in the Wage Structure Survey (*Encuesta de Estructura Salarial*). The complete version of this survey is conducted by the Spanish Statistical Office (*INE*) every four years, being the corresponding to 2010 one of the most recent ones. In intermediate years, however, only partial data are collected and the microdata are not released. If, for example, we want to explore the differences across industries on average wages by gender and type of working day in a year where the complete statistical operation is not conducted, the only information we have are at aggregate level. This situation happens, for example, in 2011, where the only available data on are the aggregates reported in Table 2, which do not allow disaggregated differences between male and female workers to be analyzed depending on the industry they belong to:

Table 2. Available information on annual wages by industry, type of working day and gender. Wage Structure Survey, 2011.

Industry	Mean Wage (EUR)
Mining and quarrying industries	29,223
Manufacturing industry	25,308
Supply of electrical energy, gas and steam	50,371
Water supply, sewerage and waste management	25,570
Construction	22,541
Trade and repair of vehicles	19,445
Transport and storage	23,347
Accommodation	14,235
Information and communications	32,491
Financial and insurance activities	41,124
Real estate activities	20,349
Professional, scientific and technical activities	25,350
Administrative and support service activities	16,199
Public administration	27,816
Education	21,565

Table 2. *Cont.*

Industry	Mean Wage (EUR)
Health and social services activities	26,058
Arts, recreation and entertainment activities	18,106
Other services	17,035
Type of Working Day and Gender	**Mean Wage (EUR)**
Full-time female	23,693
Full-time male	27,596
Part-time female	10,078
Part-time male	11,233

In such a context, if the researcher wants to study wage gender gaps across industries it would be necessary to apply an estimation procedure that produces disaggregated values for this specific year, since the official aggregated data do not allow for this type of analysis. The values in Table 2 provide the aggregates required for applying our DWP estimator. Vector y, with dimension (18×1) and elements $y_{i\cdot}$, contains the mean wage for each industry and our estimation target will be the unknown y_{ij} elements, where sub-index j refers to the type of worker (classified into four categories: full-time males, full-time female, part-time male and part-time females). The information in Table 2 is also useful for setting a regressor (x_{ij}) for our analysis. In particular, the aggregate mean wages for each type of worker ($x_{i\cdot}$, in the four bottom rows of Table 2) will be used for this purpose, assuming that $x_{i\cdot} = x_{ij}$, $j = 1, \ldots, 4$. The additional information required to define the weights (θ_{ij}) has been taken from the Spanish Labor Force Survey (EPA) corresponding to that year, where we can find information about the number of workers classified by industry, type of working day and gender. With all this information, the DWP estimator has been applied, specifying identical support vectors as those described in the previous section with the numerical simulation, and the estimates obtained are shown in Table 3:

Table 3. DWP estimates on disaggregated mean annual wages (EUR) by industry, type of working day and gender, 2011.

Industry	Full-Time, Female	Full-Time, Male	Part-Time, Female	Part-Time, Male
Mining and quarrying industries	13,338	31,307	5311	5840
Manufacturing industry	16,323	29,220	5738	5911
Supply of electrical energy, gas and steam	36,909	55,191	7445	7330
Water supply, sewerage and waste management	14,301	28,675	5419	6135
Construction	12,459	24,134	5239	5987
Trade and repair of vehicles	19,603	23,324	7453	6298
Transport and storage	14,336	26,803	5664	6248
Accommodation	15,473	17,508	6553	6230
Information and communications	23,483	39,877	6741	7326
Financial and insurance activities	38,566	46,664	7946	6078
Real estate activities	21,301	23,487	7301	6299
Professional, scientific and technical activities	25,022	29,926	7984	6565
Administrative and support service activities	17,383	20,534	9142	6290
Public administration	24,433	32,196	6269	6117
Education	25,838	21,708	8396	6720
Health and social services activities	31,832	20,406	9049	6078
Arts, recreation and entertainment activities	17,460	24,232	8094	8778
Other services	19,600	18,896	7537	6116

The aggregate information classified by industry in Table 2 displayed a high variability, ranging from slightly more than EUR 14,000 for the average worker in the Accommodation industry to almost three times higher in Financial and Insurance services. Additionally, the aggregates also showed that the male workers earned more on average than the female workers. Specifically, full-time male workers earned on average around 16% than their female counterparts, whereas this gap was around 11% in the case of part-time workers. This information, however, does not allow for checking if this gender differences on wage keep stable independently on the industry. The estimates obtained by the DWP estimator and reported on Table 3 help to shed some light on this matter.

According to the outcomes of the estimation, the gender gap for full-time workers is much larger in the case of economic branches related to mining, manufacturing or construction than in service activities. Furthermore, for the specific case of Education and Health and social services activities, we estimate significant positive difference for full-time female workers. Something similar, but to a lesser extent, happens with the case of part-time workers: the mean gender gap in favor of male workers, according to the estimates, is mainly produced by the higher wages received in mining, manufacturing and construction, but in general the activities related to services tend to alleviate this gap. Detecting these differential patterns across industries is possible due to the disaggregated information contained in the estimates, which was partially hidden in the aggregated averages. Additionally, we have explored how robust are the estimates and the patterns found by modifying the supporting vectors, which in turn impact on the priors, as depicted in equation (15). The estimates reported in Table 3 correspond to a case where the support vectors have been defined as $b' = [-100, 0, 100]$ with $M = 3$ and common for parameters α_i and β_{ij}. Appendix B reports the same estimates as in Table 3, where the support vectors are defined as $b' = [-10, 0, 10]$ (Table A1) and $b' = [-1,000, 0, 1,000]$ (Table A2) in order to check if having wider or narrower vectors impacts on the results. Despite some of the minor differences produced by the numerical simulation, the general patterns seem to be robust to this specification.

6. Conclusions

In this paper, we have tackled the problem of providing reliable estimates of a target variable in a set of small geographical areas, by showing that under certain conditions the generalized cross-entropy (GCE) solution for a matrix adjustment problem and the GME estimator of a DWR equation differ only in terms of the a priori information considered. Then, a composite estimator that combines the priors considered in both approaches is proposed and the performance among the three approaches is evaluated throughout Montecarlo experiments.

The proposed method may represent a new basis to recover estimate at a disaggregate level in presence of: (i) sampling and response errors; (ii) small samples. Within this framework, minimal distributional assumptions are necessary, and a dual loss function is used to take into account both the estimation precision and the prediction objectives. The choice of the prior is data based and endogenously determined and the method provides a simple way of introducing and evaluating prior information in the estimation process. The DWP estimation procedure seem to be a promising alternative model-based estimation technique because the implementation of the method involves minimum outlay of computing, it does not depend on any hypotheses regarding the form of the error distribution in the model, and it produces good results for small-sized samples, especially in the presence of spatial heterogeneity. Finally, theoretical and other non-sample information may be directly imposed on the DWP estimates much more easily than the classic Maximum likelihood and Bayesian estimation techniques.

The results indicate that for low values of the parameter δ (that measures the weight given to the uniform prior for each parameter), it seems preferable considering the GCE approach that does not introduce any area-specific effect and considers the indicator observed at area level as the best prediction in absence of observable information. The longer the value of this scalar, the better the relative performance of the GME-DWR approach (based on a priori uniform distributions).

The working of the proposed estimation procedure has been also illustrated by applying the procedure on the estimation of average wages for the Spanish industries in 2011, classified by gender and type of working day. Our results have shown that the DWP estimation has the potential to obtain disaggregated estimates based on minimal assumptions about the data-generating process.

Author Contributions: R.B.P. and E.F.V. designed the methodology, run the numerical simulation and the empirical application and wrote the document. E.F.V. got the data for the empirical application. All authors have read and agreed to the published version of the manuscript.

Funding: This work has been completed within the BLU-ETS project "Blue-Enterprises and Trade Statistics", a small or medium-scale focused research project funded by the Seventh Framework Programme of the European Commission, FP7-COOPERATION-SSH (*Cooperation* Work Programme: Socio-Economic Sciences and the Humanities). Additionally, this project has received funding from the European Union's Horizon 2020 research and innovation programme under grant agreement No 726950 corresponding to the IMAJINE project.

Conflicts of Interest: The authors declare no conflicts of interest.

Appendix A

To simplify the analysis, but without loss of generality, we assume that we consider a supporting vector b^β that contains $M = 2$ symmetric values, namely $-m$ and m, with respective prior probabilities q_1^β and q_2^β. Consequently:

$$\hat{y}_{ij}^0 = \left[-mq_1^\beta + mq_2^\beta\right]x_{ij} = \left[-mq_1^\beta + m(1-q_1^\beta)\right]x_{ij} \tag{A1}$$

The a priori probability distribution q^β that guaranties that $\hat{y}_{ij}^0 = x_{ij}$ is:

$$x_{ij}^0 = \left[-mq_1^\beta + m(1-q_1^\beta)\right]x_{ij} \tag{A2}$$

$$1 = \left[-mq_1^\beta + m(1-q_1^\beta)\right] = q_1^\beta(-2m) + m \tag{A3}$$

and the solution is:

$$q_1^\beta = \frac{m-1}{2m}$$
and
$$q_2^\beta = 1 - q_1^\beta = 1 - \frac{m-1}{2m} = \frac{m+1}{2m} \tag{A4}$$

whereas in the GME-DWR approach, the prior used for these parameters is $q_1^\beta = q_2^\beta = \frac{1}{2}$.

Appendix B Analysis of the Sensitivity of the Estimates

Table A1. DWP estimates on disaggregated mean annual wages (EUR) by industry, type of working day and gender, 2011. Support vectors as $b' = [-10, 0, 10]$.

Industry	Full-Time Female	Full-Time Male	Part-Time Female	Part-Time Male
Mining and quarrying industries	13,161	31,325	5231	5749
Manufacturing industry	16,206	29,259	5667	5822
Supply of electrical energy, gas and steam	22,195	58,689	5963	6238
Water supply, sewerage and waste management	14,147	28,706	5342	6051
Construction	12,275	24,150	5160	5903
Trade and repair of vehicles	19,606	23,332	7435	6222
Transport and storage	14,190	26,839	5594	6168
Accommodation	15,491	17,496	6562	6174
Information and communications	23,414	39,917	6679	7253
Financial and insurance activities	38,574	46,663	7888	5988
Real estate activities	21,325	23,476	7273	6222
Professional, scientific and technical activities	25,027	29,932	7950	6488
Administrative and support service activities	17,367	20,519	9199	6220
Public administration	24,415	32,215	6204	6031
Education	25,899	21,618	8379	6648
Health and social services activities	31,890	20,254	9023	5991
Arts, recreation and entertainment activities	17,407	24,263	8094	8771
Other services	19,651	18,825	7537	6039

Table A2. DWP estimates on disaggregated mean annual wages (EUR) by industry, type of working day and gender, 2011. Support vectors as $b' = [-1{,}000, 0, 1{,}000]$.

Industry	Full-Time Female	Full-Time Male	Part-Time Female	Part-Time Male
Mining and quarrying industries	13,276	31,313	5270	5810
Manufacturing industry	16,109	29,288	5620	5869
Supply of electrical energy, gas and steam	21,789	58,787	5866	6212
Water supply, sewerage and waste management	14,188	28,699	5358	6053
Construction	12,448	24,138	5210	5931
Trade and repair of vehicles	19,388	23,603	7070	6203
Transport and storage	14,222	26,835	5558	6146
Accommodation	15,416	17,670	6328	6149
Information and communications	22,962	40,148	6450	7046
Financial and insurance activities	37,672	47,434	7474	6015
Real estate activities	21,066	23,806	6949	6205
Professional, scientific and technical activities	24,618	30,351	7510	6427
Administrative and support service activities	17,366	20,907	8567	6210
Public administration	23,958	32,537	6070	6046
Education	25,727	22,124	7922	6579
Health and social services activities	31,809	20,810	8506	6028
Arts, recreation and entertainment activities	17,335	24,582	7618	8309
Other services	19,539	19,168	7183	6058

References

1. Freedman, D.A.; Klein, S.P.; Ostland, M.; Roberts, M.R.; King, G. A Solution to the Ecological Inference Problem. *J. Am. Stat. Assoc.* **1998**, *93*, 1518. [CrossRef]
2. Schuessler, A.A. Ecological inference. *Proc. Natl. Acad. Sci. USA* **1999**, *96*, 10578–10581. [CrossRef] [PubMed]
3. King, G.; Rosen, O.; Tanner, M.A. *Ecological Inference: New Methodological Strategies*; Cambridge University Press: Cambridge, UK, 2004.
4. Achen, C.H.; Shively, W.P. *Cross-Level Inference*; University of Chicago Press: Chicago, IL, USA, 1995.
5. Cho, W.K.T. Latent Groups and Cross-Level Inferences. *Elect. Stud.* **2001**, *20*, 243–263.
6. Rao, J.N.K. *Small Area Estimation*; Wiley: Hoboken, NJ, USA, 2003.
7. Barker, T.; Pesaran, M.H. *Disaggregation in Econometric Modelling*; Routledge: Abingdon, UK, 1990.
8. King, G. *A Solution to The Ecological Inference Problem: Reconstructing Individual Behavior from Aggregate Data*; Princeton University Press: Princeton, NJ, USA, 1997.
9. Golan, A.; Judge, G.; Robinson, S. Golan (1994) RES.pdf. *Rev. Econ. Stat. Econ. Stat.* **1994**, *76*, 541–549. [CrossRef]
10. Golan, A. A simultaneous estimation and variable selection rule. *J. Econom.* **2001**, *101*, 165–193. [CrossRef]
11. Bernardini Papalia, R. Generalized Maximum Entropy Estimation of Spatial Panel Interaction Models. *Econom. Rev.* **2008**, *27*, 596–609. [CrossRef]
12. Pukelsheim, F. The Three Sigma Rule. *Am. Stat.* **1994**, *48*, 88–91.
13. Bidani, B.; Ravallion, M. Decomposing social indicators using distributional data. *J. Econom.* **1997**, *77*, 125–139. [CrossRef]
14. Peeters, L.; Chasco, C. Ecological inference and spatial heterogeneity: An entropy-based distributionally weighted regression approach. *Pap. Reg. Sci.* **2006**, *85*, 257–276. [CrossRef]

© 2020 by the authors. Licensee MDPI, Basel, Switzerland. This article is an open access article distributed under the terms and conditions of the Creative Commons Attribution (CC BY) license (http://creativecommons.org/licenses/by/4.0/).

MDPI
St. Alban-Anlage 66
4052 Basel
Switzerland
Tel. +41 61 683 77 34
Fax +41 61 302 89 18
www.mdpi.com

Entropy Editorial Office
E-mail: entropy@mdpi.com
www.mdpi.com/journal/entropy

www.ingramcontent.com/pod-product-compliance
Lightning Source LLC
LaVergne TN
LVHW070152120526
838202LV00013BA/954